174
Advances in Polymer Science

Editorial Board:
A. Abe · A.-C. Albertsson · R. Duncan · K. Dušek · W. H. de Jeu
J. F. Joanny · H.-H. Kausch · S. Kobayashi · K.-S. Lee · L. Leibler
T. E. Long · I. Manners · M. Möller · O. Nuyken · E. M. Terentjev
B. Voit · G. Wegner

Advances in Polymer Science

Recently Published and Forthcoming Volumes

Inorganic Polymeric Nanocomposites and Membranes
Vol. 179, 2005

Polymeric and Inorganic Fibres
Vol. 178, 2005

Poly(acrylene Ethynylenes) – From Synthesis to Application
Volume Editor: Weder, C.
Vol. 177, 2005

Metathesis Polymerization
Volume Editor: Buchmeiser, M.
Vol. 176, 2005

Polymer Particles
Volume Editor: Okubo, M.
Vol. 175, 2005

Neutron Spin Echo in Polymer Systems
Volume Editors: Richter, D., Monkenbusch, M., Arbe, A., Colmenero, J.
Vol. 174, 2005

Advanced Computer Simulation Approaches for Soft Matter Sciences I
Volume Editors: Holm, C., Kremer, K.
Vol. 173, 2005

Microlithography · Molecular Imprinting
Volume Editor: Kobayashi, T.
Vol. 172, 2005

Polymer Synthesis
Vol. 171, 2004

NMR · Coordination Polymerization · Photopolymerization
Vol. 170, 2004

Long-Term Properties of Polyolefins
Volume Editor: Albertsson, A.-C.
Vol. 169, 2004

Polymers and Light
Volume Editor: Lippert, T. K.
Vol. 168, 2004

New Synthetic Methods
Vol. 167, 2004

Polyelectrolytes with Defined Molecular Architecture II
Volume Editor: Schmidt, M.
Vol. 166, 2004

Polyelectrolytes with Defined Molecular Architecture I
Volume Editors: Schmidt, M.
Vol. 165, 2004

Filler-Reinforced Elastomers · Scanning Force Microscopy
Vol. 164, 2003

Liquid Chromatography · FTIR Microspectroscopy · Microwave Assisted Synthesis
Vol. 163, 2003

Radiation Effects on Polymers for Biological Use
Volume Editor: Kausch, H.
Vol. 162, 2003

Polymers for Photonics Applications II
Nonlinear Optical, Photorefractive and Two-Photon Absorption Polymers
Volume Editor: Lee, K.-S.
Vol. 161, 2003

Filled Elastomers · Drug Delivery Systems
Vol. 160, 2002

Statistical, Gradient, Block and Graft Copolymers by Controlled/ Living Radical Polymerizations
Authors: Davis, K. A., Matyjaszewski, K.
Vol. 159, 2002

Polymers for Photonics Applications I
Nonlinear Optical and Electroluminescence Polymers
Volume Editor: Lee, K.-S.
Vol. 158, 2002

Neutron Spin Echo in Polymer Systems

By Dieter Richter, Michael Monkenbusch,
Arantxa Arbe, Juan Colmenero

 Springer

Authors

Prof. Dr. Dieter Richter
d.richter@fz-juelich.de

Dr. Michael Monkenbusch
m.monkenbusch@fz-juelich.de

Institut für Festkörperforschung
Forschungszentrum Jülich
52425 Jülich
Germany

Dr. Arantxa Arbe
Unidad Física de Materiales
(CSIC-UPV/EHU)
Apartado 1072
20080 San Sebastián
Spain
waparmea@sq.ehu.es

Prof. Dr. Juan Colmenero
Unidad Física de Materiales
(CSIC-UPV/EHU)
Apartado 1072
20080 San Sebastián
Spain

and

Departamento de Física de Materiales
(UPV/EHU) and
Donostia International Physics Center
Apartado 1072
20080 San Sebastián
Spain
wapcolej@sc.ehu.es

The series presents critical reviews of the present and future trends in polymer and biopolymer science including chemistry, physical chemistry, physics and material science. It is addressed to all scientists at universities and in industry who wish to keep abreast of advances in the topics covered.

As a rule, contributions are specially commissioned. The editors and publishers will, however, always be pleased to receive suggestions and supplementary information. Papers are accepted for "Advances in Polymer Science" in English.

In references Advances in Polymer Science is abbreviated Adv Polym Sci and is cited as a journal.

The electronic content of APS may be found springerlink.com

Library of Congress Control Number: 2004095622

ISSN 0065-3195
ISBN 3-540-22862-4 **Springer Berlin Heidelberg New York**
DOI 10.1007/b99835

This work is subject to copyright. All rights are reserved, whether the whole or part of the material is concerned, specifically the rights of translation, re-printing, re-use of illustrations, recitation, broadcasting, reproduction on microfilms or in any other ways, and storage in data banks. Duplication of this publication or parts thereof is only permitted under the provisions of the German Copyright Law of September 9, 1965, in its current version, and permission for use must always be obtained from Springer-Verlag. Violations are liable to prosecution under the German Copyright Law.

Springer is a part of Springer Science+Business Media
springeronline.com
© Springer-Verlag Berlin Heidelberg 2005
Printed in The Netherlands

The use of registered names, trademarks, etc. in this publication does not imply, even in the absence of a specific statement, that such names are exempt from the relevant protective laws and regulations and therefore free for general use.

Cover design: KünkelLopka GmbH, Heidelberg/design & production GmbH, Heidelberg
Typesetting: Fotosatz-Service Köhler GmbH, Würzburg

Printed on acid-free paper 02/3141 xv – 5 4 3 2 1 0

Editorial Board

Prof. Akihiro Abe
Department of Industrial Chemistry
Tokyo Institute of Polytechnics
1583 Iiyama, Atsugi-shi 243-02, Japan
aabe@chem.t-kougei.ac.jp

Prof. A.-C. Albertsson
Department of Polymer Technology
The Royal Institute of Technology
S-10044 Stockholm, Sweden
aila@polymer.kth.se

Prof. Ruth Duncan
Welsh School of Pharmacy
Cardiff University
Redwood Building
King Edward VII Avenue
Cardiff CF 10 3XF
United Kingdom
duncan@cf.ac.uk

Prof. Karel Dušek
Institute of Macromolecular Chemistry,
Czech
Academy of Sciences of the
Czech Republic
Heyrovský Sq. 2
16206 Prague 6, Czech Republic
dusek@imc.cas.cz

Prof. Dr. W. H. de Jeu
FOM-Institute AMOLF
Kruislaan 407
1098 SJ Amsterdam, The Netherlands
dejeu@amolf.nl
and
Dutch Polymer Institute
Eindhoven University of Technology
PO Box 513
5600 MB Eindhoven, The Netherlands

Prof. Jean-François Joanny
Physicochimie Curie
Institut Curie section recherche
26 rue d'Ulm
F-75248 Paris cedex 05, France
jean-francois.joanny@curie.fr

Prof. Hans-Henning Kausch
c/o EPFL, Science de Base (SB-ISIC)
Station 6
CH-1015 Lausanne, Switzerland
kausch.cully@bluewin.ch

Prof. S. Kobayashi
Department of Materials Chemistry
Graduate School of Engineering
Kyoto University
Kyoto 615-8510, Japan
kobayasi@mat.polym.kyoto-u.ac.jp

Prof. Dr. Kwang-Sup Lee
Department of Polymer Science &
Engineering
Institute of Hybrid Materials
for Information & Biotechnology
Hannam University
133 Ojung-Dong
Daejeon 306-791, Korea
kslee@mail.hannam.ac.kr

Prof. L. Leibler
Matière Molle et Chimie
Ecole Supèrieure de Physique
et Chimie Industrielles (ESPCI)
10 rue Vauquelin
75231 Paris Cedex 05, France
ludwik.leibler@espci.fr

Prof. Timothy E. Long
Department of Chemistry
and Research Institute
Virginia Tech
2110 Hahn Hall (0344)
Blacksburg, VA 24061, USA
telong@vt.edu

Prof. Ian Manners
Department of Chemistry
University of Toronto
80 St. George St.
M5S 3H6 Ontario, Canada
imanners@chem.utoronto.ca

Prof. Dr. Martin Möller
Deutsches Wollforschungsinstitut
an der RWTH Aachen e.V.
Pauwelsstraße 8
52056 Aachen, Germany
moeller@dwi.rwth-aachen.de

Prof. Oskar Nuyken
Lehrstuhl für Makromolekulare Stoffe
TU München
Lichtenbergstr. 4
85747 Garching, Germany
oskar.nuyken@ch.tum.de

Dr. E. M. Terentjev
Cavendish Laboratory
Madingley Road
Cambridge CB 3 OHE
United Kingdom
emt1000@cam.ac.uk

Prof. Brigitte Voit
Institut für Polymerforschung Dresden
Hohe Straße 6
01069 Dresden, Germany
voit@ipfdd.de

Prof. Gerhard Wegner
Max-Planck-Institut
für Polymerforschung
Ackermannweg 10
Postfach 3148
55128 Mainz, Germany
wegner@mpip-mainz.mpg.de

Advances in Polymer Science
Also Available Electronically

For all customers who have a standing order to Advances in Polymer Science, we offer the electronic version via SpringerLink free of charge. Please contact your librarian who can receive a password for free access to the full articles by registering at:

springerlink.com

If you do not have a subscription, you can still view the tables of contents of the volumes and the abstract of each article by going to the SpringerLink Homepage, clicking on "Browse by Online Libraries", then "Chemical Sciences", and finally choose Advances in Polymer Science.

You will find information about the

– Editorial Board
– Aims and Scope
– Instructions for Authors
– Sample Contribution

at springeronline.com using the search function.

Contents

1	**Introduction**	2
2	**Neutron Scattering and Neutron Spin Echo**	9
2.1	Neutron Scattering Principles	9
2.1.1	Coherent and Incoherent Scattering	10
2.1.2	Coherent Scattering and Coarse Graining	11
2.1.3	Contrast Variation	11
2.2	The Neutron Spin-Echo Method	12
2.2.1	Limitations	15
2.2.2	Technical Principles	16
2.2.3	Zero Field Spin-Echo Technique	18
2.2.4	Accessible High Resolution NSE Spectrometers	21
2.2.5	Related Methods: Pulsed Field Gradient NMR and Dynamic Light Scattering	21
3	**Large Scale Dynamics of Homopolymers**	24
3.1	Entropy Driven Dynamics – the Rouse Model	25
3.1.1	Gaussian Chains	25
3.1.2	The Rouse Model	26
3.1.3	Neutron Spin Echo Results	31
3.1.4	Rouse Model: Rheological Measurements and NSE	35
3.1.5	Rouse Model: Computer Simulation and NSE	37
3.2	Reptation	41
3.2.1	Self-Correlation Function	45
3.2.2	Single-Chain Dynamic Structure Factor	45
3.2.3	Neutron Spin Echo Results on Chain Confinement	48
3.2.4	Reptation, NSE and Computer Simulation	55
3.2.5	Contour Length Fluctuations	61
4	**Local Dynamics and the Glass Transition**	67
4.1	α-Relaxation	72
4.1.1	Structural α-Relaxation: Dynamic Structure Factor ..	72
4.1.2	Self-Atomic Motions of Protons	82
4.2	β-Relaxation	96
4.2.1	Model for β-Relaxation	99

4.3	$\alpha\beta$-Merging	105
4.4	Mode Coupling Theory	112

5 Intermediate Length Scales Dynamics ... 117
5.1	Chain-Rigidity and Rotational Transitions as Rouse Limiting Processes	117
5.1.1	Mode Description of Chain Statistics	118
5.1.2	Effect of Bending Forces	120
5.1.3	Internal Viscosity Effects	121
5.1.4	Hydrodynamic Interactions and Internal Viscosity Effects	123
5.1.5	NSE Results on Chain Specific Effects Limiting the Rouse Dynamics	125
5.1.6	Internal Viscosity	128
5.1.7	Polymer Solutions	130
5.1.8	Intrachain Viscosity Analysis of the PIB-Data	133
5.2	Collective Motions	136
5.2.1	Application of Mode Coupling Theory	141
5.3	Self-Motion	142
5.4	Scenario for the Intermediate Length Scale Dynamics	147
5.4.1	Direct Comparison of the Correlators	147
5.4.2	Stressing the Internal Viscosity Approach	151

6 The Dynamics of More Complex Polymer Systems ... 153
6.1	Polymer Blends	153
6.1.1	Neutron Scattering and Polymer Blend Dynamics	155
6.1.2	Neutron Spin Echo Results in Polymer Blends	156
6.2	Diblock copolymers	162
6.2.1	Dynamic Random Phase Approximation (RPA)	163
6.2.2	Static Scattering Function	165
6.2.3	Dynamic Structure Factor	165
6.2.4	Neutron Scattering Results	167
6.2.5	Static Structure Factor	167
6.2.6	Collective Dynamics	170
6.2.7	Single Chain Dynamics	173
6.3	Gels	181
6.4	Micelles, Stars and Dendrimers	184
6.4.1	Micelles	184
6.4.2	Stars	185
6.4.3	Dendrimers	186
6.5	Rubbery Electrolytes	188
6.6	Polymer Solutions	193
6.6.1	Semidilute Solutions	195
6.6.2	Theta Solution/Two Length Scales	197
6.6.3	Multicomponent Solutions	199
6.7	Biological Macromolecules	200

7	**Conclusion and Outlook**	207
8	**References**	209

Abbreviations and Symbols 219

Author Index Volumes 101–174 223

Subject Index 243

Neutron Spin Echo in Polymer Systems

D. Richter[1] · M. Monkenbusch[1] · A. Arbe[2] · J. Colmenero[2,3]

[1] Institut für Festkörperforschung, Forschungszentrum Jülich, D-52425 Jülich, Germany
d.richter@fz-juelich.de, m.monkenbusch@fz-juelich.de
[2] Unidad Física de Materiales (CSIC-UPV/EHU), Apartado 1072, 20080 San Sebastián, Spain
waparmea@sq.ehu.es, wapcolej@sq.ehu.es
[3] Departmento de Física de Materiales (UPV/EHU) and Donostia International Physics Center, Apartado 1072, 20080 San Sebastián, Spain

Abstract Neutron spin echo spectroscopy (NSE) provides the unique opportunity to unravel the molecular dynamics of polymer chains in space and time, covering most of the relevant length and time scales. This article reviews in a comprehensive form recent advances in the application of NSE to problems in polymer physics and describes in terms of examples expected future trends. The review commences with a description of NSE covering both the generic longitudinal field set-up as well as the resonance technique. Then, NSE results for homopolymers chains are presented, covering all length scales from the very local secondary β-relaxation to large scale reptation. This overview is the core of the review. Thereafter the dynamics of more complex systems is addressed. Starting from polymer blends, diblock copolymers, gels, micelles, stars and dendrimers, rubbery electrolytes and biological macromolecules are discussed. Wherever possible the review relates the NSE findings to the results of other techniques, in particular emphasizing computer simulations.

Keywords Neutron spin echo spectroscopy · Polymer dynamics · Reptation · Glassy relaxation

1
Introduction

Among the experimental techniques for studying the structure and dynamics of polymers, neutron scattering plays a unique role for several reasons:

i. The suitability of the length and time scales. These are accessed in particular by small angle neutron scattering (SANS) and neutron spin echo (NSE) and allow the exploration of large scale properties – for instance the conformation of a large macromolecule, its diffusion in the embedding medium and its entropy driven dynamics – as well as features characteristic for more local scales, e.g. the inter- and intrachain correlations in a glass-forming polymer and their time evolution, the rotational motion of methyl groups, the vibrations and so on.
ii. By variation of the contrast between the structural units or molecular groups, complex systems may be selectively studied. In particular, the large contrast achieved by isotopic substitution of hydrogen – one of the main components of polymers – by deuterium constitutes the most powerful tool for deciphering complex structures and dynamic processes in these materials.
iii. Neutron reflectometry constitutes a unique technique for the investigation of surfaces and interfaces in polymeric systems.
iv. The high penetration of neutrons in matter allows the study of the influence of external fields, e.g. shear or pressure or the evolution of the system under processing conditions.
v. The space-time resolution of these techniques reveals the molecular motions leading to the viscoelastic and mechanical properties of polymeric systems. This knowledge is of great importance for scientific reasons and is also a basis for the design of tailor-made materials.

The unique power of neutron scattering for revealing essential features in the field of polymer science can be exemplified by two pioneering experiments that can already be considered as "classic". The first is the experimental proof of the random coil conformation of polymer chains in the melt or in the glassy state, as proposed in the 1950s by Flory [1]. Its confirmation was only possible in the 1970s [2] with the development of SANS. Since in the bulk a given macromolecule is surrounded by identical units, Flory's proposition could only be demonstrated by using contrast variation and deuterating single molecules. This measurement of a single chain form factor by SANS was one of the first applications of neutron scattering to polymer science. Its dynamic counterpart could only be realized 25 years later. Neutron spin echo investigations on the long time chain dynamics recently allowed the confirmation of de Gennes' predictions [3] on the mechanism of tube-like confinement and reptation in polymer melts and dense systems [4].

In this review we will concentrate on the dynamic aspects of the polymer ensemble and describe as comprehensively as possible what has been achieved

1 Introduction

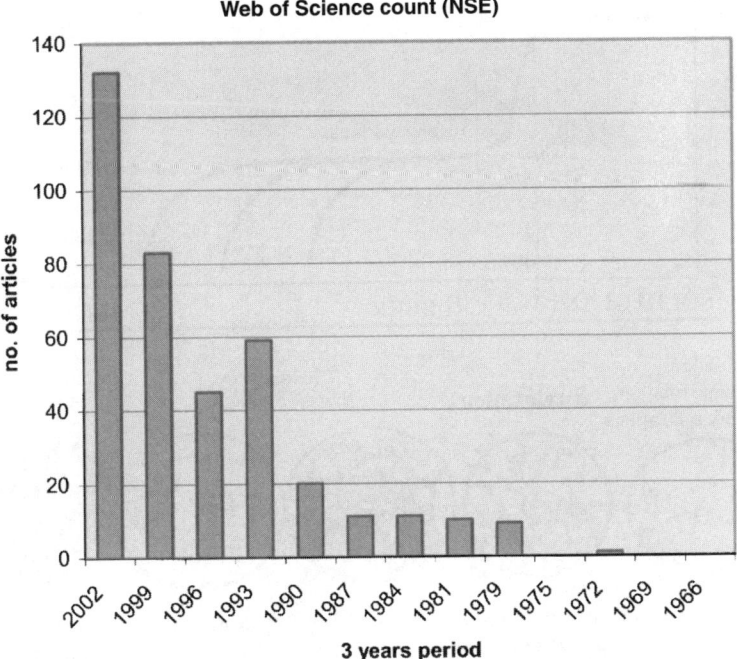

Fig. 1.1 Development of the number of NSE-related articles in 3-year periods from now to the invention of the method

so far by neutron spin echo spectroscopy. The last major review in this field appeared in 1997 [5]. Since then, a strong growth of publications on NSE results may be observed (Fig. 1.1). This figure displays the total number of publications on neutron spin echo results in three years intervals. It is evident that during recent years the publication rate on NSE results has increased dramatically. This increase is due to two reasons. First, apparently more and more scientists are discovering the power of NSE for facilitating the observation of slow dynamics in condensed matter, and, second, the number of NSE instruments available for public use has increased significantly over the last 10 years. Table 2.1 in Chap. 2 gives an overview on the NSE spectrometers that are available today for neutron users at user facilities worldwide. The table includes e-mail addresses to provide the reader access to information on these instruments.

Dynamic processes in polymers occur over a wide range of length and time scales(see Fig. 1.2 and Fig. 1.3). Figure 1.2 relates the dynamic modulus as it may be observed on a polymer melt with the length and time scales of molecular motion underlying the rheological behaviour. Our example deals with an amorphous polymer system excluding any crystallization processes. A typical relaxation map for this kind of systems is that displayed in Fig. 1.3 for the archetypal polymer polyisoprene. It is clear that we can distinguish several different regimes:

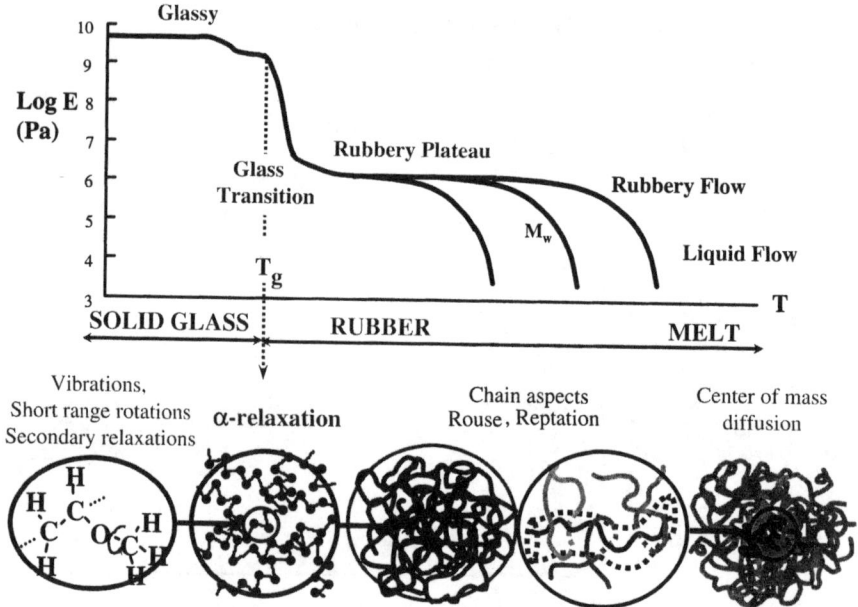

Fig. 1.2 Richness of dynamic modulus in a bulk polymer and its molecular origin. The associated length scales vary from the typical bond length (≈Å) at low temperatures to interchain distances (≈10 Å) around the glass transition. In the plateau regime of the modulus typical scales involve distances between "entanglements" of the order of 50–100 Å. In the flow regime the relevant length scale is determined by the proper chain dimensions

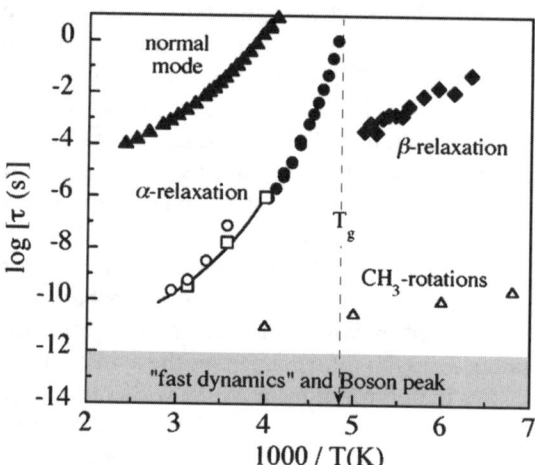

Fig. 1.3 Relaxation map of polyisoprene: results from dielectric spectroscopy (inverse of maximum loss frequency *full symbols*), rheological shift factors (*solid line*) [7], and neutron scattering: pair correlation ($\langle \tau(Q=1.44\ \text{Å}^{-1})\rangle$ *empty square*) [8] and self correlation ($\langle \tau(Q=0.88\ \text{Å}^{-1})\rangle$ *empty circle*) [9], methyl group rotation (*empty triangle*) [10]. The *shadowed area* indicates the time scales corresponding to the so-called fast dynamics [11]

i. At low temperature the material is in the glassy state and only small amplitude motions like vibrations, short range rotations or secondary relaxations are possible. Below the glass transition temperature T_g the secondary β-relaxation as observed by dielectric spectroscopy and the methyl group rotations may be observed. In addition, at high frequencies the vibrational dynamics, in particular the so called Boson peak, characterizes the dynamic behaviour of amorphous polyisoprene. The secondary relaxations cause the first small step in the dynamic modulus of such a polymer system.

ii. At the glass transition temperature T_g the primary relaxation (α-relaxation) becomes active allowing the system to flow. The temperature dependence of its characteristic relaxation time is displayed in Fig. 1.3 combining dielectric, rheological and neutron scattering experiments. The time range over which this relaxation takes place easily covers more than ten orders of magnitude. This implies the necessity to combine different experimental techniques to fully characterize this process. As shall be demonstrated in this review, the length scale associated with α-relaxation is the typical interchain distance between two polymer chains. In the dynamic modulus, the α-relaxation causes a significant step of typically three orders of magnitude in strength.

iii. The following rubbery plateau in the modulus relates to large scale motions within a polymer chain. Two aspects stand out. The first is the entropy-driven relaxation of fluctuations out of equilibrium. Secondly, these relaxations are limited by confinement effects caused by the mutually interpenetrating chains. This confinement is modelled most successfully in terms of the reptation model by de Gennes [3] and Doi and Edwards [6]. There, the confinement effects are described in terms of a tube following the coarse grained chain profile. Motion is only allowed along the tube profile leading to the reptation process – the snake-like motion of a polymer chain.

iv. When a chain has lost the memory of its initial state, rubbery flow sets in. The associated characteristic relaxation time is displayed in Fig. 1.3 in terms of the "normal mode" (polyisoprene displays an electric dipole moment in the direction of the chain) and thus dielectric spectroscopy is able to measure the relaxation of the end-to-end vector of a given chain. The rubbery flow passes over to liquid flow, which is characterized by the translational diffusion coefficient of the chain. Depending on the molecular weight, the characteristic length scales from the motion of a single bond to the overall chain diffusion may cover about three orders of magnitude, while the associated time scales easily may be stretched over ten or more orders.

In this review we will present the outcome of NSE studies on polymer systems covering results beyond those reported in an earlier review in *Advances in Polymer Science* [5] eight years ago. Table 1.1 shows the chemical structure and information on the chain dimensions of the systems considered here. In Chap. 2 we will commence with a brief description of neutron scattering principles and a discussion of the two different ways neutron spin echo may be implemented – the traditional NSE approach with precession coils and the neu-

Table 1.1 Names, acronyms and structure of the repeat unit of the polymers that appear in this work. The ratio between the average end-to-end distance $\langle R^2 \rangle_0$ and the molecular weight M at 413 K is also shown [12]

Common name	Acronym	Structure of the repeat unit	$\langle R^2 \rangle_0/M$ (Å^{2}mol/g)
Polyisoprene	PI	$[-CH_2-CH=C(CH_3)-CH_2-]_n$	0.625
Polydimethyl siloxane	PDMS	$[-Si(CH_3)_2-O-]_n$	0.457
Polyethylene	PE	$[-CH_2-CH_2-]_n$	1.21[a]
Poly(ethyl ethylene)	PEE	$[-CH-CH_2-]_n$ with CH_2-CH_3 branch	0.507
Poly(ethylene propylene)[b]	PEP	$[-CH_2-CH_2-CH(CH_3)-CH_2-]_n$	0.834
1,4-Polybutadiene	PB	$[-CH_2-CH=CH-CH_2-]_n$	0.876
Polyisobutylene	PIB	$[-CH_2-C(CH_3)_2-]_n$	0.570
Atactic polypropylene	aPP	$[-CH(CH_3)-CH_2-]_n$	0.670
Polyurethane	PU	$HO[-CH-CH_2-O]_n-\overset{CH_2-CH_3}{\underset{[O-CH_2-CH-]_nOH}{C}}-[O-CH_2-CH-]_nOH$	
Poly(vinyl chloride)	PVC	$[-CH_2-CClH-]_n$	
Poly(vinyl ethylene)	PVE	$[-CH-CH_2-]_n$ with $CH=CH_2$ branch	0.664[c]
Poly(vinyl methyl ether)	PVME	$[-CH_2-CH-]_n$ with $O-CH_3$ branch	
Polyethyl-methyl-siloxane	PEMS	$[-Si(CH_3)-O-]_n$ with CH_2-CH_3 branch	
Poly(ethylene oxide)	PEO	$[-CH_2-CH_2-O-]_n$	0.805

Table 1.1 (continued)

Common name	Acronym	Structure of the repeat unit	$\langle R^2 \rangle_0/M$ (Å^{2}mol/g)
Poly-(propylene oxide)	PPO	$[-C(CH_3)-CH_2-O-]_n$	0.741^d
Poly-(N-isopropyl-acryl-amide)	PNIPA	$[-CH_2-CH-]_n$ $\quad\quad\quad\mid$ $\quad\quad CO-NH-CH(CH_3)_2$	
Poly-fluoro-silicone)	PFS	$[-Si(CH_3)_2-O-]_n$ $\quad\quad\mid$ $CH_2-CH_2-CF_3$	
Poly-styrene	PS	$[-CH-CH_2-]_n$ $\quad\mid$ Ph	
Polyamido-amine	PAMAM	$[-CH_2-CH_2-CO-NH-CH_2-CH_2-N]_n$	

^{a} PE is obtained by anionic polymerization from hydrogenating a precursor 1,4-PB polymer. Since 1,4-PB contains a small fraction of 1,2-units the PE studied here has two ethylene side branches per 100 backbone carbons.
^{b} alt-PEP: Essentially alternating poly(ethylene co-1-butene).
^{c} T=298 K.
^{d} T=493 K.

tron spin echo resonance technique. Then, in Chap. 3 we will turn to the standard model of polymer dynamics, the Rouse model. This model is described in some detail in order to display the principles of the stochastic dynamics and the way in which a dynamic structure factor may be calculated from a model for polymer motion. We will present NSE results both for the self- as well as for the pair correlation function and compare these space- and time-dependent neutron data with computer simulation results. Thereafter we will address confinement effects and discuss reptation and other competing models, compare them with neutron spin echo results, discuss the limiting mechanism for reptation and, finally, look at computer simulation results for comparison.

Chapter 4 deals with the local dynamics of polymer melts and the glass transition. NSE results on the self- and the pair correlation function relating to the primary and secondary relaxation will be discussed. We will show that the macroscopic flow manifests itself on the nearest neighbour scale and relate the secondary relaxations to intrachain dynamics. The question of the spatial heterogeneity of the α-process will be another important issue. NSE observations demonstrate a sublinear diffusion regime underlying the atomic motions during the structural α-relaxation.

Chapter 5 considers the connection between the universal large scale dynamics discussed first and the local specific dynamics discussed in the second step. The dynamics at intermediate length scales bridges the two and we will address the leading mechanism limiting the universal dynamics in flexible polymers.

Most of the experiments reported so far have been performed on linear homopolymer systems. In Chap. 6 we discuss what has been achieved so far beyond such simple materials. We begin with the discussion of neutron spin echo data on miscible polymer blends, where the main issue is the "dynamic miscibility". There are two questions: Firstly, on what length and time scales and to what extent does a heterogeneous material like a blend exhibit homogeneous dynamics? Secondly, how does it relate to the corresponding homopolymer properties?

Then we address the dynamics of diblock copolymer melts. There we discuss the single chain dynamics, the collective dynamics as well as the dynamics of the interfaces in microphase separated systems. The next degree of complication is reached when we discuss the dynamic of gels (Chap. 6.3) and that of polymer aggregates like micelles or polymers with complex architecture such as stars and dendrimers. Chapter 6.5 addresses the first measurements on a rubbery electrolyte. Some new results on polymer solutions are discussed in Chap. 6.6 with particular emphasis on theta solvents and hydrodynamic screening. Chapter 6.7 finally addresses experiments that have been performed on biological macromolecules.

Finally, in the conclusion we will give some outlook on future developments and opportunities.

2
Neutron Scattering and Neutron Spin Echo

2.1
Neutron Scattering Principles

Thermal and cold neutrons have de Broglie wavelengths from $\lambda=0.1-2$ nm corresponding to velocities of $v=4000$ m/s down to 200 m/s. The wavelength range covers that of X-ray and synchrotron radiation diffraction instruments. However, in contrast to electromagnetic radiation the neutron velocity has the same order of magnitude as the atomic velocities in the sample, the kinetic energy of the neutrons – 82 meV down to 0.2 meV in contrast to the typical X-ray energies of several KeV – compares with the excitation energies of atomic or molecular motions. Therefore, even the slow relaxational motions in soft condensed matter are detectable by a velocity change of the neutron. The spatial character of the motion can be inferred from the angular distribution of the scattered neutrons.

In general, scattering of thermal neutrons yields information on the sample by measurement and analysis of the double differential cross section:

$$\frac{d\sigma(\theta)}{d\Omega dE} = \frac{k_f}{k_i} \frac{1}{N} \sum_{i,j} \langle b_i b_j \rangle S_{i,j}(Q,\omega) \tag{2.1}$$

i.e. the intensity of scattered neutrons with energy E_f into a given direction θ. The energy transfer, i.e. the difference of kinetic energy before and after the scattering, $\Delta E = E_f - E_i$ relates to $\hbar\omega = \Delta E$.

The momentum transfer $\hbar Q$, respectively the wave vector, is given by $Q = \underline{k}_i - \underline{k}_f$ where $\underline{k}_i$ and $\underline{k}_f$ are the wave vectors of the incoming and outgoing (scattered) neutrons. They relate to the neutron wavelength $k_{i,f} = 2\pi/\lambda_{i,f}$. The neutron momenta are $\underline{p}_{i,f} = m_n \underline{v}_{i,f} = \hbar \underline{k}_{if}$. Therefore:

$$\omega = \frac{\Delta E}{\hbar} = \frac{(p_f^2 - p_i^2)}{(2\hbar m_n)} \tag{2.2}$$

The energy transfer ΔE and ω can be determined by measurement of the neutron velocities v_i and v_f. Note that for all problems discussed in this article $|\underline{k}_i| \sim |\underline{k}_f|$ and therefore:

$$Q = \frac{4\pi}{\lambda_i} \sin\left(\frac{\theta}{2}\right) = 2|k_i| \sin\left(\frac{\theta}{2}\right) \tag{2.3}$$

can be assumed. Finally b_i denotes the scattering length of atom nucleus "i" and $\langle\ldots\rangle$ is the ensemble average.

The unique features of neutrons that render them into the powerful tool for the investigation of "soft matter" are:

1. The isotope and spin dependence of b_i
2. Typical wavelengths of cold and thermal neutrons that match molecular and atomic distances
3. Even slow motions of molecules cause neutron velocity changes that are large enough to be detectable, in particular NSE is able to resolve changes Δv of the order of $10^{-5}\, v_i$.

To proceed further we introduce the intermediate scattering function as the Fourier transform of $S(Q,\omega)$:

$$S(Q,t) = \int_{-\infty}^{\infty} S(Q,\omega)\, e^{i\omega t}\, d\omega \qquad (2.4)$$

The intermediate scattering function directly depends on the (time-dependent) atomic positions:

$$S_{ij}(Q,t) = \left\langle \sum_{n,m} e^{iQ\cdot[\underline{R}_n^i(t) - \underline{R}_m^j(0)]} \right\rangle \qquad (2.5)$$

Note that in general the position of an atom "n" of type "i" $\underline{R}_n^i(t)$ is a quantum mechanical operator rather than a simple time-dependent coordinate and $S(Q,\omega)=X(Q,\omega)S(Q,-\omega)$, with $X(Q,\omega)$ taking account for detailed balance and e.g. recoil effects (see e.g. [13]).

Again for soft matter problems in the (Q,ω)-range discussed here $\hbar\omega \ll k_BT$, $T \approx 250$–500 K and $E_i \ll E_{bond}$, conditions for which $\underline{R}_n^i(t)$ may safely treated as classical coordinate and $S(Q,\omega) \approx S(Q,-\omega)$.

2.1.1
Coherent and Incoherent Scattering

Considering the ensemble average of Eq. 2.1 we have to observe the fact that chemically equivalent atoms may have a number of different scattering lengths that are randomly distributed over the ensemble of all atoms of the same kind in the sample. Most important in the present context is the variation due to the spin-dependent component of the proton scattering length. Whereas the average value $\langle b_i \rangle$ leads to coherent scattering, the fluctuating part $b_i - \langle b_i \rangle$ leads to incoherent scattering that yields an additional contribution of the atom–atom self-correlation, i.e.

$$S_i^{self}(Q,t) = \left\langle e^{iQ\cdot(\underline{R}_n^i(t) - \underline{R}_n^j(0))} \right\rangle \qquad (2.6)$$

Applying the Gaussian approximation, i.e. assuming that the atomic displacement distribution function are Gaussian Eq. 2.6 transforms into:

$$S^{self}(Q,t) = \exp\left(-\frac{Q^2}{6}\langle R_i^2(t)\rangle\right) \qquad (2.7)$$

where $\langle R_i^2(t)\rangle$ is the mean-square displacement of atom "i".

The spin-incoherent scattering (prominent for protons) involves a change of the spin-state of the scattered neutrons (spin-flip scattering) with a probability of $2/3$.

2.1.2
Coherent Scattering and Coarse Graining

Many polymer problems – including those discussed in this article – depend on structure and dynamics in a mesoscopic regime. Here a description in terms of individual atom coordinates $\underline{R}$ is not adequate. Rather a coarse grained description in terms of scattering length density $\Delta\rho(\underline{r},t)$ is used. To do so, a molecular unit of type "j" (e.g. a polymer segment, monomer or a whole smaller molecule) is selected and the sum of the scattering lengths of the contained atoms is related to the effective volume V of this unit, $\rho_j = \Sigma_{i\in j} b_i / V$. The same has to be done with the embedding matrix, e.g. the solvent. The scattering in the low Q-regime only depends on the scattering length density difference, the contrast $\Delta\rho(\underline{r},t) = \rho_{polymer} - \rho_{matrix}$. The proper effective volume corresponds to the volume increase of the system when one molecular unit is added to it. To yield a valid description of the scattering the extension of the molecular unit L should be smaller than $L < 1/Q_{max}$. The related scattering function then is:

$$S(Q,t) = \int \langle \Delta\rho(\underline{r},t) \cdot \Delta\rho(\underline{r}',0) \rangle e^{iQ\cdot(r-r')} d^3\underline{r}. \tag{2.8}$$

The corresponding small angle neutron scattering (SANS) intensity is proportional to $S(Q,t=0)$.

2.1.3
Contrast Variation

The above description implies that contrast variation and matching can be employed to enhance or suppress the contribution of a relaxation signal from selected subunits of a system. Only the motion of those structures that contribute to the SANS intensity are seen in the corresponding NSE experiment. However, for NSE a few restrictions are to be observed:

1. The majority of the sample should be deuterated to optimize the intensity to background ratio, in particular at $Q > 1-2$ nm^{-1} where for typical soft matter samples the coherent intensity drops below the incoherent contribution
2. If solvents consisting of larger organic molecules such as e.g. dodecane are used, h/d-mixing (h, d refer to protonated and deuterated systems, respectively) leads to a sizeable additional scattering due to isotope incoherence. If d- and h-solvents are mixed the scattering length within the individual molecules – containing either only H or D atoms – add and the isotopic mixture effectively consists of these units. As a consequence, the inelastic signal

will contain the solvent diffusion which may extend into the time range of a few ns.

Prominent examples of successful application of contrast variation are the investigations of the single chain dynamics of polymers in melts. Here a mixture of about 10% h-polymer in a matrix of d-polymer is used. Further details are obtained by investigating d-polymers that contain only a h-labelled section, i.e. at the ends, at branching points or at its centre in a fully deuterated matrix.

2.2
The Neutron Spin-Echo Method

Relaxation motions at mesoscopic scale as encountered in polymer samples yield energy transfers of μeV and less which cannot be analysed at any reasonable intensity level by direct filtering. The neutron spin-echo (NSE) method fills the gap between dynamic light scattering and "conventional" neutron spectroscopy and aims at much slower processes with typical times up to several 100 ns[1] as they occur in polymer samples on mesoscopic length scales. A resolution necessary to observe processes at time scales of 100 ns and above requires the detection of neutron velocity changes of 1 in 10^4 to 1 in 10^5. Preparation of a suitable incoming beam with $\Delta v/v < 10^{-5}$ would only be possible by removing all neutrons with unwanted directions and velocities leaving only a few 10^3 neutrons/cm²s at the sample, from which only a tiny fraction will be scattered to a final momentum passing into the detector.

The essence of the NSE techniques consists of a method to decouple the detectability of tiny velocity changes caused by the scattering process from the width of the incoming velocity distribution. NSE instruments at reactors routinely run with 10–20% width of the velocity distribution. This yields about 10^4 times more neutrons in the primary beam than a direct selection of neutrons with velocities within a band of 10^{-5} relative width. Basically the NSE trick consists of replacing the velocity filter transmitting only one extremely narrow band by a filter with a cosine modulated transmission. A velocity increment Δv_c of 10^{-5} between adjacent maxima and minima of such a filter enables the detection of a 10^{-5} velocity change, whereas the number of neutrons in a velocity band that is several 10^3–10^4 times wider than the detected velocity increment ensures high intensity. The complete information on the distribution of velocity changes during scattering – here as its Fourier transform – is obtained by scanning a parameter that controls the period of the cosine filter. Any reference to the neutron spin and its motion relates only to the above described velocity filter. The spin quoted in the name of the method only refers

---

[1] Such long times usually appear when the observed objects are large compared to single atoms, but are still mesoscopic, e.g. polymer molecules or aggregates of smaller molecules, or when dynamic processes at all length scales slow down in the vicinity of phase transitions or due to a glass transition.

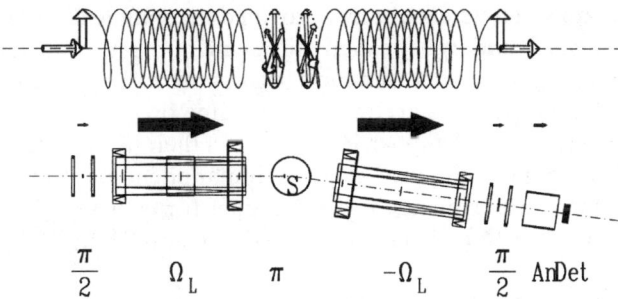

Fig. 2.1 Spin history leading to the formation of the spin-echo. Longitudinally polarized neutrons enter from the left. *Upper part* spin motion. *Lower part* NSE setup, π/2-flipper between belonging current rings, primary main precession solenoid ω_L with symmetry scan windings in the middle and stray-field compensating loops (as used in the FZ-Jülich design) at both ends, π-flipper near the sample S, symmetric arrangement on the secondary side followed by analyser An and detector Det. The *arrows* indicate the strength of the longitudinal field. (Reprinted with permission from [18]. Copyright 1997 Elsevier)

to an ingenious monochromatization method and not at all to the phenomena to be investigated by the NSE instrument. However, the neutron spin is the key element in realizing the filter function. The basic NSE instrument invented by F. Mezei [14] works as follows (see Fig. 2.1).

A beam of longitudinally polarized neutrons, i.e. neutrons with spins pointing into the beam direction, enters the instrument and traverses a π/2-flipper located in a low longitudinal magnetic field. During the passage of this flipper the neutron spins are rotated by 90°≡π/2 and are then perpendicular to the beam, e.g. the spins are all pointing upwards. Immediately after leaving the flipper they start to precess around the longitudinal field generated by the primary precession solenoid. As they proceed into the precession coil the Larmor frequency ω_L, which is proportional to the field, increases up to several MHz (i.e. about one turn per 0.1 mm length of path) in the middle of the solenoid. The field and ω_L decrease to low values again on the way to the sample. Arriving at the sample the neutron may have performed a total of up to several ten to hundred thousand precessions, however, different neutrons with different velocities from the incoming 10–20% distribution have total precession angles that differ proportionally. For that reason the ensemble of neutron spins at the sample contains any spin direction on a disc perpendicular to the longitudinal field with virtual equal probability. Nevertheless each single neutron has a tag of its velocity by the individual precession angle (modulo 2π) of its spin – which may be viewed as the neutron's own stopwatch. Now at (i.e. near) the sample position there is a π-flipper turning the spins during neutron passage by 180°≡π around an e.g. upward pointing axis. Thereby transforming a spin with precession angle α mod 2π to $-\alpha$ mod 2π=$-\alpha$ mod (-2π), effectively reversing the sense of the "spin-clocks". Then the neutrons enter the secondary part of the spectrometer, which is symmetric to the primary part. During the passage of the second main solenoid each spin – provided the sample did not change the neutron's velocity – undergoes exactly the same number of precessions as in the primary part. Due to the reversing action of the π-flipper this leads to the result that all neutron spins arrive with the same precession angle, pointing upwards, at the second π/2-flipper, irrespective of their individual velocity. This effect is called the spin-echo in analogy to similar phenomena in conventional nuclear resonance experiments. The π/2-flipper turns these spins by 90° into the longitudinal direction, thereby freezing the acquired longitudinal spin polarization.

The analyser transmits only neutrons with spin components parallel (or antiparallel) to the axis, thus the spin projection to the axis determines the probability that the neutrons reach the detector. If the neutrons now undergo a velocity change Δv_s in the course of the scattering by the sample, the final spin direction is no longer upwards. Spins pointing downwards are rotated to the axially anti-parallel direction by the π/2-flipper, and are therefore blocked by the analyser and do not reach the detector, i.e. the echo signal is reduced. The cosine of the final precession angle determines the analyser transmission, which finally leads a cosine modulating filter. The filter period Δv_c is controlled by the magnetic field inside the main precession solenoids. The influence of the sample on the neutrons is described in terms of the scattering function $S(Q,\omega)$. The frequency ω is proportional to the energy transfer between neutron and sample

$$\frac{\omega}{2\pi} = \frac{m_n}{2h}[v^2 - (v+\Delta v_s)^2] \tag{2.9}$$

For the very small Δv_s values under consideration the above expression may be linearized, $\omega \sim \Delta v_s$. Due to the cosine modulating filter function the NSE instrument measures the cosine transform of $S(Q,\omega)$. The detector output of the (ideal) NSE instrument at exact symmetry is:

$$I_{Det} \propto \frac{1}{2}\left[S(Q) \pm \int \cos\left(J\lambda^3\gamma\frac{m_n^2}{2\pi h^2}\omega\right) S(Q,\omega)d\omega\right] \tag{2.10}$$

where $J=\int_{path}|\underline{B}|dl$ is the integral of the magnetic induction along the flight path of the neutron from π/2-flipper to the sample, and $\gamma=1.83033\times10^8$ radian/sT. The sign of the integral depends on the type of analyser and on a choice of the sign of the flipping angle of the secondary π/2-flipper. The time parameter is $t=J\lambda^3\gamma\frac{m_n^2}{2\pi h^2}$. It may be easily scanned by varying the main solenoid current to which J is (approximately) proportional. Note that the maximum achievable time t, e.g. the resolution, depends linearly on the (maximum) field integral and on the cube of the neutron wavelength λ^3! The fact that the instrument signal represents the intermediate scattering function $S(Q,t)$ directly and not $S(Q,\omega)$ makes it especially useful for relaxation type scattering because:

1. The relaxation function is measured directly as a function of time. Therefore, any instrumental resolution correction consists of a simple point by point division by the result of a measurement of a resolution sample, instead of a tedious deconvolution that would be required for $S(Q,\omega)$ measured at a real – finite resolution – instrument.
2. The large dynamic range – for a neutron instrument – of up to 1:1000 with one instrumental setting plus another two orders of magnitude if the wavelength is changed.

2.2.1
Limitations

In contrast to direct spectral filtering where only those neutrons with a well-defined energy $E_f=E_i+\hbar\omega\pm\delta E$ reach the detector, the Fourier transformation property of the measuring principle implies that at any spectrometer setting virtually 50% of the neutrons that are scattered into the detector direction are counted. As a consequence weak spectral features are buried under the counting noise caused by the ω-integrated neutron flux. Reasonable performance is expected only for relaxation type scattering contributing at least several percent of the total scattering in the considered direction (Q-value).

The NSE principle – as described above – only works if the neutron spin is not affected by the scattering process (some exceptions like complete spin-flip could be tolerated and would just replace the π-flipper). A related problem occurs if the scattering nuclei have a non-zero spin and their scattering power depends significantly on the relative orientation of nuclear and neutron spin.

A prominent example is the so-called incoherent scattering from hydrogen, which has a very large cross section and is often used to measure the proton one-particle self-correlation function using other types of neutron spectroscopy. This type of scattering consists of $2/3$ spin-flip and $1/3$ non-flip processes. The average signal left is therefore only $1/3$ with reversed polarization on top of $2/3$ of the intensity being converted to background, which renders this type of investigations difficult and time consuming, but not impossible. A virtue of the spin-flip scattering is that multiple scattering is efficiently suppressed since two subsequent scattering events only leave $1/9$ of the signal and more events further reduce the contribution to the echo amplitude, i.e. multiple scattering has a strong tendency to become depolarized and thereby only increases the background which is – except for the size of statistical errors – neutral to the extraction of the desired scattering functions. The overall intensity is much smaller (10–20 times) than the coherent small angle scattering (SANS) from chain-labelled samples. Due to the low intensity, background scattering from the sample container and spectrometer components contribute significantly to the spin-echo signal. The latter effect is especially severe at low Q, i.e. low scattering angles require very careful studies of the experimental background.

2.2.2
Technical Principles

The NSE instruments are operated at a neutron beam emerging from a cold source (liquid H_2 or D_2 or other hydrogenous materials held at temperatures in the 20 K region) located near the core of a research reactor. The neutrons propagate through a neutron guide (an evacuated channel with walls that totally reflect neutrons that hit them at a low angle) over a distance of several 10 m to the instrument. Polarization is performed by reflection at (transmission through) magnetic multilayer mirrors [16]. The preparation of the 10–20% velocity band is done by a mechanical velocity selector (helically bent channels on the circumference of a rotating cylinder give free passage for neutrons with velocities equal to the speed of the open channel position along the rotation axis). These elements are located somewhere in the neutron guide track. As soon as the neutron beam is polarized it needs a guide field 1–2 orders of magnitude higher than any stray fields to prevent the latter from turning the aligned spins in an undefined manner.[2]

All the spin manipulations are done by a tailored magnetic field that the neutron spin feels during its passage through the instrument. The motion of the expectation value of the neutron spin $\langle \underline{s} \rangle$ obeys the Bloch equation:

$$\frac{d\underline{s}}{dt} = \underline{s} \times \gamma \underline{B}\{\underline{x}(t)\} \tag{2.11}$$

$\underline{x}(t)$ being the position of the neutron. At constant $\underline{B}$ this equation implies the precession of $\langle \underline{s} \rangle$ around an axis parallel $\underline{B}_0$ with a frequency $\omega_L = \gamma |\underline{B}|$ at fixed angle between $\langle \underline{s} \rangle$ and $\underline{B}$, e.g. $\langle \underline{s} \rangle$ describes a cone around $\underline{B}$. The field felt by the neutron spin depends implicitly on time due to the position-dependence of $\underline{B}$ along the neutron path, $\underline{B}(t) = \underline{B}\{\underline{x}(t)\}$. If $\omega_L \gg$ *rotation speed of the direction of* $\underline{B}(t)$, the angle of the precession cone, *angle* $(\underline{s}, \underline{B})$, remains virtually unchanged. The design and operation of an NSE instrument has to ensure this so-called adiabatic condition everywhere outside the flippers.

Flipping is performed by exposing the neutron spin to an abrupt (non-adiabatic) change of the direction of $\underline{B}$. Technically this is realized by a pair of current sheets, being the broad sides of a one layer coil of a neutron transparent wire (Al) wound around a rectangular cross section. They are oriented such that the neutron beam points perpendicular to the broad sides and enters and leaves the flipper through them, see Fig. 2.2. The spin keeps its orientation during the infinitesimal short passage time through the current sheet, then starts precession around the "new" field inside the flipper. There it performs a certain

---

[2] A true zero field region would also work, however, is much more difficult to realize. To avoid significant spin rotation an upper field limit of about 10^{-8} T must be observed and much attention must be paid to the transition from field- to zero-field-regions. The zero-field or resonance NSE instruments work with corresponding flight paths [17].

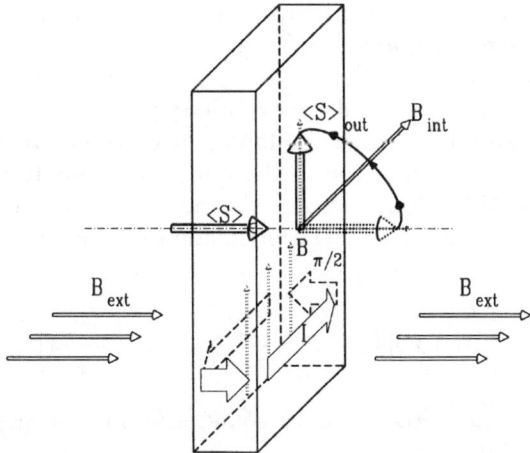

Fig. 2.2 A π/2-flipper effecively rotating the spin direction from longitudinal $\langle S \rangle$ $\langle S_{out} \rangle$ to perpendicular to the path and to the external field. The vertical field $B_{\pi/2}$ generated inside the flipper adds to the embedding external field B_{ext}

rotation angle until it leaves the flipper through the second sheet. The executed rotation is the flipping action, of which the magnitude depends on the flipper thickness, the neutron velocity and the value of the inner field.

The key elements of the instrument are the two main precession solenoids providing a high field along the neutron path. They represent rotation devices for spins perpendicular to the field with a total precession angle inversely proportional to the individual neutron velocity. The primary and secondary parts have to be perfectly symmetric, which is ensured by identical construction and by electrical operation of the coils in series. The maximum Fourier time depends on the path integral of the field depending on the length as well as on the average inner field of the solenoid.

The spin-echo signal is lost if neutrons taking different paths through the solenoids show precession angle dephasing of more than 180°. To prevent this, $\int |\underline{B}|\, dl$ along all paths within the neutron beam of finite width and divergence have to be equal within about 1:(10×*maximum number of precessions*), e.g. ≈1:10^5–10^6 for state-of-the-art instruments. The corresponding ratio for a plain solenoid and a neutron beam both with reasonable parameters is about 1:10^3. The required homogeneity of $\int |\underline{B}|\, dl$ can only be achieved by insertion of correction elements, so-called "Fresnel-coils" [14, 18]. Flippers need a defined embedding field, typically about 10^{-3} T, irrespective of the field in the main solenoids. For this purpose, axial coils are used to compensate or enhance the stray field of the main solenoids at the flipper positions. The lower the external flipper (sample) fields are chosen the less sensitive $\int |\underline{B}|\, dl$ is on the flipper position and sample size. Beyond the second π/2-flipper the neutrons enter an analyser composed of magnetic multilayer mirrors. Only neutrons of one longitudinal spin direction are transmitted to the detector. The secondary part of the instrument may be rotated around the sample position in order to allow for a variation of the momentum transfer Q.

The measurement of the echo signal at given (Q,t) requires scanning of one of the precession fields around the symmetry point by varying the current through some extra turns wound on top of the corresponding solenoid. If an echo is present this scanning rotates the spin ensemble in front of the second $\pi/2$-flipper around the longitudinal axis and results in a cosine modulated intensity output of the detector. The modulation amplitude is $\sim S(Q,t)$ whereas the average value is $\sim S(Q,0)$

$$I_{Det}[J,\Delta J] \propto \frac{1}{2}\left[S(Q) \pm \eta \int \cos\left(J\lambda^3\gamma\frac{m_n^2}{2\pi h^2}\omega + \Delta J\gamma\frac{m_n}{h}\lambda\right)S(Q,\omega)d\omega\right] \quad (2.12)$$

Assuming that $S(Q,\omega)=S(Q,-\omega)$ and accounting for the finite width wavelength distribution $w(\lambda)$ this leads to:

$I_{Det}[J,\Delta J] \propto$

$$\frac{1}{2}\int w(\lambda)\left[S(Q) \pm \eta \cos\left(\Delta J\gamma\frac{m_n}{h}\lambda\right)\int \cos\left(J\lambda^3\gamma\frac{m_n^2}{2\pi h^2}\omega\right)S(Q,\omega)d\omega\right]d\lambda \quad (2.13)$$

where $\eta=\eta(J,Q,\lambda)\approx 0.5-1.0$ accounts for resolution effects of the spectrometer. The main effect of a small asymmetry is the $\cos\left(\Delta J\gamma\frac{m_n}{h}\right)$ modulation of the contribution $\sim \eta S(Q,t)$. Therefore a symmetry scan across at least one half period of the modulation allows for the extraction of $\eta S(Q,t)$ in terms of the oscillation amplitude.

Under the assumption that $A=\int\cos(t[\lambda]\omega)S(Q[\lambda],\omega)d\omega$ depends only weakly on λ, the intensity variation as function of the asymmetry ΔJ is $I\sim A\int w(\lambda)\cos\left(\Delta J\lambda\frac{\gamma m_n}{h}\right)d\lambda$. The resulting cosine modulation is indicated in Fig. 2.3. The period depends on the nominal wavelength and the width of the envelope is inversely proportional to the width of the incoming wavelength band. The symmetry scan is used to extract the amplitude A which carries the desired information of $S(Q,t)$.

2.2.3
Zero Field Spin-Echo Technique

The most "expensive" parts of a conventional NSE instrument are the main solenoids providing the precession field. A closer look at Bloch's equation of motion for the spins (Eq. 2.11) shows that in a coordinate system that rotates with the precession frequency around $\underline{B}_0$ the spin is stationary, the coordinate system rotation is equivalent to the addition of $-\underline{B}_0$ to all magnetic fields. By this means the large precession field inside the main coils may be transformed to zero ($\to$ *zero field* spin-echo). The flippers are viewed as elements rotating

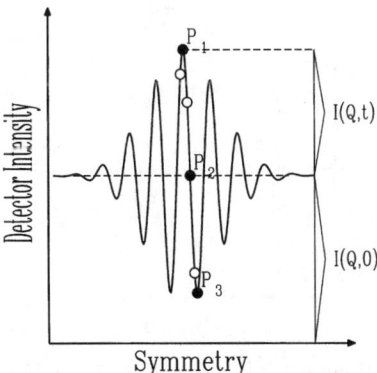

Fig. 2.3 Oscillation of detected intensity as function of the symmetry reveals the echo amplitude. At least three points must be measured to determine the exact symmetry point and the amplitude. However, a more complete symmetry scan is more robust against external disturbances

with the precession frequency. Consequently the NSE described above can be replaced by a zero field flight path limited by flippers. Their inner field, previously very low, needs to be transformed to a value comparable to the old precession field.

Whereas the above argument yields a rough idea of how to realize a zero-field or resonance NSE instrument, only a somewhat different technical realization makes it feasible [17]. All magnetic fields are transverse and confined in the flippers that are used to code the time of neutron passage in terms of a time varying spin rotation angle. In a first pair of flippers separated by a zero field region of length L the neutron time of flight is coded into a net spin rotation angle $\Delta\Psi = \Omega \frac{L}{v}$ (depending on the type of flipper $\Omega = n\omega_L$ with $n=2,4$). This angle is preserved in the zero field region around the sample. In a symmetric flight path between two RF-flippers (RF: radio frequency) in the scattering arm a rotation angle $\Delta\Psi = \Omega \frac{L}{v+\Delta v}$ is acquired and the final polarization is detected by an analyser – as in the conventional NSE. The symmetry scan is performed by mechanical movement of one of the flippers ($L_1=L_2\pm\Delta L$), the zero field regions are realized with the help of μ-metal screens around the flight paths. The flippers have to combine a confined static magnetic field with a RF-field with magnitude equivalent to the flipper fields in a conventional NSE instrument. The static field has the same magnitude as the main solenoid field in a conventional NSE spectrometer divided by n. The resonance RF-frequency corresponds to the Larmor frequency ω_L of neutrons in the DC-field. Due to a geometry as shown in Fig. 2.4 the resulting spin coding rotation Ω is $2\omega_L$. A "sandwich" of two such flippers in a so-called bootstrap combination even yields $\Omega = 4\omega_L$ [19].

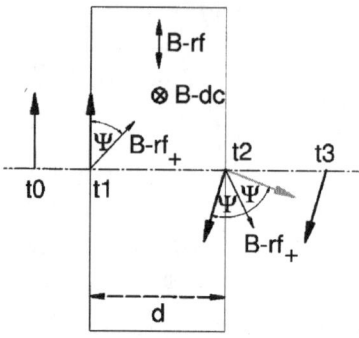

Fig. 2.4 Spin coding in a RF-flipper of thickness d as used in zero field NSE. Before the neutron enters the flipper, e.g. at t_0, it moves in a zero field region and the spin (*arrow*) keeps its direction. Inside the flipper a DC field B-dc perpendicular to the drawing plane is present inside the flipper. It causes the neutron spins to precess with ω_L, as a smaller RF-field B-rf with frequency ω_L acts in the plane. The RF-field may be decomposed into two counter rotating fields, B-rf=B-rf$_+$+B-rf$_-$ only the + component moves synchronously with the neutron spins in the DC-field and thereby can change its state (resonance). The time coding is now affected by the phase (field rotation angle of B-rf$_+$) at t_1 when the neutron enters the flipper, i.e. $\Psi=\Psi_0+\omega_L t_1$. Inside the flipper the neutron spin rotates with B-rf$_+$ and the magnitude of the RF-field is adjusted such that at t_2 the spin is rotated by π around the B-rf$_+$. The effect is an extra rotation angle of 2Ψ outside the flipper. At t_3 the field is again zero and the acquired spin state is preserved. An analogous second flipper, placed after a distance L but fed with the same RF-current, serves as an effective readout of the time delay between coding and decoding (in terms of spin orientation)

The complete instrument consists of two flippers with distance L before the sample and a symmetric set of two flippers after the sample. All flight paths between the first and the fourth (last) flipper are surrounded by magnetic shielding to yield the zero field condition that preserves the spin orientation. All RF flippers are operated synchronously, i.e. with the same current. Their field direction (i.e. rotation) is +, −, −, +. A neutron that enters the first flipper at a time t_1^1 suffers a spin rotation of Ωt_1^1. If the other flippers are passed by the neutron at t_1^2, t_1^3 and t_1^4, the net spin rotation is $\Phi=\Omega[t_1^1-t_1^2-t_1^3+t_1^4]=\Omega[(t_1^1-t_1^2)-(t_1^3+t_1^4)]$, i.e. the time difference the neutron needs to travel the first distance between flippers and that to travel the second distance. At perfect symmetry both distances are equal to L and the time difference is zero, except for a velocity change of the neutron. $\Delta t=L/v_2-L/v_1=L/(v+\Delta v)-L/v\approx(L/v)\times\Delta v/v$ and therefore the final net spin rotation $\Phi=\Omega(L/v)\times\Delta v/v$. This implies that ΩL directly corresponds to the field integral J in normal NSE. After the final polarization analysis and detection the signals are analogous, the symmetry scan is performed by translation of the flippers.

An interesting variant of the resonance NSE is the so-called MIEZE technique [19]. Using two RF-flippers that operate at different frequencies a neutron beam is prepared such that a special correlation between a time varying spin rotation $\omega=(\Omega_1-\Omega_2)$ and the velocity of the neutrons is achieved. An analyser after the second RF-flipper translates the spin rotation into an intensity modulation. The correlation of spin rotation (i.e. analyser transmission) with the velocity is such that, after a distance $L_D = L\Omega_1/(\Omega_1 - \Omega_2)$ behind the second RF-flipper, the intensity modulation for all incoming neutron velocities is in phase and a detector at that position with time-of-flight (TOF) channels synchronized to the modulation frequency reveals the RF-intensity modulation. If the velocities are changed during scattering at the sample the contrast of the modulation is reduced analogous to the echo amplitude reduction in the other NSE methods. For high resolution, however, the flight paths must be very well defined in length. This means that one needs very thin samples in a special geometry, a detector with a very thin detecting depth and a beam divergence that is rather narrow. The benefit of this method is the decoupling of the spin-analysis from the influence the sample has on the neutrons spin state. For instance, spin-incoherent scattering will not lead to signal reduction and sign reversal of the echo signal as in the normal NSE spectrometers.

There are some technical limitations that currently prevent the zero field NSE techniques from reaching the highest resolution and high effective intensity by the use of large solid angles. These are the difficulties in creating the large DC-field inside the RF-flippers and simultaneously keeping the neutron transmission and the mechanical accuracy high. Also, except for untested propositions [19, 20], there is no working scheme to correct for path length differences in a divergent beam.

2.2.4
Accessible High Resolution NSE Spectrometers

Currently worldwide six conventional (generic IN11 type) high resolution instruments are in operation and one is under construction. Two zero-field instruments that are suitable for polymer investigation are available. In addition, several installations for other purposes such as large angle scattering or phonon line width analysis exist (see Table 2.1 for more details).

2.2.5
Related Methods: Pulsed Field Gradient NMR and Dynamic Light Scattering

Pulsed field gradient NMR (PFGNMR) and dynamic light scattering (DLS) are methods that yield information that is related to or similar to the NSE data but on larger length and time scale. PFGNMR is analogous to incoherent neutron scattering, whereas DLS corresponds to coherent scattering with the index of refraction replacing the scattering length density.

PFGNMR measures the ratio Ψ of an NMR spin-echo signal A, which is observed after two field gradient pulses g are applied in a pulse sequence $\pi/2$-g-$\pi/2$, compared to the signal A_0 from the same sequence without gradient pulses:

$$\Psi = \frac{A}{A_0} = \int P(\underline{r}_0) \int P(\underline{r}_0 | \underline{r}, \Delta) \exp(-i\gamma \delta g[\underline{r} - \underline{r}_0]) \, dV \, dV_0 \qquad (2.14)$$

Table 2.1 Existing high resolution NSE instruments that are open for neutron users

Instrument	Type	Facility	Ref.
IN11	Generic IN11	ILL, Grenoble, France	[14, 21]
IN15	Generic IN11	ILL/FZJ/HMI	[22, 23]
FRJ2-NSE	Generic IN11	FZ-Jülich, Germany	[18]
NIST-NSE	Generic IN11	NIST, Gaithersburg, USA	[24]
MESS	Generic IN11	LLB, Saclay, France	www.llb.cea.fr/spectros/spectros_e.html
C2-2	Generic IN11	JAERI, Tokai, Japan	[25]
SNS-NSE	Generic IN11	SNS, Oak Ridge, USA/FZJ	under construction [26]
MUSES	Zero-field	LLB, Saclay, France	[27]
RESEDA	Zero-field	FRM2, TUM, Munich, Germany	www.frm2.tu-muenchen.de

The gradient g is applied during a short time interval δ and with a time delay Δ between both pulses, γ is the gyromagnetic ratio of the observed nuclei. The combination $\gamma\delta g$ has the same meaning as the momentum transfer vector $\underline{Q}$ in a scattering experiment. $P(\underline{r}_0)$ denotes the probability to find a nuclear spin at $\underline{r}_0$ and $P(\underline{r}_0|\underline{r},t)$ is the conditional probability to find the spin which was at $\underline{r}_0$ at position $\underline{r}$ after time t which is analogous to the self-correlation of positions of the spin carrying nuclei [28, 29]. Thus $\Psi \equiv S_{\mathrm{inc}}(Q=\gamma\delta g, t=\Delta)$ if longitudinal spin-relaxation effects can be neglected. The maximum value of Q is in the order of 10^{-2} nm^{-1}, i.e. about 1–2 orders of magnitude smaller than the smallest Q-values of NSE experiments. The time parameter Δ ranges in the order of ms to seconds.

This effective Q,t-range overlaps with that of DLS. DLS measures the dynamics of density or concentration fluctuations by autocorrelation of the scattered laser light intensity in time. The intensity fluctuations result from a change of the random interference pattern (speckle) from a small observation volume. The size of the observation volume and the width of the detector opening determine the contrast factor C of the fluctuations (coherence factor). The normalized intensity autocorrelation function $g^2(Q,t)$ relates to the field amplitude correlation function $g^1(Q,t)$ in a simple way: $g^2(t)=1+C|g^1(t)|^2$ if Gaussian statistics holds [30]. $g^1(Q,t)$ represents the correlation function of the fluctuating local index of refraction as $S(Q,t)/S(Q)$ measures the scattering length density correlation in the NSE case. For solutions this is analogous to a NSE spectrum from a molecule with a sizeable scattering contrast to the solvent, if there is also a sizeable difference of refraction index between solvent and molecule in the case of DLS. Isotopic labelling is not possible, i.e. single chain scattering function from melts are inaccessible by DLS. The time range of DLS ranges from fractions of a µs to seconds, and $Q=10^{-3}$ to 0.02 nm^{-1}.

The extension of DLS into the X-ray regime (X-ray photon correlation spectroscopy, XPS) has become available with the third generation synchrotron sources. The contrast is that of X-ray scattering and therefore sets strict limitation on the correlation functions that can be obtained from polymer samples. In particular, neither isotopic labelling nor the equivalent of incoherent scattering is accessible. However, block-copolymers or polymer blends, for example, may provide enough contrast between A and B polymer-rich domains that yield a detectable scattering signal, as is shown in [31] with a shortest correlation time of 5 s. The signal intensity determines the shortest correlation time that can be measured before shot-noise sets a limitation. For low scattering samples highly efficient counting is achieved by area detection with a CCD device, the shortest time then is in the range of ms [32]. The coherence factor may stay in the 10% region. It has also to be observed that radiation damage of the sample starts to become an issue. Only extremely intense scattering (e.g. as nearly specular reflections from free standing films [33]) enables the extension down to the ns regime. In cases where the contrast is appropriate and radiation damage can be limited XPS offers a Q-range that corresponds to NSE at much larger correlation times – comparable to those of DLS.

Both methods yield the time information directly from a real time applied in the experiment, namely the time Δ between gradient pulses or the time shift in the correlator. PFGNMR uses a coding scheme (similar to that used in NSE to code velocity) to encode displacements of individual nuclei and thereby generates spatial information, whereas DLS acts as any scattering method and yields a scattering angle-dependent value of Q. Where contrast allows, these methods enable the extension of incoherent and coherent scattering functions as obtained by NSE into the macroscopic regime.

3
Large Scale Dynamics of Homopolymers

In general, flexible long chain polymers easily form rotational isomers at the bonds of the chain along the backbone. Therefore such polymers possess a very large number of internal degrees of freedom, which contribute importantly to the entropy part of the molecular free energy. At length scales somewhat larger than the size of the monomer the detailed chemical structure of the chain building blocks is not important so that the very general properties determined by statistical mechanics of the chain prevail, e.g. the conformational entropy follows from the number of possible arrangements of chain sequences in space. According to the central limit theorem the most probable arrangement is that of a Gaussian coil, i.e. the polymer chain performs a random walk in space. If pieces of a chain perform fluctuations out of equilibrium, an entropic force arising from the derivative of the free energy acts on these segments endeavouring to restore them to the most probable contorted state. Such forces are the basis of rubber elasticity [34–37]. This is the content of the so-called Rouse model [38], which is discussed in the first part of this chapter. We derive the conformational free energy and go through the derivation of the Rouse model in detail, including calculation of the appropriate dynamic structure factors [6]. Then we present NSE results on Rouse chains, discussing both the single chain dynamic structure factor as well as the self-motion. We compare the NSE rheological results with atomistic molecular dynamic simulations. This last comparison tells us about the limitations of the Rouse approach at short length scales.

Apart from entropic forces, the confinement of long chains due to their mutual interpenetration in the melt determines their dynamic behaviour to a large extent. It is well known that the dynamic modulus of long chain linear polymer melts displays a plateau regime that expands with growing chain length [6]. In this plateau region the stress is proportional to strain – although liquid, the polymer melt reacts elastically. The material yields for a long time and viscous behaviour takes over.

In the second part of this chapter we address the confinement effects and in particular discuss the reptation model of de Gennes [3, 37], which pictures this confinement as a localization tube following the coarse grained chain profile. We describe NSE experiments unravelling the single chain and the collective dynamics of confined chains; we show the basic correctness of the reptation concept and discuss the constraint release mechanism, the leading process limiting reptation at early times. We demonstrate that NSE experiments confirm quantitatively both the regime of local reptation as well as the idea of contour length fluctuation. Finally, we compare the results with computer simulations addressing, in particular, fully atomistic MD simulations.

3 Large Scale Dynamics of Homopolymers

3.1
Entropy Driven Dynamics – the Rouse Model

The dynamics of a generic linear, ideal Gaussian chain – as described in the Rouse model [38] – is the starting point and standard description for the Brownian dynamics in polymer melts. In this model the conformational entropy of a chain acts as a resource for restoring forces for chain conformations deviating from thermal equilibrium. First, we attempt to exemplify the mathematical treatment of chain dynamics problems. Therefore, we have detailed the description such that it may be followed in all steps. In the discussion of further models we have given references to the relevant literature.

3.1.1
Gaussian Chains

The conformation of a flexible linear polymer chain is – on scales somewhat larger than the main chain bond length ℓ_0 – close to a random walk. The conformations of such a chain are described by a set of segment vectors $\{\underline{r}(n)\}$ with $\{\underline{r}(n)\}=(\underline{R}(n)-\underline{R}(n-1))$, where $\underline{R}(n)$ is the position vector of segment n. Following the central limit theorem the length distribution of a vector $\underline{R}$ connecting segments that have a topological distance of n steps is Gaussian:

$$\phi(R,n) = \left(\frac{3}{2\pi n\ell^2}\right)^{3/2} \exp\left(-\frac{3R^2}{2n\ell^2}\right) \tag{3.1}$$

For the segment length we take $\ell=\ell_0\sqrt{C_\infty}$, where the characteristic ratio C_∞ accounts for the local stiffness arising from the non-random bond angle distribution of the bonds of length ℓ_0. The Rouse model is based on a further idealization of the chain statistics assuming that the bond vector $\underline{r}$ of hypothetical connecting points of the chain has a Gaussian distribution of length:

$$\psi(r) = \left(\frac{3}{2\pi b^2}\right)^{3/2} \exp\left(-\frac{3r^2}{2b^2}\right) \tag{3.2}$$

yielding $\langle r^2\rangle=b^2$. For simplicity throughout the paper we will take $b^2=\ell^2$ keeping in mind that the building block of a Gaussian chain may well contain a larger number of main chain bonds.[1]

The conformational probability of a conformation $\{\underline{r}(n)\}$ follows as:

$$\text{Prob}(\{\underline{r}(n)\}) = \prod_{n=1}^{N}\psi(\underline{r}_n) = \left(\frac{3}{2\pi\ell^2}\right)^{\frac{3}{2}N} \exp\left[-\sum_{n=1}^{N}\frac{3[\underline{R}(n)-\underline{R}(n-1)]^2}{2\ell^2}\right] \tag{3.3}$$

---

[1] In fact, in a recent work on poly(vinylethylene), the size of the Gaussian blob has been found to correspond to about 20 bonds [39].

with N counting the number of segments of the chain. The free energy of a Gaussian chain is entirely described by its conformational entropy:

$$S = k_B \ln Prob(\{\underline{r}(n)\}) \tag{3.4}$$

The Gaussian chain model yields a spring constant even for a single "bond" $k=3k_B T/\ell^2$, where k_B is the Boltzmann constant. From Eq. 3.3 the chain extension between arbitrary points along the chain may be computed to $\langle(\underline{R}(n) - \underline{R}(m))^2\rangle = |n-m|\ell^2$.

3.1.2
The Rouse Model

The Rouse model starts from such a Gaussian chain representing a coarse-grained polymer model, where springs represent the entropic forces between hypothetic beads [6] (Fig. 3.1).

We are interested in the motion of segments on a length scale $\ell < r < R_E$, where $R_E^2 = N\ell^2$ is the end-to-end distance of the chain. The segments are subject to an entropic force resulting from Eq. 3.4 (x-components):

$$\frac{\partial}{\partial x(n)} k_B T \ln Prob(\{x(n)\}) = \frac{3k_B T}{\ell^2}[x(n+1) - 2x(n) + x(n-1)] \tag{3.5}$$

and a stochastic force $f_x(n,t)$ which fulfills $\langle f_x(n,t)\rangle = 0$ and $\langle f_\alpha(n,t) f_\beta(m,0)\rangle = 2k_B T \zeta_0 \delta_{nm} \delta_{\alpha\beta} \delta(t)$; ζ_0 denotes a friction coefficient and α, β the Cartesian components. With this the Langevin equation for segment motion assumes the form:

$$\zeta_0 \frac{\partial x(n)}{\partial t} = \frac{3k_B T}{\ell^2}[x(n+1) - 2x(n) + x(n-1)] + f_x(n,t) \tag{3.6}$$

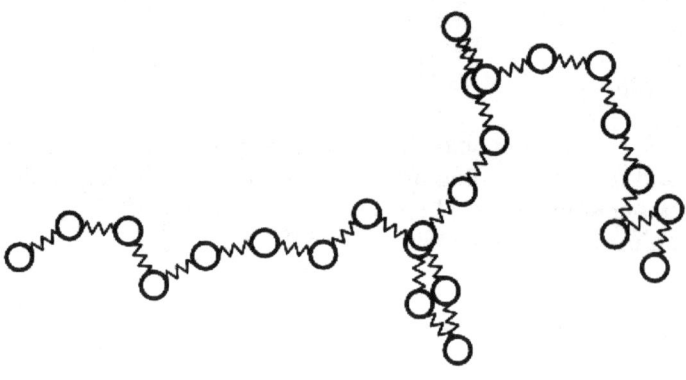

Fig. 3.1 Bead-spring-bead model of a Gaussian chain as assumed in the Rouse model. The beads are connected by "entropic springs" and are subject to a frictional force $\zeta_0 v$, where v is the bead velocity and ζ_0 the bead friction coefficient

Regarding the index n as a continuous variable we arrive at:

$$\zeta_0 \frac{\partial x}{\partial t} = \frac{3k_B T}{\ell^2} \frac{\partial^2 x(n)}{\partial n^2} + f_x(n,t) \tag{3.7}$$

The boundary condition of force-free ends requires $\left.\frac{\partial x(n)}{\partial n}\right|_{n=0,N} = 0$. The partial differential equation is solved by cos-Fourier transformation to normal coordinates fulfilling the boundary conditions:

$$\tilde{x}(p,t) = \frac{1}{N} \int_0^N dn \cos\left(\frac{p\pi n}{N}\right) x(n,t) \tag{3.8}$$

In normal coordinates the Langevin Eq. 3.7 becomes:

$$\zeta_p \frac{\partial}{\partial t} \tilde{x}(p,t) = -k_p \tilde{x}(p,t) + \tilde{f}(p,t) \tag{3.9}$$

where $k_p = \frac{2\pi^2 p^2}{N} k$, $\zeta_p = 2N\zeta_0$. For the stochastic forces we have:

$$\langle f_\alpha(p,t) f_\beta(q,t') \rangle = 2\zeta_p k_B T \delta_{pq} \delta_{\alpha\beta} \delta(t-t') \tag{3.10}$$

Equation 3.9 is readily solved by a single exponential:

$$\tilde{x}(p,t) = \frac{1}{\zeta_p} \int_{-\infty}^t dt \exp\left((t-t')/\tau_p\right) \tilde{f}(p,t') \tag{3.11}$$

where the mode relaxation time τ_p is given by:

$$\tau_p = \frac{\zeta_p}{k_p} = \frac{\zeta_0 N^2 \ell^2}{3\pi^2 k_B T p^2} = \frac{\tau_R}{p^2} = \frac{N^2}{W\pi^2 p^2}; \quad W = \frac{3k_B T}{\ell^2 \zeta_0} \tag{3.12}$$

Here τ_R is the Rouse time – the longest time in the relaxation spectrum – and W is the elementary Rouse rate. The correlation function $\langle x(p,t) x(p,0) \rangle$ of the normal coordinates is finally obtained by:

$$\langle \tilde{x}(p,t) \tilde{x}(p,0) \rangle = \frac{1}{\zeta_p^2} \int_{-\infty}^t dt_1 \int_{-\infty}^0 dt_2 \exp[-(t-t_1-t_2)/\tau_p] \langle f(p,t_1) f(p,t_2) \rangle$$

$$= \frac{1}{\zeta_p^2} \int_{-\infty}^t dt_1 \int_{-\infty}^0 dt_2 \exp[-(t-t_1-t_2)/\tau_p] 2\zeta_p k_B T \delta(t_1-t_2)$$

$$= \frac{k_B T}{k_p} \exp(-t/\tau_p) = \frac{N\ell^2}{6\pi^2 p^2} \exp\left(-\frac{t}{\tau_p}\right) \tag{3.13}$$

For the centre of mass coordinate one finds:

$$\langle \tilde{x}(0,t)\, \tilde{x}(0,0) \rangle = \frac{2k_B T}{N \zeta_0} t \qquad (3.14)$$

Scattering experiments are sensitive to the mean-square segment correlation functions:

$$B(n,m,t) = 3\langle [x(n,t) - x(m,0)]^2 \rangle \qquad (3.15)$$

They are obtained by back-transformation of the normal coordinates (inverse of Eq. 3.8):

$$B(n,m,t) = \langle r^2(t) \rangle = 6D_R t + |m-n|\ell^2 + \frac{4N\ell^2}{\pi^2}$$

$$\times \sum_{p=1}^{N} \frac{1}{p^2} \cos\left(\frac{p\pi m}{N}\right)\cos\left(\frac{p\pi n}{N}\right)\left[1 - \exp\left(-\frac{tp^2}{\tau_R}\right)\right] \qquad (3.16)$$

In order to arrive at Eq. 3.16, we have used $D_R = \frac{1}{6t}\langle (x(0,t) - x(0,0))^2 \rangle$ (D_R is the Rouse diffusion coefficient) and $\sum_{p=1}^{\infty} \frac{1}{p^2}\left[\cos\left(\frac{p\pi m}{N}\right) - \cos\left(\frac{p\pi n}{N}\right)\right]^2 = \frac{\pi^2}{2N}|n-m|$.

For the special case of the self-correlation function ($n=m$) $B(n,n,t)$ reveals the mean-square displacement of a polymer segment. For large p the $\cos^2$ in Eq. 3.16 is a rapidly oscillating function which may be replaced by the mean-value $1/2$. With this approximation we can convert the sum into an integral and obtain:

$$B(n,n,t) = \langle r^2(t) \rangle = \frac{4N\ell^2}{\pi^2} \int_0^\infty dp \frac{1}{p^2}\frac{1}{2}\left(1 - \exp\left(-\frac{tp^2}{\tau_R}\right)\right) + 6D_R t$$

$$= 2N\ell^2 \left(\frac{t}{\pi^3 \tau_R}\right)^{1/2} = 2\ell^2 \left(\frac{3k_B T t}{\pi \zeta \ell^2}\right)^{1/2} + 6D_R t \qquad (3.17)$$

In contrast to normal diffusion, in the segmental regime the mean-square displacement does not grow linearly, but with the square route of time. For the translational diffusion coefficient $D_R = k_B T/N\zeta_0 = \frac{W\ell^4}{3N\ell^2} = \frac{W\ell^4}{3R_E^2}$ is obtained; D_R is inversely proportional to the number of friction performing segments.

The self-correlation function relates directly to the mean-square displacement of the diffusing segments. Inserting Eq. 3.17 into Eq. 2.6 for $t < \tau_R$ we have:

$$S_{\text{self}}(Q,t) = \exp[-Q^2 D_R t]\exp\left\{-\left(\frac{t}{\tau_{\text{self}}(Q)}\right)^{1/2}\right\} \qquad (3.18)$$

with $\tau_{self}(Q) = \left(\dfrac{9\pi}{W\ell^4 Q^4}\right)$. In the case of coherent scattering, which observes the pair correlation function, interferences from scattering waves emanating for various segments complicate the scattering function. Inserting Eq. 3.16 into Eq. 2.5 we obtain:

$$S_{chain}(Q,t) = \frac{1}{N} \exp[-Q^2 D_R t] \sum_{n,m} \exp\left\{-\frac{1}{6}|n-m|Q^2\ell^2\right\} \times$$
$$\exp\left\{-\frac{2}{3}\frac{R_E^2 Q^2}{\pi^2} \sum_p \frac{1}{p^2} \left\{\cos\left(\frac{p\pi m}{N}\right)\cos\left(\frac{p\pi n}{N}\right)\left(1 - \exp\left(-\frac{tp^2}{\tau_R}\right)\right)\right\}\right\}$$
(3.19)

For small Q ($QR_E < 1$) the second and third terms are negligible and $S_{chain}(Q,t)$ describes the centre of mass diffusion of the chain:

$$S_{self}(Q,t) = \frac{1}{N} S_{chain}(Q,t) = \exp[-D_R Q^2 t] \qquad (3.20)$$

For $QR_E > 1$ and $t < \tau_R$ the internal relaxations dominate. For $t=0$ we have $S(Q,t) = S(Q)$; i.e. the structure factor corresponds to a snapshot of the chain structure:

$$S_{chain}(Q) = \frac{1}{N} \sum_{n,m} \exp\left(-\frac{1}{6} Q^2 |n-m|\ell^2\right) \qquad (3.21)$$

Replacing the summations by integrals and observing the relation $R_g^2 = \dfrac{1}{6} N\ell^2$ for the radius of gyration Eq. 3.21 immediately leads to the well known Debye function:

$$S_{chain}(Q) = Nf_{Debye}(Q^2 R_g^2) \qquad (3.22)$$

$$f_{Debye}(x) = \frac{2}{x^2}(e^{-x} - 1 + x) \qquad (3.23)$$

The shape of the Debye function corresponds to the uppermost curve in Fig. 3.2.

Important scaling properties are revealed by an approximate computation of the high-Q behaviour of $S_{chain}(Q,t)$ [41]. Replacing sums by integrals and performing some simplifications:

$$S_{chain}(Q,t) = \frac{12}{Q^2\ell^2} \int_0^\infty du \, \exp\{-u - (\Omega_R t)^{1/2} h(u(\Omega_R t)^{-1/2})\} \qquad (3.24)$$

with $\quad h(y) = \dfrac{2}{\pi} \int_0^\infty dx \, \dfrac{\cos(xy)}{x^2}(1 - \exp(-x^2))$

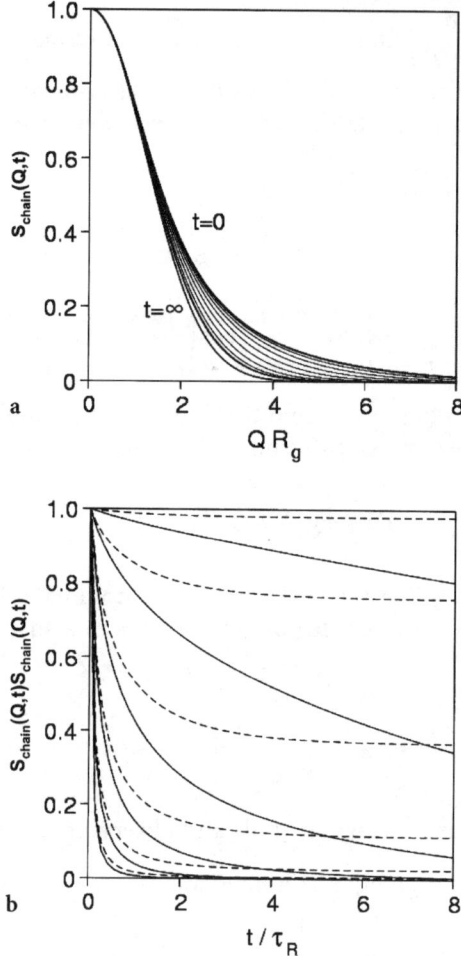

Fig. 3.2 Development of $S_{chain}(Q,t)$ for different times (**a**) and the normalized relaxation function $S_{chain}(Q,t)/S_{chain}(Q)$ (**b**) for QR_g=1, 2, ... 6. The *dashed lines* contain only the intra-chain relaxation whereas the *solid lines* include the centre-of-mass diffusion. Note that for short chains and for small Q the diffusion dominates the observed dynamics (Reprinted with permission from [40]. Copyright 2003 Springer, Berlin Heidelberg New York)

is obtained. Note that Eq. 3.24 only depends on one variable, the Rouse variable:

$$(\Omega_R t)^{1/2} = \frac{Q^2}{6}\sqrt{\frac{3k_B T \ell^2 t}{\zeta_0}} = \frac{Q^2 \ell^2}{6}\sqrt{Wt} \qquad (3.25)$$

Since the Rouse model does not contain an explicit length scale, for different momentum transfers the dynamic structure factors are predicted to collapse to one master curve, if they are represented as a function of the Rouse variable.

Further note that for $t=0$ Eq. 3.24 does not resemble the Debye function but yields its high Q-limiting behaviour $\propto Q^{-2}$, i.e. it is only valid for $QR_g \gg 1$. In that regime the form of Ω_R immediately reveals that the intra-chain relaxation increases $\propto Q^4$ in contrast to normal diffusion $\propto Q^2$. Finally, Fig. 3.2 illustrates the time development of the structure factor.

3.1.3
Neutron Spin Echo Results

The self-correlation function of a Rouse chain was first observed on polydimethylsiloxane (PDMS) [42]. Since a straightforward study of the incoherent scattering by NSE is very difficult – due to spin flip scattering a severe loss of polarization occurs leading to very weak signals – the measurements of the self-correlation function were performed on high molecular weight deuterated PDMS chains that contained short protonated labels at random positions. In such a sample the scattering essentially originates from the contrast between the protonated sequence and a deuterated environment and therefore is coherent. On the other hand the sequences are randomly distributed, so there is no constructive interference of partial waves arising from different sequences. Under these conditions the scattering experiments measures the self-correlation function.

In Fig. 3.3 the corresponding NSE spectra are plotted against the scaling variable of the Rouse model. As predicted by Eq. 3.18 the results for the different momentum transfers follow a common straight line.

Recently it also became possible to observe directly the incoherent cross section of a protonated chain. As eluded to in Sect. 2.2, incoherent scattering

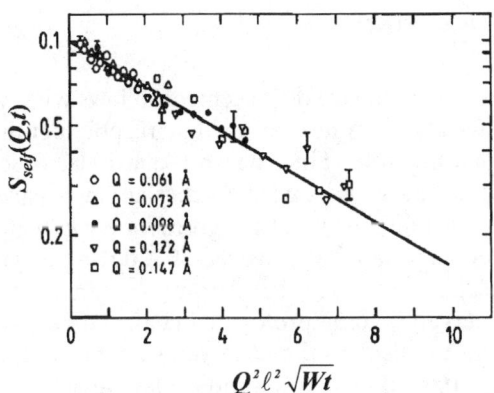

Fig. 3.3 Self-correlation for a PDMS melt $T=100$ °C. The data at different momentum transfers are plotted vs. the scaling variable of the Rouse model. (Reprinted with permission from [42]. Copyright 1989 The American Physical Society)

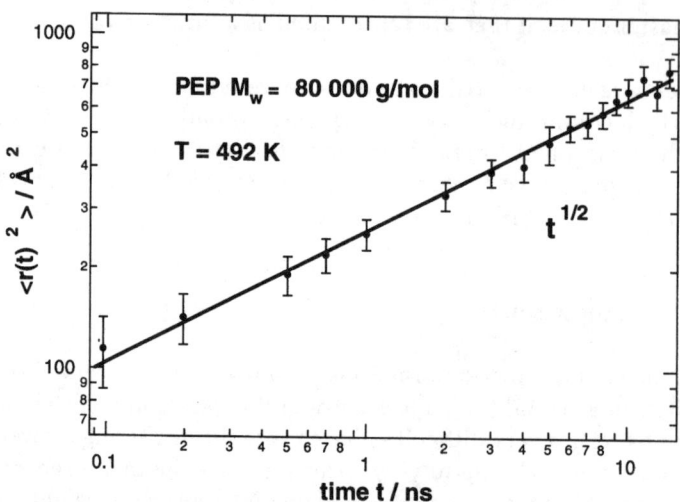

Fig. 3.4 Time-dependent mean-square displacement of a PEP segment in the melt at 492 K. The *solid line* indicates the prediction of the Rouse model. (Reprinted with permission from [43]. Copyright 2003 The American Physical Society)

experiments at an NSE instrument pose some challenges due to the reduction of the echo amplitude to $-1/3$ and the significantly increased background due to spin flip scattering. Figure 3.4 displays the time-dependent mean-square displacement $\langle r^2(t) \rangle$ obtained from a high molecular weight (M_w=80,000) monodisperse polyethylene-propylene (PEP) melt at 492 K. The sample was synthesized by the anionic polymerization of polyisoprene (PI) and converted to PEP by hydrogenation [43]. $\langle r^2(t) \rangle$ has been calculated, in Gaussian approximation Eq. 2.7 as:

$$\langle r^2(t) \rangle = -\frac{6}{Q^2} \ell n \, S_{\text{self}}(Q,t) \tag{3.26}$$

The thus obtained mean-square displacement follows with high accuracy the predicted square root law in time (for the high M_w polymer the translation diffusion does not play any role). Since neutron quasi-elastic scattering resolves dynamic processes in space *and* time, these measurements give direct information about the segment displacement at a given time, e.g. at 10 ns, $<r^2(10\,\text{ns})>$= 620 Å^2, i.e. the average proton has travelled 25 Å during this time interval.

The pair correlation function arising from the segment motion within one given chain is observed if some protonated chains are dissolved in a deuterated matrix. Figure 3.5 displays the observed spectra from polyethylethylene (PEE) (90% dPEE, 10% hPEE) with a molecular weight of M_w^h=21.5 kg/mol; M_w^d=24.5 kg/mol and a narrow molecular weight distribution [44]. The solid lines give the prediction of the dynamic structure factor of Eq. 3.19. Obviously very good agreement is achieved.

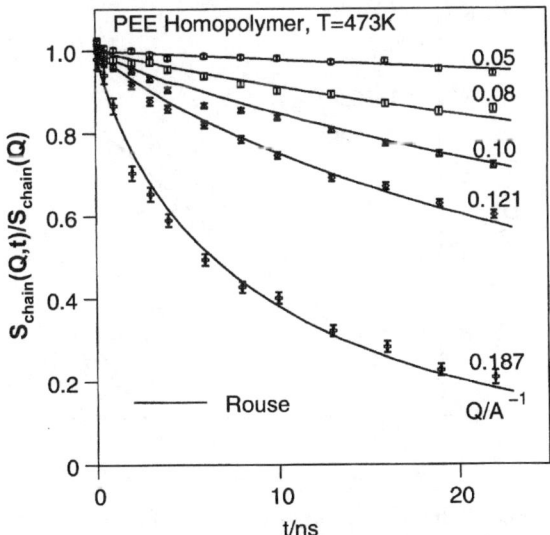

Fig. 3.5 Single chain structure factor from a PEE melt at 473 K. The numbers along the curves represent the experimental Q-values in [Å^{-1}]. The *solid lines* are a joint fit with the Rouse model (Eq. 3.19). (Reprinted with permission from [44]. Copyright 1999 American Institute of Physics)

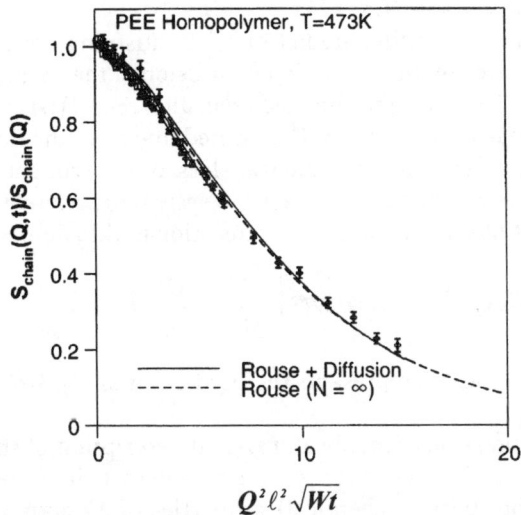

Fig. 3.6 Single chain structure factor from PEE melts as a function of the Rouse scaling variable. The dashed line displays the Rouse prediction for infinite chains, the solid lines incorporate the effect of translational diffusion. The different symbols relate to the spectra displayed in Fig. 3.5. (Reprinted with permission from [40]. Copyright 2003 Springer, Berlin)

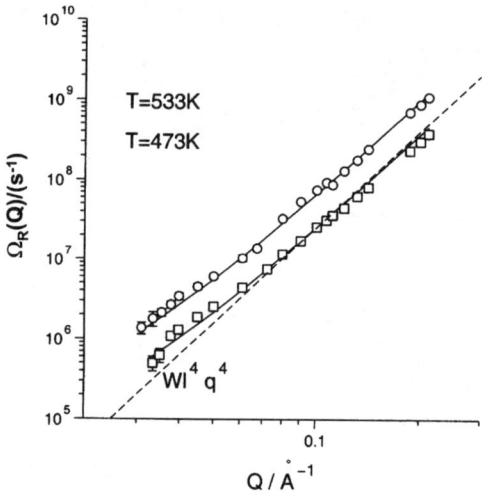

Fig. 3.7 Relaxation rates Ω_R from PEE melts vs. Q for two different temperatures. The *dashed line* represents the $\Gamma \sim Q^4$ prediction of the Rouse model. The *solid lines* include the contribution from translational diffusion (Eq. 3.27). (Reprinted with permission from [40]. Copyright 2003 Springer, Berlin Heidelberg New York)

We now use these data, in order to investigate the scaling prediction inherent in Eq. 3.24. Figure 3.6 presents a plot of the data of Fig. 3.5 as a function of the Rouse scaling variable (Eq. 3.25).

The data follow the scaling prediction with satisfying precision. The small deviations are related to the translational diffusion of the chains. This becomes evident from Fig. 3.7, where the obtained relaxation rates $\Gamma(Q)$ are plotted versus Q in a double logarithmic fashion. The dashed line gives the Rouse prediction $\Gamma \propto W\ell^4 Q^4$. While at larger momentum transfers the experimental results follow this prediction very well, towards lower Q a systematic relative increase of the relaxation rate is observed. Including translational diffusion, we have:

$$\Gamma(Q) = Q^2 \left[D + Q^2 \frac{W\ell^4}{6} \right] = Q^2 W\ell^2 \left[\frac{\ell^2}{3R_E^2} + \frac{Q\ell^2}{6} \right]^2 \tag{3.27}$$

The solid lines in Fig. 3.7 represents the prediction of Eq. 3.27. Perfect agreement is obtained.

The above expressions provide a universal description of the dynamics of a Gaussian chain and are valid for real linear polymer chains on intermediate length scales. The specific (chemical) properties of a polymer enter only in terms of two parameters $N\ell^2 = R_E^2$ and ℓ^2/ζ_0. The friction parameter is governing the Rouse variable in terms of $W\ell^4 = \dfrac{3k_B T \ell^2}{\zeta_0}$. As eluded to in Eq. 3.27 the centre of mass diffusion coefficient may also be expressed in these terms $D_R = k_B T \ell^2 / (\zeta_0 R_E^2)$. Since the Rouse model does not contain an inherent length scale, the parameter N (chain length) and ℓ^2 (segment length squared) are somewhat arbitrary as long as the physical values of ℓ^2/ζ_0 and R_E^2 are kept constant. The NSE experiments measure directly the friction coefficient/length squared!

3.1.4
Rouse Model: Rheological Measurements and NSE

Setting the time scale, the monomeric friction coefficient is a basic quantity in all rheological measurements. This quantity is inferred indirectly in two ways:

i. From measurements of the dynamic modulus $G(\omega)$ in the first relaxation regime. Here in terms of the Rouse model the relaxation spectrum H is given by [34]:

$$H = \frac{\sqrt{\langle \ell_{mon}^2 \rangle} \rho N_A}{2\pi M_0} \left(\frac{\zeta k_B T}{6}\right)^{1/2} \omega^{1/2} \qquad (3.28)$$

M_0 is the monomer mass, $\sqrt{\langle \ell_{mon}^2 \rangle} = \sqrt{nC_\infty \ell_0^2}$ is the size of the monomer, n the number of bonds/monomer, ρ the polymer density, N_A the Avogadro number, ζ the friction coefficient related to the monomer size and ω the relaxation frequency.

ii. From viscosity measurements, where in the Rouse regime:

$$\eta = \frac{N_A}{36} \frac{\rho}{M_0} \zeta \ell_{mon}^2 N_{mon} \qquad (3.29)$$

is predicted with $N_{mon} = N/n$. Very often viscosity studies at high molecular weight are also used. Then the effect of entanglements has to be considered in an empirical fashion [45]. The NSE experiments in the Rouse regime provide direct microscopic access to ζ. From $W\ell^4 = \dfrac{3k_B T \ell^2}{\zeta_0}$ we obtain:

$$\zeta = \left(\frac{3k_B T}{W\ell^4}\right) \ell_{mon}^2 \qquad (3.30)$$

ℓ_{mon}^2 is obtained from measurements of the chain end-to-end distance by $\ell_{mon}^2 = \dfrac{R_E^2}{M} M_0$. A detailed tabulation is given for example in [46]. NSE experiments on polymer melts in the Rouse regime have been performed by now on seven different polymers including two studies of a systematic temperature dependence [42, 44, 47–50]. Table 3.1 presents the monomeric friction coefficient originating from these measurements and compares them with rheological data taken mostly from Ferry's book [34]. While in some instances very good

Table 3.1 Monomeric friction coefficients from NSE experiments compared to rheological data from the compilation of Ferry [34]

Polymer	Temperature (K)	ζ^{NSE} (10^{11} Ns/m)	ζ^{rheo} (10^{11} Ns/m)	Ref.
PVE	408	1.59	1.26	[47]
PI	408	0.569	0.19	[47]
PB	353	0.60	3.1	[48]
PEP	373	3.82	0.98 (6.3[a])	[49, 50]
PE	415	0.21	0.255[a]	[49, 50]
PDMS	373	0.298	0.282	[42]
PEE	473	0.695		[44]

[a] [51].

agreement is evident [PDMS, Poly(vinylethylene) (PVE)] in other cases substantial deviations are visible. Also the method by which the rheological results were obtained seems to significantly affect the rheological friction coefficient. For example, in the case of PEP a rescaling of the high molecular weight viscosity at 373 K gives $\zeta^{VIS}=6.3\cdot10^{-11}$ Ns/m, while an evaluation from the dynamic modulus yields $\zeta^{Mod}=0.98\cdot10^{-11}$ Ns/m. The microscopic value of 3.8×10^{-11} Ns/m is just in between.

For polyethylene (PE) and PEP systematic temperature-dependent measurements in the Rouse regime were performed. Furthermore data on the temperature-dependent end-to-end chain distance (R_E^2) also exist. Table 3.2 displays

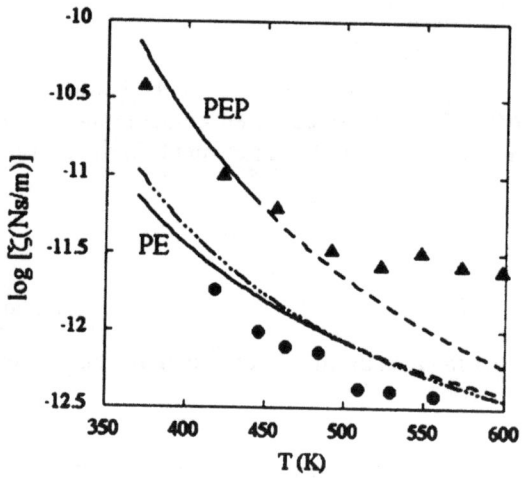

Fig. 3.8 Temperature dependence of the monomeric friction coefficients for PEP and PE. The symbols present the NSE results (*filled triangle* PEP, *filled circle* PE). The *solid lines* display the respective rheological predictions; extrapolations are shown as *dashed lines* (*solid lines*: prediction [51], *point dashed line*: prediction [34] for PEP)

Table 3.2 Rouse rates $W\ell^4$ mean-square monomer length $\ell^2_{mon}=nC_\infty\ell^2_0$ and monomeric friction coefficients ζ for PE and PEP for various temperatures [50]

Temperature (K)	$W\ell^4$ (nm^4/ns)	ℓ^2_{mon} (nm^2)	ζ (10^{11} Ns/m)
PEP			
373	0.236±0.022	59.6	3.9
423	0.95±0.07	56.5	1.04
457	1.65±0.1	55.4	0.64
492	3.26±0.11	54.4	0.34
523	4.30±0.16	53.6	0.27
548	3.68±0.37	53.1	0.33
573	4.65±0.99	52.8	0.27
598	5.3±0.9	52.5	0.245
PE			
418	1.47±0.2	17.65	0.21
446	2.81±0.4	17.12	0.11
463	3.6±0.5	16.76	0.089
484	3.95±0.4	16.38	0.083
509	7.0±0.7	15.9	0.048
529	7.3±0.7	15.54	0.047
556	8.0±0.5	15.09	0.043

the measured monomeric friction coefficients. Figure 3.8 compares them with rheological data according to [51] and [34]. First we note that within the range of validity, the PEP data are very well reproduced by the rheological results of [51]. Ferry's prediction (point dashed line) on the other hand is off and predicts significantly smaller friction coefficients. At high temperatures the experimental data display a much weaker temperature dependence compared to what one would infer from an extrapolation of the rheological data. For PE the rheological extrapolation is again quite good, though at all temperatures a somewhat too high friction coefficient was evaluated.

3.1.5
Rouse Model: Computer Simulation and NSE

In order to learn more about the Rouse model and its limits a detailed quantitative comparison was recently performed of molecular dynamics (MD) computer simulations on a 100 C-atom PE chain with NSE experiments on PE chains of similar molecular weight [52]. Both the experiment and the simulation were carried out at T=509 K. Simulations were undertaken, both for an explicit (*EA*) as well as for an united (*UA*) atom model. In the latter the H-atoms are not explicitly taken into account but reinserted when calculating the dynamic structure factor. The potential parameters for the MD-simulation were either based on quantum chemical calculations or taken from literature. No adjusting

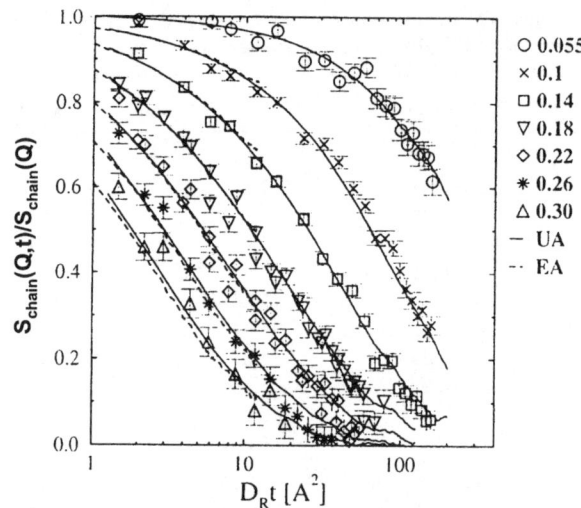

Fig. 3.9 Dynamic structure factor for a 100 monomer PE chain in the melt at 509 K vs. scaled time for the experiment (*symbols*), the united atom model (*full curves*) and the explicit atom model (*dashed curves*). (Reprinted with permission from [52]. Copyright 1998 American Institute of Physics)

parameter was introduced. Figure 3.9 compares the results from the MD-simulation (solid and broken lines) with the NSE-spectra. In order to correct for the slightly different overall time scales of experiment and simulation, the time axis is scaled with the centre of mass diffusion coefficient. This procedure amounted to a 20% rescaling of the time axis. After this correction from Fig. 3.9 quantitative agreement between both results is evident.

Figure 3.10 compares the same experimental data, with a best fit to the Rouse model (Eq. 3.19). Here a good description is observed for small Q-values ($Q \leq 0.14$ Å^{-1}), while at higher Q important deviations appear. Similarly, the simulations cannot be fitted in detail with a Rouse structure factor. Recently this result was confirmed by an atomistic computer simulation on PE molecules of different lengths. Again, at high Q the Rouse model predicts a too-fast decay for $S_{\text{pair}}(Q,t)$ [53].

Having obtained a very good agreement between experiment and simulation, the simulations containing the complete information about the atomic trajectories may be further exploited in order to rationalize the origin of the discrepancies with the Rouse model. A number of deviations from Rouse behaviour evolve.

According to the Rouse model the mode correlators (Eq. 3.14) should decay in a single exponential fashion. A direct evaluation from the atomic trajectories shows that the three major contributing Rouse modes decay with stretched exponentials displaying stretching exponents β of (1: β=0.96 and 2, 3: β=0.86). We note, however, that there is no evidence for the extreme stretching of in-

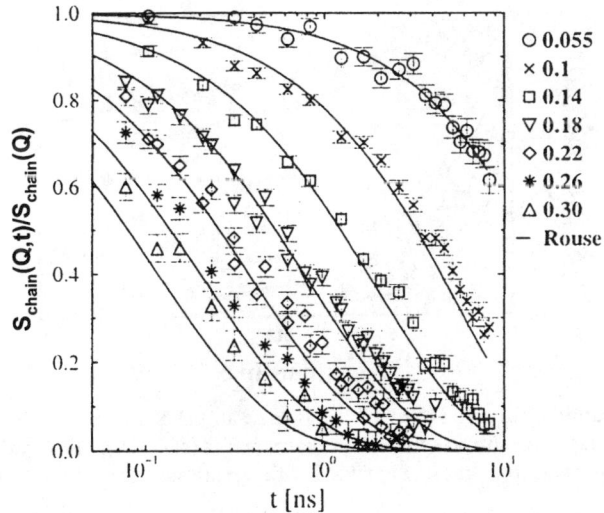

Fig. 3.10 NSE dynamic structure factor for ($C_{100}H_{202}$) for momentum transfers indicated in the legend. The *full lines* are the Rouse prediction. (Reprinted with permission from [52]. Copyright 1998 American Institute of Physics)

termediate Rouse modes (β down to about 0.5) as recently predicted by a coarse grained simulation on PE [54].

A detailed scrutiny of the Gaussian approximation (Eq. 2.7) reveals that for $t<\tau_R$ deviations occur. This was studied later in more detail for the case of polybutadiene (PB) [55]. These simulations demonstrated that the deviations from the Gaussian approximation relate to intermolecular correlations that are not included in any of the analytical models at hand.

While the Rouse model predicts a linear time evolution of the mean-square centre of mass coordinate (Eq. 3.14), within the time window of the simulation ($t<9$ ns) a sublinear diffusion in form of a stretched exponential with a stretching exponent of $\beta=0.83$ is found. A detailed inspection of the time-dependent mean-squared amplitudes reveals that the sublinear diffusion mainly originates from motions at short times $t<\tau_R=2$ ns.

The prediction of a time-dependent centre of mass diffusion coefficient has recently been corroborated by a combined atomistic simulation and an NSE approach on PB ([55]). The dynamic structure factor from simulation and experiment obtained at 353 K are displayed in Fig. 3.11.

In order to adjust the overall time scales a 25% scaling of the experimental times needed to be performed. In the low Q-regime, by rearranging Eq. 3.20 in the way of Eq. 3.26 the centre of mass mean-square displacement $\langle R_{cm}^2(t)\rangle_{app}$ may be obtained. Here, $\langle R_{cm}^2(t)\rangle_{app}$ indicates an apparent mean-square centre of mass displacement, since Eq. 3.20 yields the true centre of mass displacement only in the limit $Q \ll 2\pi/R_g$. $\langle R_{cm}^2(t)\rangle_{app}$ is obtained from application of Eq. 3.26 to the experimental $S_{pair}(Q,t)$ data for $Q=0.05, 0.08$ and 0.10 Å^{-1} and is shown

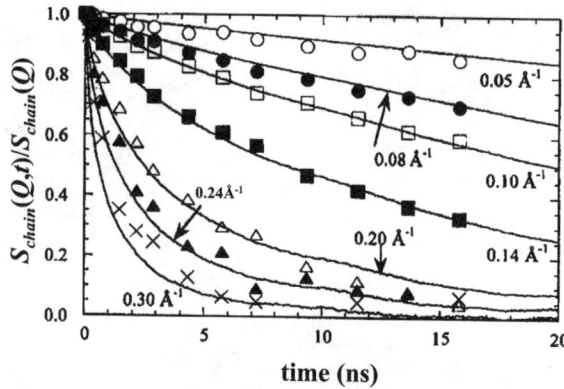

Fig. 3.11 Dynamic structure factor for a PB melt at 353 K (M_w=1,600) obtained from simulation (*lines*) and neutron spin echo measurements (*symbols*). The Q-values are given adjacent to the respective lines. (Reprinted with permission from [55]. Copyright 2000 Elsevier)

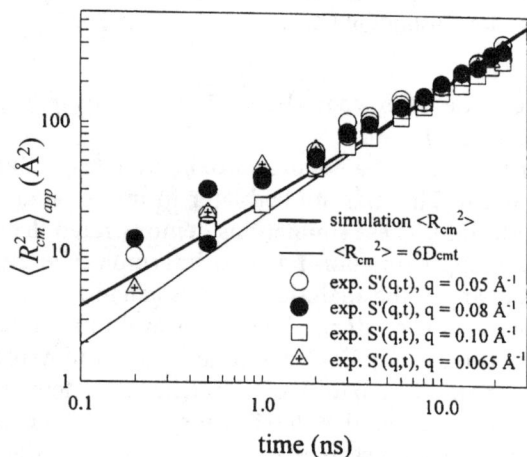

Fig. 3.12 Mean-square centre of mass displacement for PB chains in the melt at 353 K obtained from $\langle r^2(t) \rangle = -6/Q^2 \ln S(Q,t)$. *Solid line*: simulation result; *thin line*: $\langle r^2(t) \rangle = D_R t$. (Reprinted with permission from [55]. Copyright 2000 Elsevier)

in Fig. 3.12 along with $\langle R_{cm}^2(t) \rangle$ obtained directly from simulation. Also shown is $\langle R_{cm}^2(t) \rangle$ obtained from Eq. 3.20 using the polymer self-diffusion coefficient obtained from simulation. For times longer than the chain Rouse time τ_R=15 ns, the centre of mass motion is clearly diffusive and is well described by Eq. 3.20. However, for time less than τ_R, the motion is sub-diffusive. In this time regime, the mean-square centre of mass motion is well described by a power law relationship $\langle R_{cm}^2(t) \rangle \sim t^\beta$, where $\beta \approx 0.8$. Also a recent theoretical study found that interactions of a chain in the melt with $\sqrt{N}$ other chains leads to sub-diffusive behaviour for times less than τ_R [56].

The overall picture emerging from this combined simulation and experimental effort is, that:

i. At lower Q-values the Rouse model gives a rather good description of the experimental and simulation results.
ii. In detail, the model is fulfilled quantitative only at time scales of the order of the Rouse time or larger and therefore on length scales of the order of the radius of gyration of the chains or larger and in the regime where the chains actually show Fickian diffusion.
iii. The self-diffusion behaviour for times smaller than the Rouse time and the relaxation of the internal modes of the chains show small but systematic deviations from the Rouse prediction. The origin of these discrepancies is traced to interchain interactions.
iv. The deviations observed at higher Q relate to a large extent to intrachain interactions, as will be seen later.

3.2
Reptation

The dynamic modulus of a polymer melt is characterized by a plateau in frequency which broadens with increasing chain length. In this plateau regime the polymer melt acts like a rubber, where the elastic properties are derived from the entropy elasticity of the chains between permanent cross links. The modulus of a rubber is inversely proportional to the mesh size and proportional to the temperature. In analogy, it is suggestive to assume that the entanglement or topological interactions between the chains in a melt lead to the formation of a temporary network, which then displays rubber elastic properties. Other than in the rubber, the chains may slowly disentangle and the melt flow. Using the analogy to the modulus of a rubber, we may estimate the distances between entanglement points from the plateau modulus G_N^0 [6]. For different polymers they come out to be between 30 Å and 100 Å. On the basis of such assumptions, a number of theories of viscoelasticity have been developed [6, 41, 57, 58]. The most famous among them is the reptation model of de Gennes [3] and Doi and Edwards [58]. In this model, the dominating chain motion is a reptile-like creep along the chain profile. The lateral restrictions are modelled by a tube with a diameter d parallel to the chain profile. d relates to the plateau modulus of the melt:

$$d^2 = \frac{4}{5} \frac{R_E^2}{M} \frac{k_B T}{G_N^0} \tag{3.31}$$

The restrictions of the motions by the presence of other chains are not effective on a monomer scale but rather permit lateral freedom on intermediate length scales. The experimental observations for viscosity and diffusion can be made directly comprehensible in this simple model.

The viscosity relates to the longest relaxation time in a system. If we consider Rouse diffusion along the tube with a Rouse diffusion coefficient $D_R \approx 1/(N\zeta_0)$ then an initial tube configuration is completely forgotten when the mean-square displacement along the tube fulfils $\langle r^2(t) \rangle_{\text{tube}} = $ (contour length $L)^2$. Thus, for the longest relaxation time, we obtain:

$$\tau_\eta \approx \frac{L^2}{D_R} \approx N^3 \zeta_0 \qquad (3.32)$$

The diffusion coefficient is found by considering that during this time in real space the mean-square displacement just amounts to the end-to-end distance of the chain squared. Thus, we have:

$$D_{\text{Rep}} = \frac{R_E^2}{\tau_\eta} \approx \frac{N}{N^3 \zeta_0} \approx \frac{1}{N^2 \zeta_0} \qquad (3.33)$$

The reptation model predicts that the viscosity of a melt scales with the chain length to the third power while the diffusion coefficient decreases with the second power of the chain length.

We now consider the predictions of the reptation model for the mean-square displacement of the chain segments. Figure 3.13 gives an overview.

i. For short times, when the chain segment has not yet realized the topological constraints, i.e. for distances smaller than the tube diameter ($r<d$), we expect unrestricted Rouse motion $\langle \Delta r^2(t) \rangle \approx t^{1/2}$ (Eq. 3.17). Experimentally this was the case for PEP (Fig. 3.4), where Rouse dynamics were observed for an entangled chain for times up to 20 ns and displacement up to about 30 Å. If the mean-square displacement reaches the order of the tube diameter, then motional restrictions are expected.

We will take a somewhat different but equivalent criterion in order to describe the crossover. As the crossover time τ_e, we take the Rouse relaxation time of a polymer section, spanning the tube diameter:

$$\tau_e = \frac{1}{3\pi^2} \frac{d^4 \zeta_0}{k_B T \ell^2} = \frac{1}{\pi^2} \frac{d^4}{W\ell^4} \qquad (3.34)$$

ii. For times $t > \tau_e$ one dimensional curvilinear Rouse motion along the tube has to be considered. For displacements $s_n(t)$ along the tube Eq. 3.16 yields:

$$\langle (s_n(t) - s_n(0))^2 \rangle = \langle (Y_0(t) - Y_0(0))^2 \rangle \qquad (3.35)$$

$$+ \frac{4N\ell^2}{3\pi^2} \sum_p \cos^2\left(\frac{p\pi n}{N}\right) \times \left(1 - \exp\left(-\frac{tp^2}{\tau_R}\right)\right)$$

3 Large Scale Dynamics of Homopolymers

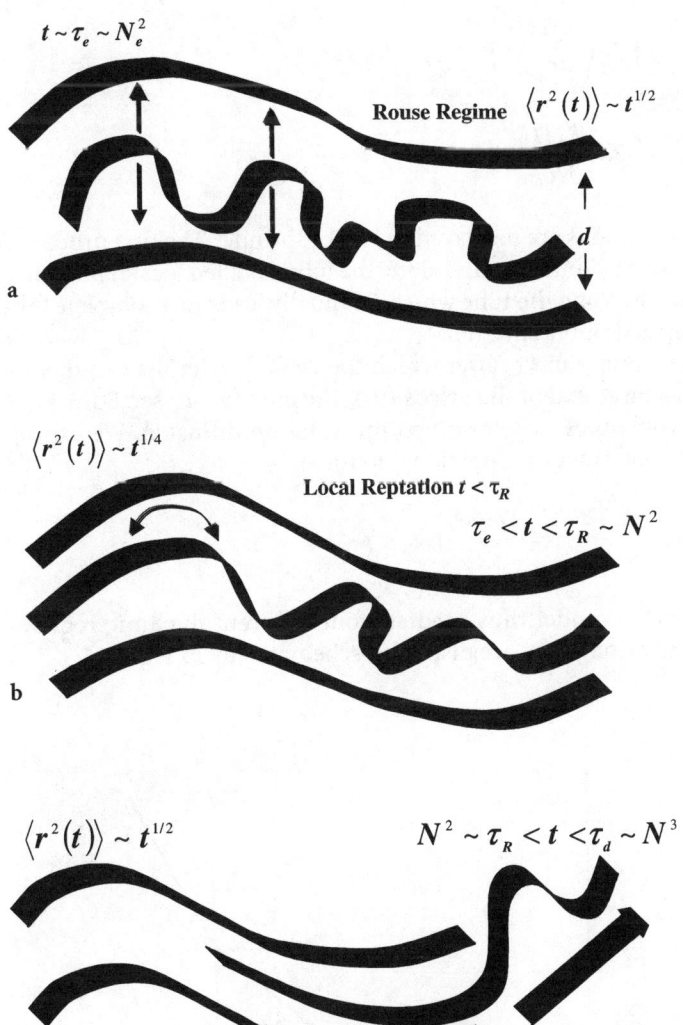

Fig. 3.13 Schematic sketch of the different time regimes of reptation: **a** unrestricted Rouse motion for $\langle r^2(t)\rangle \lesssim d^2/4$, **b** local reptation, i.e. Rouse relaxation along the confining tube, and **c** pure reptation, i.e. diffusion along the tube leading to tube renewal

Y_0 is the centre of mass coordinate for the curvilinear motion. For times $t<\tau_R$ we again get the well known $t^{1/2}$ law, while for $t>\tau_R$ the first term dominates. In this case we deal with curvilinear-diffusion along the tube $\langle (s_n(t)-s_n(0))^2\rangle = (k_B T/N\zeta_0)\,t$.

If a segment is displaced along the tube by $\langle (s_n(t)-s_n(0))^2\rangle$ then the mean-square displacement in 3D real space is $d\langle (s_n(t)-s_n(0))^2\rangle^{1/2}$. With that we obtain:

$$\langle r^2(t) \rangle = \begin{cases} 2d \left(\dfrac{k_B T \ell^2 t}{\zeta_0 \pi} \right)^{1/4} & \tau_e < t < \tau_R \\ 2d \left(\dfrac{k_B T t}{N \zeta_0} \right)^{1/2} & \tau_R < t < \tau_d \end{cases} \qquad (3.36)$$

The two situations are displayed in Fig.3.13b and c. The first process, where the chain performs Rouse motion along the tube, is called local reptation; the creep-like diffusion along the tube which eventually leads to a complete tube renewal is also termed pure reptation.

The terminal time τ_d after which the chain has left its original tube determines to a large extent the viscosity of the melt ($\tau_d \approx \tau_\eta$; see Eq. 3.32). For times $t > \tau_d$ the dynamics is determined by reptation diffusion. We expect a mean-square displacement proportional to time:

$$\langle r^2(t) \rangle = \frac{k_B T d^2}{N^2 \ell^2 \zeta_0} t = \frac{W \ell^4 d^2}{3 R_E^4} t \qquad (3.37)$$

The reptation model thus predicts four different dynamic regimes for segmental diffusion. They are displayed schematically in Fig. 3.14.

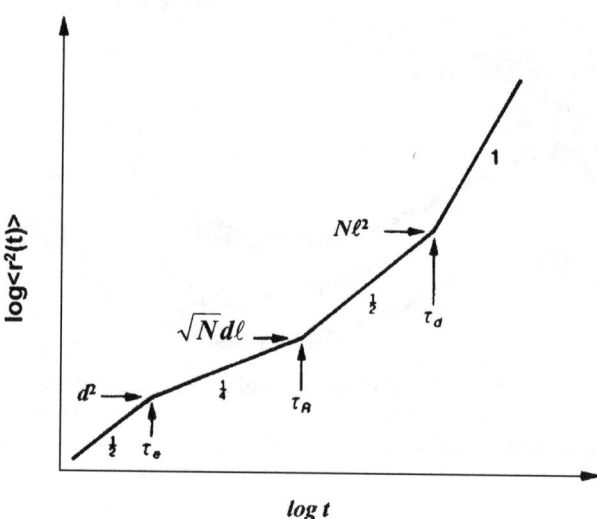

Fig. 3.14 Mean-square displacement of a chain segment in the reptation model. The exponents of the different power law regimes are noted along the respective lines

3.2.1
Self-Correlation Function

In Gaussian approximation the self-correlation function of a reptating chain would directly relate to the above calculated mean-square displacements (see Eq. 2.7). This supposes that after a diffusion time t the Gaussian width of the single segment distribution along the 1-dimensional tube contour path coordinates may be taken for the time-dependent displacement. Projecting this on the Gaussian contorted tube then would again correspond to a Gaussian sublinear diffusion in real space.

However, the real process has to be modelled by projecting the *segment probability distribution* due to curvilinear Rouse motion with the linear coordinate s onto the random walk-like contour path of the contorted tube. This leads to a non-Gaussian probability distribution of the segment at times $t > \tau_e$. The necessity of performing the proper averaging was first shown by Fatkullin and Kimmich in the context of interpretation of field-gradient NMR diffusometry data [59], yielding results that are analogous to the incoherent neutron scattering functions (see Sect. 2.2). In these experiments the corresponding to time and space regime $\tau_R < t < \tau_d$ was covered. Their result:

$$S_{\text{self}}(Q, t > \tau_e) = \exp\left[\frac{Q^4 d^2}{72} \frac{\langle r^2(t) \rangle}{3}\right] \mathrm{erfc}\left[\frac{Q^2 d}{6\sqrt{2}} \sqrt{\frac{\langle r^2(t) \rangle}{3}}\right] \quad (3.38)$$

invalidates the Gaussian approximation for times above τ_e. We note that Eq. 3.38 is strictly valid only for $t \gg \tau_e$ when $\langle r^2(t) \rangle \gg d^2$. The effect on the scattering function is that if (wrongly) interpreted in terms of the Gaussian approximation the crossover to local reptation appears to occur at significantly lower values of τ_e. The general asymptotic $t^{1/4}$ law remains untouched.

3.2.2
Single-Chain Dynamic Structure Factor

Now we turn to the single-chain dynamic structure factor, which is also strongly affected by the topological tube constraints. Qualitatively we would expect the following behaviour:

i. At short times $t < \tau_e$ the chain will perform unrestricted Rouse motion and the dynamic structure factor of Eq. 3.19 or Eq. 3.24 should well describe the dynamics. This has been exemplified in early measurements of the Rouse dynamic structure factor of entangled PDMS melts – in these materials d is large [60].
ii. In the regime of local reptation the chain has already explored the tube laterally and further density fluctuations of the labelled chain will only be possible via Rouse relaxation along the tube. Under such circumstances the structure factor to a first approximation will mirror the form factor of the

tube $S_{chain}(Q,\tau_R>t>\tau_e)/S_{chain}(Q) \approx \exp(-Q^2d^2/36)$. In this regime the experiment should reveal the size of the topological constraints without applying any detailed model.

iii. In the creep regime $t>\tau_R$ the memory of the tube confinement will be gradually lost and the dynamic structure factor should reveal the fraction of the still confined polymer segments.

iv. Finally, in the diffusive regime at very small $Q(QR_g\ll 1)$ an equivalent to Eq. 3.20 will be valid, where now the reptation diffusion coefficient will be measured.

De Gennes [61] and Doi and Edwards [6] have formulated a tractable analytic expression for the dynamic structure factor. They neglected the initial Rouse regime, i.e. the derived expression is valid for $t>\tau_e$ once confinement effects become important. The dynamic structure factor is composed from two contributions $S^{loc}(Q,t)$ and $S^{esc}(Q,t)$ reflecting local reptation and escape processes (creep motion) from the tube:

$$\frac{S_{chain}(Q,t)}{S_{chain}(Q)} = \left[1 - \exp\left(-\frac{Q^2d^2}{36}\right)\right] S^{loc}(Q,t) + \exp\left(-\frac{Q^2d^2}{36}\right) S^{esc}(Q,t) \quad (3.39)$$

The local reptation part was calculated as:

$$S^{loc}(Q,t) = \exp\left(\frac{t}{\tau_0}\right) erfc\left(\sqrt{\frac{t}{\tau_0}}\right) \quad (3.40)$$

where $\tau_0 = 36/(W\ell^4 Q^4)$ and $erfc$ is the complementary error function $erfc(x) = 2/\sqrt{\pi} \int_x^\infty e^{-t^2} dt$. The general expression for $S^{esc}(Q,t)$ due to pure reptation was given by Doi and Edwards:

$$S^{esc}(Q,t) = \sum_{p=1}^{\infty} \frac{2N\mu}{\alpha_p^2(\mu^2 + \alpha_p^2 + \mu)} \sin^2(\alpha_p) \exp\left(-\frac{4t\alpha_p^2}{\pi^2 \tau_d}\right) \quad (3.41)$$

where $\mu = Q^2N\ell^2/12$ and α_p are the solutions of equation $\alpha_p \tan(\alpha_p) = \mu$. $\tau_d = \frac{3R_E^6}{\pi^2 W\ell^4 d^2}$ is the reptation time. For $QR_E \gg 1$ the escape term simplifies to:

$$S^{esc}(Q,t) = \frac{8}{\pi^2} \exp\left(-\frac{Q^2d^2}{36}\right) \sum_{n\,odd} \frac{1}{n^2} \exp\left(-\frac{n^2 t}{\tau_d}\right) \quad (3.42)$$

Without the tube form factor $\exp\left(-\dfrac{Q^2d^2}{36}\right)$ Eq. 3.42 gives directly the total tube survival probability $\mu_{DE}(t) = \dfrac{8}{\pi^2} \sum_{n\,\text{odd}} \dfrac{1}{n^2} \exp\left(-\dfrac{tn^2}{\tau_d}\right)$ which also determines the time and frequency dependence of the dynamic modulus $G(t) \sim \mu_{DE}(t)$. For $t < \tau_d$ or $\omega > 1/\tau_d$ a behaviour $G''(\omega) \sim \omega^{-1/2}$ is predicted [62].

For short times $S_{\text{chain}}(Q,t)$ decays mainly due to local reptation (first term of Eq. 3.39), while for longer times (and low Q) the second term resulting from the creep motion dominates. The ratio of the two relevant time scales τ_0 and τ_d is proportional to N^3. Therefore, for long chains at intermediate times $\tau_e < t < \tau_d$ a pronounced plateau in $S_{\text{chain}}(Q,t)$ is predicted. Such a plateau is a signature for confined motion and is present also in other models for confined chain motion. Besides the reptation model other entanglement models have also been brought forward. We will discuss them briefly by categories.

In generalized Rouse models, the effect of topological hindrance is described by a memory function. In the border line case of long chains the dynamic structure factor can be explicitly calculated in the time domain of the NSE experiment. A simple analytic expression for the case of local confinement evolves from a treatment of Ronca [63]. In the transition regime from unrestricted Rouse motion to confinement effects he finds:

$$\frac{S(Q,t)_{\text{chain}}}{S(Q)_{\text{chain}}} = \frac{Q^2d^2}{24} \int_0^\infty \exp\left[-\frac{Q^2d^2}{48} - g\left(z, \frac{16W\ell^4 t}{d^4}\right)\right] dz \qquad (3.43)$$

with $\quad g(x,y) = 2x - \exp(x)\,\text{erfc}\left(\sqrt{y} + \dfrac{x}{\sqrt{2y}}\right) + \exp(-x)\,\text{erfc}\left(\dfrac{x}{\sqrt{2y}} - \sqrt{y}\right)$

Like the dynamic structure factor for local reptation (Eq. 3.40). Eq. 3.43 develops a plateau region with a plateau height that depends on Qd. Further generalized Rouse models were brought forward by e.g. Hess, Chaterjee and Loring [64, 65].

Rubber-like models take entanglements as local stress points acting as temporary cross links. De Cloizeaux [66] has proposed such a model, where he considers infinite chains with spatially fixed entanglement points at intermediate times. Under the condition of fixed entanglements, which are distributed according to a Poisson distribution, the chains perform Rouse motion. This rubber-like model is closest to the idea of a temporary network. The resulting dynamic structure factor has the form:

$$\frac{S_{\text{chain}}(Q,t)}{S_{\text{chain}}(Q)} = \frac{1}{Z}\ell n(1+Z) + \int_0^\infty dp\, e^{-p}\, F(X, pZ) \qquad (3.44)$$

$Z = q^2 d^2/6 \qquad X = \pi^{1/2}\, t/\tau_0$

$$F(X, pZ) = pZ \int_0^1 dz \int_0^1 dv \exp - \left\{ pZz + \pi^{-1/2} \int_0^x dw \, B(z, v, w, pZ) \right\}$$

$$B(z, v, w, pZ) = \sum_{p=-\infty}^{\infty} \exp\left[-\frac{(z-2p)^2 Z^2}{w^2} \right] \exp\left[-\frac{(v-2p)^2 Z^2}{w^2} \right]$$

Qualitatively the rubber-like and the generalized Rouse model of Ronca arrive at similar results for $S_{\text{chain}}(Q,t)$ in the transition region. Quantitatively the Q-dispersion of the plateau heights, however, is more pronounced in the Ronca model.

In a mode coupling approach, a microscopic theory describing the polymer motion in entangled melts has recently been developed. While these theories describe well the different time regimes for segmental motion, unfortunately as a consequence of the necessary approximations a dynamic structure factor has not yet been derived [67, 68].

3.2.3
Neutron Spin Echo Results on Chain Confinement

Figure 3.15 compares the dynamic structure factors from two polyethylene melts both measured at 509 K for two different molecular weights (2 kg/mol [69, 70] and 12.4 kg/mol [71]). The solid lines in Fig. 3.15a display a fit with the Rouse dynamic structure factor according to Eq. 3.19. Very good agreement is achieved. Figure 3.15b presents equivalent results from the higher M_w melt with the solid lines again showing the prediction of the Rouse model. Please note that the time scale is extended by one order of magnitude compared to the short chain case. While for the short chain melts, the Rouse model describes well the experimental observations, for the longer chains it fails completely. The initial decay of the dynamic structure factor is depicted only in the short time regime, while for longer times the relaxation behaviour is strongly retarded, signifying confinement effects.

In Fig. 3.16 dynamic structure factor data from a M_w=36 kg/mol PE melt are displayed as a function of the Rouse variable $Q^2 \ell^2 \sqrt{Wt}$ (Eq. 3.25) [4]. In Fig. 3.6 the scaled data followed a common master curve but here they split into different branches which come close together only at small values of the scaling variable. This splitting is a consequence of the existing dynamic length scale, which invalidates the Rouse scaling properties. We note that this length is of purely dynamic character and cannot be observed in static equilibrium experiments.

In the spirit of Eq. 3.39 and neglecting the ongoing decay of $S_{\text{chain}}(Q,t)$ due to local reptation, from the heights of the achieved plateaus we may obtain a first estimate for the amount of confinement. Identifying the plateau levels with a Debye-Waller factor describing the confinement we get d=44 Å, a value that is a lower estimate for the true tube diameter since $S^{\text{loc}}(Q,t)$ is not fully relaxed. The horizontal bars in Fig. 3.16 are the predictions from this Debye-Waller factor estimate.

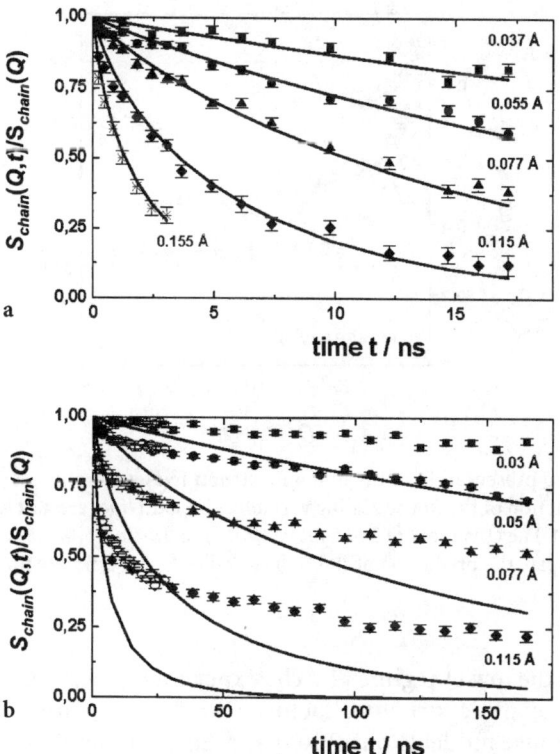

Fig. 3.15 Dynamic structure factors from PE melts at 509 K: **a** M_w=2,000 [69, 70] and **b** M_w=12,400 [71]. The *solid lines* display the predictions of the Rouse model. The Q-values are noted adjacent to the respective lines. Note that the time frame in **b** is extended by an order of magnitude compared to **a**. (**a** Reprinted with permission from [69]. Copyright 1993 The American Physical Society)

First detailed studies of the dynamic structure factor of highly entangled polymer melts were reported at the beginning of the 90's. At that time the attainable Fourier times were limited to about 40 ns. On this time scale, NSE has already played a crucial role in helping to understand the dynamics of polymeric systems [49, 50, 72, 73]. At that time the existence of an entanglement length scale in a linear polymer was been proven [72]. However, with the available time resolution, it was not possible to separate the predictions of the different confinement models addressed above.

In order to proceed further, a significant step in instrument development was needed, which was achieved with the IN15 instrument at the Institut Laue Langevin in Grenoble [22, 23]. This instrument routinely accesses time scales in the order of 200 ns and opens opportunities for exploring dynamic phenomena that were previously inaccessible. Figure 3.17 and Fig. 3.18 display experimental results from a M_w=36 kg/mol PE sample on the dynamic single chain struc-

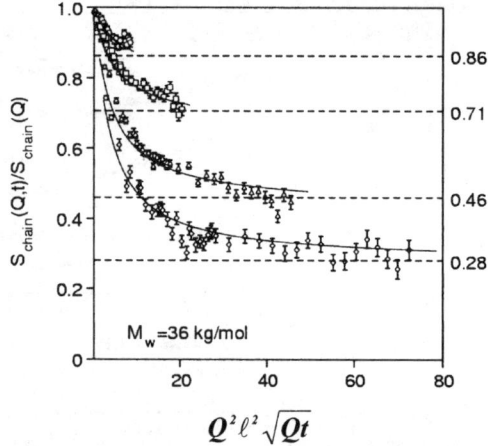

Fig. 3.16 Scaling presentation of the dynamic structure factor from a M_w=36,000 PE melt at 509 K as a function of the Rouse scaling variable. The *solid lines* are a fit with the reptation model (Eq. 3.39). The Q-values are from above Q=0.05, 0.077, 0.115, 0.145 Å^{-1}. The *horizontal dashed lines* display the prediction of the Debye-Waller factor estimate for the confinement size (see text)

ture factor in the low Q-regime [4]. The experimental results were fitted with the different analytic structure factors (Eq. 3.39: reptation [61], Eq. 3.43: generalized Rouse model (Ronca) [63] and Eq. 3.44: rubber like model [66]). The entanglement distances emerging from the different fits are given in Table 3.3.

From Fig. 3.17 and Fig. 3.18 it is apparent that these data clearly favour the reptation model. The reptation model is the only model for which the dynamic structure factor has been calculated and which is in quantitative agreement with these results. We observe that the models of Ronca [63] (see Fig. 3.17) and des Cloizeaux [66] (see Fig. 3.18) produce a plateau which is too flat. On the other hand, the model of Chaterjee and Loring relaxes too quickly and does not form a plateau (see Fig. 3 in [65]).

Table 3.3 Fit results for the entanglement distance d in PE for various models. The reduced χ^2 is also indicated

Model	Ref.	d(Å)	Reduced χ^2	Equation
Reptation	[61]	46.0±0.1	3.03	3.39
Local reptation	[61]	46.5±0.1	3.21	3.39[a]
Des Cloizeaux	[66]	59.8±0.2	7.19	3.44
Ronca	[63]	47.4±0.1	12.2	3.43
Reptation (Doi–Edwards)	[6, 61]	47.4	3.0	3.41

[a] Here only the time dependence of the $S^{loc}(Q,t)$ was considered; $S^{esc}(Q,t)$=const.

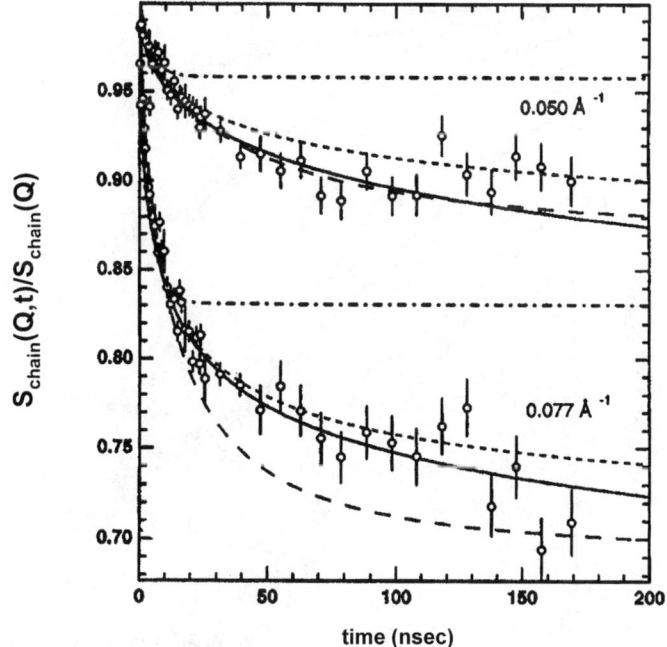

Fig. 3.17 Plot of $S_{chain}(Q,t)$ vs. t measured at $Q=0.050$ and 0.077 Å^{-1} with a comparison between the predictions of reptation (*solid lines*) [61], local reptation (*dotted lines*) [61], the model of des Cloizeaux [66] (*dashed lines*) and the Ronca model [63] (*dot-dashed lines*). (Reprinted with permission from [4]. Copyright 1998 The American Physical Society)

Moreover, from Fig. 3.18 it is apparent that the model of des Cloizeaux also suffers from an incorrect Q-dependence of $S(Q,t)$ in the plateau region, which is most apparent at the highest Q measured. It is important to note that the fits with the reptation model were done with only one free parameter, the entanglement distance d. The Rouse rate $W\ell^4$ was determined earlier through NSE data taken for $t<\tau_e$. With this one free parameter, quantitative agreement over the whole range of Q and t using the reptation model with $d=46.0\pm1.0$ Å was found.

In Fig. 3.17 dotted lines for local reptation and solid lines for the full reptation process are distinguished. A later re-evaluation of these data showed that these differences are a consequence of the fit to high Q asymptotic Eq. 3.42 instead of the correct Eq. 3.39. If Eq. 3.39 is applied then (i) no influence of the creep term exists and (ii) the local reptation term by itself also describes the low Q-data with high precision [74]. The value for the tube diameter comes out slightly higher at $d=47.4$ Å (see also Table 3.3). At this resolution level highly entangled PEP was also studied [43]. Figure 3.19 shows the result. The solid lines in this figure represent a fit with Eq. 3.39 simultaneously to all Q-values. The tube diameter d was varied as the only free parameter, resulting in $d=60$ Å corresponding to a crossover time $\tau_e=40$ ns (Eq. 3.34).

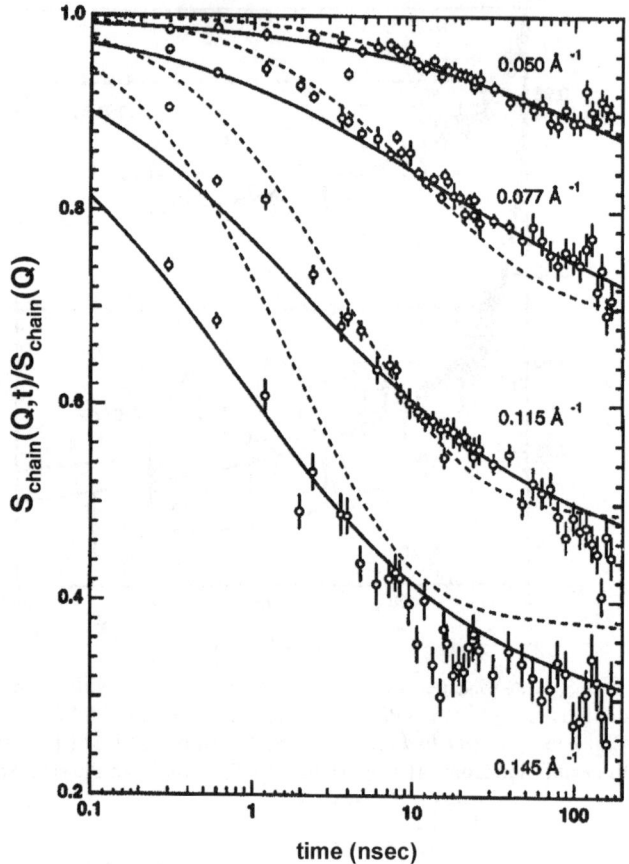

Fig. 3.18 Semi-log plot of the time-dependent single chain dynamic structure factor from a PE melt at T=509 K (M_w=36 kg/mol) for various Q. The *solid lines* are the fit of the reptation model (Eq. 3.39). The *dashed lines* are a fit using the model of des Cloizeaux (Eq. 3.44). (Reprinted with permission from [4]. Copyright 1998 The American Physical Society)

We now turn to the mean-square displacement of a chain segment, which may be observed by incoherent quasi-elastic scattering. In the local reptation regime we expect to observe a cross over of the time-dependent mean-square displacement from a $t^{1/2}$- to a $t^{1/4}$-law (Eq. 3.36). Experiments were performed on PE and PEP samples at the same temperatures at which dynamic structure factors were studied [43]. Figure 3.20 displays the incoherent spectra obtained from the hydrogenated PE at 509 K. For such measurements the experimental background has to be controlled with great accuracy and care. Note that after background subtraction with the proper transmission ratios and without further corrections the short time behaviour follows the Rouse prediction (solid lines). Figure 3.21 converts these data via Eq. 3.26 towards mean-square displacement. Thereby the Gaussian assumption is implied.

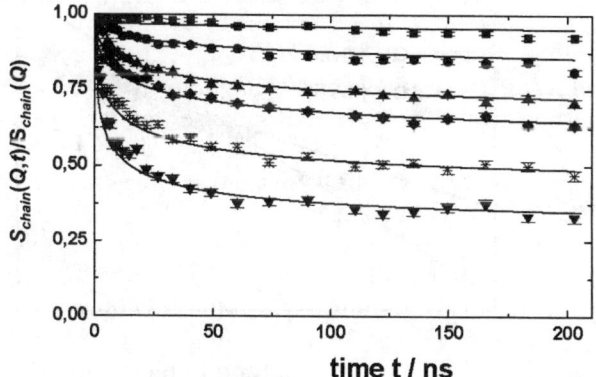

Fig. 3.19 Single chain dynamic structure factor $S_{chain}(Q,t)/S(Q)_{chain}$ from a M_w=80 kg/mol PEP melt at T=492 K for the following scattering wave vectors: Q=0.03 Å^{-1}, Q=0.05 Å^{-1}, Q=0.068 Å^{-1}, Q=0.077 Å^{-1} (from above), Q=0.09 Å^{-1}, Q=0.115 Å^{-1}. The *solid lines* represents a fit with Eq. 3.39. (Reprinted with permission from [43]. Copyright 2003 The American Physical Society)

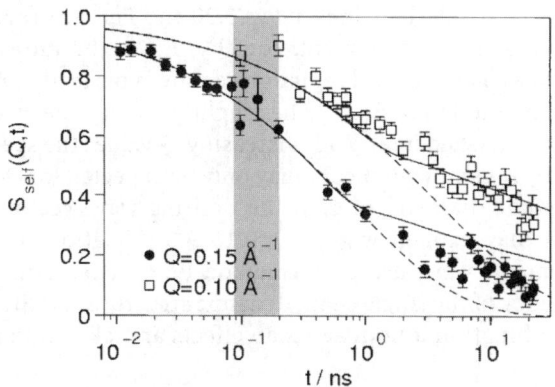

Fig. 3.20 NSE data obtained from the incoherent scattering from a fully protonated PE melt at 509 K (M_w=190 kg/mol). The data close to the 0.1 ns boundary (*grey bar*) are, due to technical difficulties at the range boundaries of the two spectrometer configurations, more uncertain than the bulk of the data points, as seen by the size of the error bars. *Lines*: see text. (Reprinted with permission from [43]. Copyright 2003 The American Physical Society)

Inserting the Rouse rate $W\ell^4$(509 K)=(7±0.7)×10^4 Å^4/ns (Table 3.2) obtained from single chain structure factor measurement into Eq. 3.18 the solid line $\sim t^{1/2}$ is obtained. It quantitatively corroborates the correctness of the Rouse description at short times. The data also reveal clearly a transition to a $t^{1/4}$ law, though Eq. 3.36 would predict the dotted line. The discrepancy explains itself in considering the non-Gaussian character of the curve-linear Rouse motion (Eq. 3.38). Fixing $W\ell^4$ and d to the values obtained from the single chain struc-

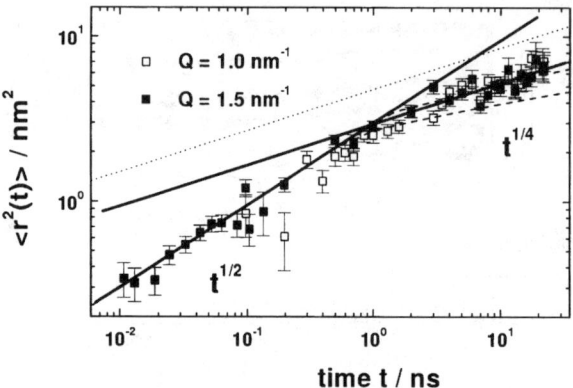

Fig. 3.21 Same data as shown in Fig. 3.20 in a representation of $-6\ell n[S_{self}(Q,t)]/Q^2$, which is the mean-square displacement $\langle r^2(t)\rangle$ as long as the Gaussian approximation holds. *Solid lines* describe the asymptotic power laws $\langle r^2(t)\rangle \alpha t^{1/2}, t^{1/4}$. *Dotted lines*: prediction from the Gaussian approximation, *dashed lines*: see text. (Reprinted with permission from [43]. Copyright 2003 The American Physical Society)

ture measurement, the dashed lines in Fig. 3.20 and Fig. 3.21 reveal the prediction of the non-Gaussian treatment. For $Q=0.1$ Å^{-1}, the unrestricted Rouse regime ($t<\tau_e$) as well as the local reptation regime is perfectly reproduced. For $Q=0.15$ Å^{-1} the prediction of Eq. 3.38 lies slightly outside the error band of the data points (lower dashed line). With increasing Q-values the spatial resolution increases and agreement with theory may only be expected for Q-values $<2\pi/d$. For PE with a tube diameter of 47 Å the limiting wave vector would be $Q=0.13$ Å^{-1}, which may explain why at $Q=0.15$ Å^{-1} deviations become visible. Similar experiments were also performed on PEP melts, resulting again in a perfect consistency of the single chain dynamic structure and the results for the self-correlation function if non-Gaussian effects are taken into account properly [43].

Moreover, we note that recently in reconstructing relaxation times via the time-temperature superposition principle using double quantum nuclear magnetic resonance (*DQ*-NMR) the $t^{1/2}$ and $t^{1/4}$ power laws were invoked without giving the spatial information of NSE [75].

Summarizing, from the NSE results reported so far on the dynamics of long highly entangled polymer melts, a consistent picture evolves, where both pair and self-correlation functions are quantitatively described with identical parameters in terms of the structure factors for reptation. The data now cover a substantial region of the time domain where the reptation concept is in principle applicable. Compared with other phenomenological entanglement models, reptation is now the only approach that provides a consistent description of all the NSE data. This is compelling evidence that reptation is indeed the principle relaxation mechanism in entangled linear-chain systems.

Finally, we would like to remark on two results from rheology:

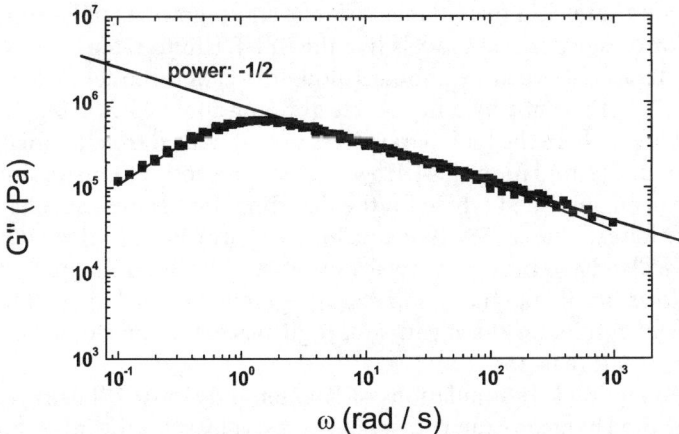

Fig. 3.22 Frequency-dependent dynamic modulus $G''(\omega)$ from a PE chain of M_w=800 kg/mol at 509 K. The *solid line* gives the reptation prediction of $G''(\omega) \sim \omega^{-1/2}$. The *peak* here may not be confused with the α-relaxation of the glass dynamics. It immediately follows from the Fourier transform of $\mu_{DE}(t)$ and strongly depends on molecular weight. The glass relaxation would be many orders of magnitude faster and merge with the local Rouse dynamics on the ns-timescales at the measurement temperature T=509 K

i. For $\omega > 1/\tau_d$ the dynamic modulus $G''(\omega)$ mirrors the total tube survival probability $\mu_{DE}(t)$ and in general in this regime $G''(\omega) \sim \omega^{-1/4}$ is observed in contradiction to the reptation prediction of $G''(\omega) \sim \omega^{-1/2}$ [62]. $\mu_{DE}(t)$ is directly reflected by $S^{esc}(Q,t)$, but for long chains it cannot be probed by NSE due to resolution restrictions. Very recently the reptation prediction was verified beautifully by measuring $G(\omega)$ on PE-chains of M_w=800 kg/mol and very low polydispersity M_w/M_n=1.02 (M_n=number averaged molecular weight) at the NSE temperature of 509 K (Fig. 3.22) [76]. As we will discuss later, the deviations leading to weaker power law decays relate to reptation limiting mechanisms, which are avoided by the very high M_w PE chain.

ii. The dimensions of local confinement observed at the proper mesoscopic scales are significantly larger than those deduced from rheological tube diameter determination: d^{PE}=47 Å vs. d^{PE}=32 Å [77] and d^{PEP}=60 Å vs. d^{PEP}= 43 Å (rheology) [7]. Obviously the local freedom for segmental motion is larger than anticipated so far from rheology. The result casts some doubts on the determination of the plateau modulus from rheological data in terms of the reptation model.

3.2.4
Reptation, NSE and Computer Simulation

Compared to the available clear-cut comparisons between NSE results and simulations for short Rouse chains, the situation in the entangled regime is less satisfactory. The molecular dynamics (MD) simulation of ensembles of sufficiently long chains is a formidable task and up to now has only been accomplished for coarse grained models like the FENE (finite extensible non-linear elastic potential between neighbours along the chain) model by Kremer and Grest [78, 79] or recently by other coarse graining approaches using, e.g. coarse grained blobs, where the blob parameters are calculated from atomistic molecular dynamics simulations [54]. However, very recently atomistic simulations have also been published, which for the first time show significant chain length effects for longer chains [53]. Aside from this route Monte Carlo (MC) simulations have also been brought forward using, e.g. the bond fluctuation model, which allows for the fluctuation of chain segments on the lattice [80]. We will first inspect results on the mean-square displacement and then turn to the dynamic structure factor.

The pioneering MD-simulations of Kremer and Grest [78] using the FENE model resulted in mean-square displacements, at least for the innermost chain segments, which indicated the different power laws predicted by reptation (see Eq. 3.36 and Eq. 3.37). Recently Pütz et al. [79] extended such calculations to chains with up to 10,000 beads. Figure 3.23 displays the outcome for two correlation functions. $g_{2,i}$ gives the mean-square displacements relative to the centre of mass motion $g_{2,i}(t) = \langle (\underline{r}_i(t) - \underline{R}_{cm}(t) - \underline{r}_i(0) + \underline{R}_{cm}(0))^2 \rangle$, where i is the segment number. $g_3(t)$ describes the centre of mass displacement. The figure displays results for three different chain lengths: $N=350$, 700 and 10,000 segments.

While the simulation times for $N=350$ were long enough to reach the diffusive regime, the data for $N=700$ and 10,000 just reach far into the predicted reptation regime. After an initial Rouse-like motion $g_2(t) \propto t^{1/2}$ for inner chain

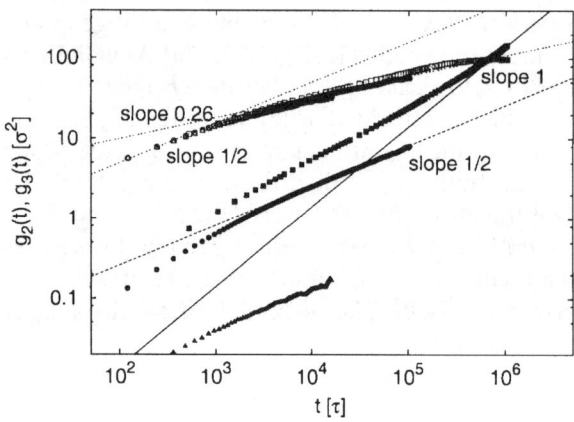

Fig. 3.23 Mean-square displacements $g_2(t)$ (*open symbols*) and $g_3(t)$ (*closed symbols*) for different chain lengths: $N=350$ (*empty square*), $N=700$ (*empty circle*) and $N=10,000$ (*empty triangle*). The *straight lines* show some power law behaviours to guide the eye. The local reptation power laws $g_2(t) \propto t^{1/4}$ and $g_3(t) \propto t^{1/2}$ are verified with remarkable clarity. (Reprinted with permission from [79]. Copyright 2000 EDP Sciences)

3 Large Scale Dynamics of Homopolymers

segments up to a time $\tau_e=(1{,}420\pm100)\tau$ [$\tau_e=(1{,}100\pm100)\tau$] for $N=700$ [$N=10{,}000$] the motion slows down and is proportional to $t^{0.26(1)}$, which is in remarkable agreement with the reptation model (τ is the basic unit of time in the simulation [79]). From the crossover times using Eq. 3.34 values for the tube diameter are inferred. Even between the $N=700$ and $N=10{,}000$ chains a change of d by about 7% can be observed. With $d^2=\ell^2 N_e$ the entanglement distances of 32 or 28 segments, respectively, evolve indicating significant finite size effects even for chains with an overall length of 22 entanglements ($N=700$ chains). This result is in line with recent MC-simulations on the basis of the bond fluctuation model. There it turned out that the cross over from non-entangled to entangled dynamics is very gradual. Even the longest considered chain, which had an overall length of 14 entanglement lengths, could only be considered as slightly entangled when analysed in terms of the asymptotic power laws of reptation theory [81].

On the other hand, the group of Briels [54] recently proposed a coarse graining in terms of blobs with coarse grained properties determined beforehand by atomistic simulations. The mean-square displacement of these blobs displays only a very narrow Rouse regime and then is slowed down further, but never reaches the $t^{1/4}$ power law of reptation. The authors relate this observation to the shortness of the chains (~8 entanglement lengths) and their averaging over all blobs (not only the innermost!).

We conclude this brief discussion of recent results on the segment mean-square displacement (msd) with a brief look at the latest atomistic simulation put forward by Harmandaris et al. [53]. In this simulation fully atomistic PE

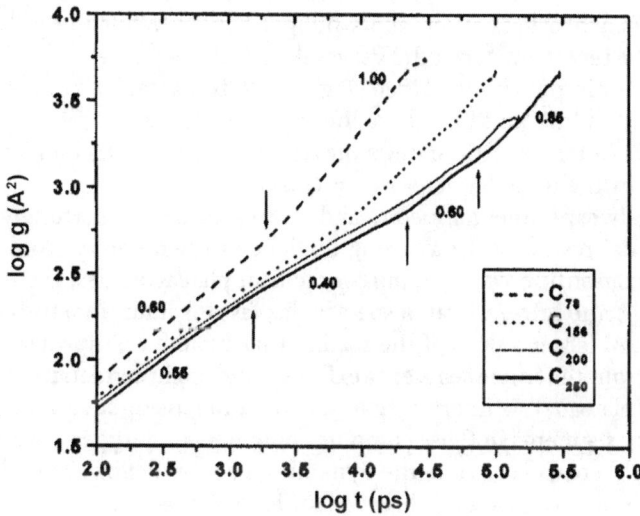

Fig. 3.24 Mean-square displacement $g(t)$ of the innermost atomistic chain segments vs. time t in a log-log plot for the PE melt systems of different chain length (see *insert*). (Reprinted with permission from [53]. Copyright 2003 American Chemical Society)

chains up to a length of 250 monomers were simulated at 450 K. Figure 3.24 displays the msd of the nine innermost segments from this simulation obtained for various chain lengths.

The dashed line refers to the C_{78} PE melt and is clearly seen to obey the scaling laws predicted by the Rouse model: a short time behaviour where $g \sim t^{1/2}$ and the long time (Fickian diffusion) behaviour where $g \sim t^1$. In fact, due to the non-negligible contribution of the diffusion term, the slope of the short-time part of the curve is slightly higher than 0.5 and close to 0.6. It is also interesting to note this short-time behaviour ends at $t \approx 2$ ns, which is practically the Rouse time of the C_{78} melt.

Also shown in Fig. 3.24 are the results corresponding to the C_{156} (dotted), the C_{200} (long-dashed), and the C_{250} (solid) PE melts. Although the first two curves do not exhibit any pronounced structure, the curve corresponding to the C_{250} PE melt appears to indicate some of the power law regimes of reptation. However, the exponents (0.55, 0.4, 0.60, 0.85) are still significantly apart from the reptation prediction (0.5, 0.25, 0.5, 1).

Since the reptation limiting mechanisms, which we will discuss in the next section, are very strong for such short chains, a better agreement could not have been expected. Obviously atomistic simulations with longer PE melts are needed to compare theory and experiment. Finally, in order to improve the comparison between the incoherent scattering and computer simulation it would be advantageous to extract $S_{self}(Q,t)$ from the simulation results – with negligible extra effort – where presently only the msd are computed. Such results could be immediately compared to NSE since they also would incorporate the non-Gaussian effects discussed above.

We now address the dynamic structure factor which incorporates all time-dependent correlations of segments along the chain. While the early MD-simulations of Kremer and Grest did very well with the msd, the dynamic structure factor was only poorly described. Figure 3.25b displays a comparison with NSE results on PEP and PE, where the simulation results were mapped to the experiment in terms of time units measured by τ_e ($\tilde{t} = t/\tau_e$) and length scales measured by the tube diameter d ($\tilde{Q} = Q/d$) [50].

Obviously, experimental spectra and MD results disagree strongly. While the experimental results show a strong tendency to bend away from the initial Rouse relaxation, thereby crossing over to near plateau values, and to split with Q (Fig. 3.16), the MD-generated spectra display only marginal traces of such a behaviour. At larger values of the scaling variable the MD spectra in Fig. 3.25 resulting from the inner segments exhibit some slight tendencies to split and to reach a plateau. It is interesting to note that the mean-square displacement of the inner segments in these computer polymers undergoes a sharp well-defined $t^{1/2}$–$t^{1/4}$ crossover, while the dynamic structure factor even for the inner segments is far from what is observed for long chains.

Similarly the experimental dynamic structure factor also seems to be in disagreement with the latest and most extensive coarse grained MD-simulations by Pütz et al. [79]. Figure 3.26 compares the computed structure factors

3 Large Scale Dynamics of Homopolymers

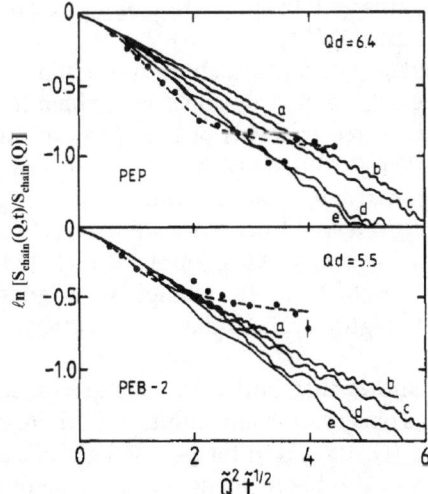

Fig. 3.25 Comparison of the experimental dynamic single chain structure factors for PEP at $Q=0.135$ ($Qd=6.4$) and PE at $Q=0.128$ Å^{-1} ($Qd=5.5$) with the dynamic structure factors from the computer polymer. The various *full lines* represent MD results for different $Qd=3.1$ (a), 3.9 (b), 4.6 (c), 6.2 (d), and 7.7 (e). In the *upper part* the computer results are the structure factors from a fully labelled chain, while in the *lower part* only the centre 35 monomers are labelled. (Reprinted with permission from [49]. Copyright 1992 American Chemical Society)

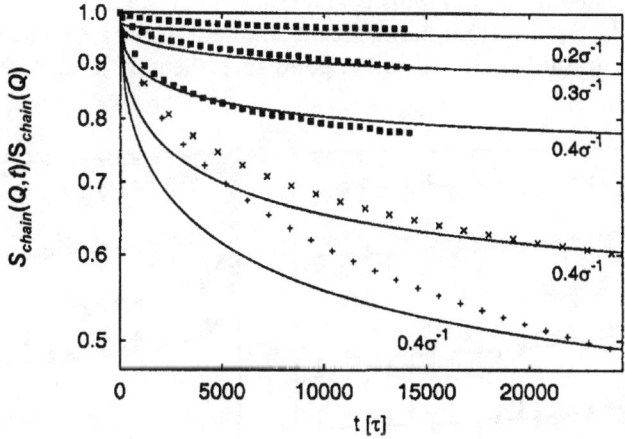

Fig. 3.26 Simulated single chain dynamic structure factor $S_{\mathrm{chain}}(Q,t)$ for different chain lengths $N=350$ (*pluses*), 700 (*crosses*), and 10,000 (*filled squares*) for various Q-values [79] (Q is given in terms of bead size σ). *Solid lines* are fits to Eq. 3.39 and Eq. 3.42. For equal Q-values the plateaus show a strong N-dependence. (Reprinted with permission from [79]. Copyright 2000 EDP Sciences)

for different chain lengths with the prediction of Eq. 3.39 using the high Q form of the escape term (Eq. 3.42) [74].

The agreement with the structure factor is only good for the very long chain and dramatic changes of the plateau values with chain length indicate severe finite size effects. Expressed in terms of the apparent tube diameter, with a decrease from $d=15.76$ ($N=350$), $d=12.76$ ($N=700$) to $d=8.56$ ($N=10,000$) bead sizes evolve. Only an extrapolation to infinite chain length leads to a good agreement with results inferred from the msd ($d_\infty=7.656$). These results stimulated an NSE study on a $M_w=190$ kg/mol PE melt [74]. This experiment increased the chain length from 15 entanglement length ($M_w=36$ kg/mol, Figs. 3.17, 3.18) to 80. Figure 3.27 displays $S_{chain}(Q,t)/S_{chain}(Q)$ for both polymers.

The comparison shows that both data sets are identical at comparable Q-values, i.e. from this observation any influence of finite-chain-length effects on the measured $S_{chain}(Q,t)/S_{chain}(Q)$ for $M_w=36$ kg/mol can be excluded. Taking as reference the very good agreement between NSE results and theoretical predictions, the same comparison to the simulations is rather poor, pointing towards some problems in the simulation of such large chain ensembles.

Finally, we remark on the plateau moduli. The simulation computed G_0^N from the normal stress decay. Depending on the evaluation method, moduli evolve that are much smaller than would be expected from Eq. 3.31. While in the experiment the local freedom of a chain to move was found to be about 50% larger than predicted from the modulus (PE: $d^{rheo}=32$ Å [77]; $d^{NSE}=48$ Å, PEP: $d^{rheo}=43$ Å [7]; $d^{NSE}=60$ Å) the simulation appears to tell the opposite: the tube diameter related to the modulus is about 50% larger than that measured microscopically. On the other hand, recent simulations for PE melts based on a blob coarse graining revealed rather good agreement with Eq. 3.39 for chains

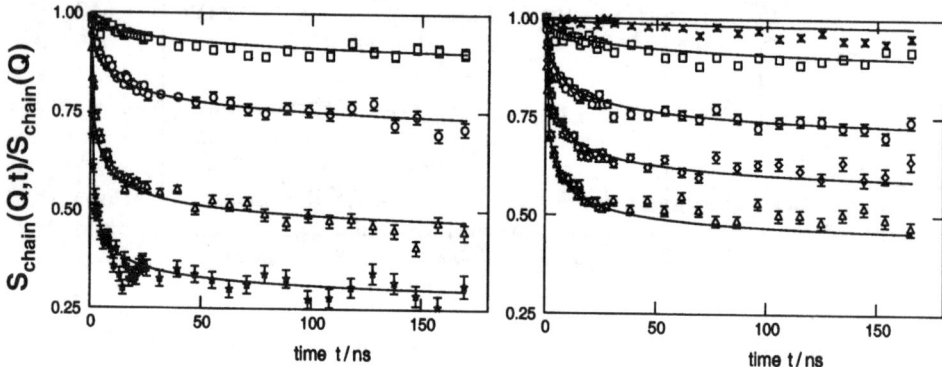

Fig. 3.27 NSE data from PE melts with a molecular weight of $M_w=36,000$ g/mol (*left*) and 190,000 g/mol (*right*) ($M_w/M_n<1.02$) for $Q=0.03$ (*crosses*), 0.05 (*squares*), 0.077 (*circles*), 0.096 (*diamonds*), 0.115 (*triangles*) and 0.145 Å^{-1} (*stars*). For the common Q-values the data are identical. The *lines* represent a fit with the reptation model (using $W\ell^4=7\times10^4$ Å^4/ns for the Rouse rate). (Reprinted with permission from [74]. Copyright 2000 EDP Sciences)

with lengths of about 5–8 entanglement length. For the longest chain $d=54$ Å was found, which agrees well with NSE measurements of similar M_w (see next section). Finally, the fully atomistic simulations on PE chains [53] for the longest chains display a significant retardation of relaxation compared to the Rouse model. A detailed analysis in view of existing NSE data still remains to be done.

Thus, while the experimental situation has matured significantly with consistent results for pair and self correlation functions, in the area of simulations fully atomistic MD-calculations on long-enough chains are needed in order to resolve the existing discrepancies between different simulation approaches and experiment.

3.2.5
Contour Length Fluctuations

Though in general the reptation predictions Eq. 3.22 and Eq. 3.33 for a number of experimental observables like viscosity or the translation diffusion coefficient are qualitatively in good agreement with experiment, deviations are evident. For the viscosity, instead of the $\eta \sim N^3$ power law, higher exponents in the order of 3.4 are generally found. In a similar way, it became evident recently that the translational diffusion coefficient does not follow Eq. 3.33 but displays a stronger molecular weight dependence. Figure 3.28 presents self-diffusion coefficients for PB melt at a reference temperature of 175 °C [82].

Outside the experimental error these data show that, similar to the viscosity result, the pure reptation prediction is violated ($D \sim N^{-2.30 \pm 0.05}$ instead of $D \sim N^{-2}$).

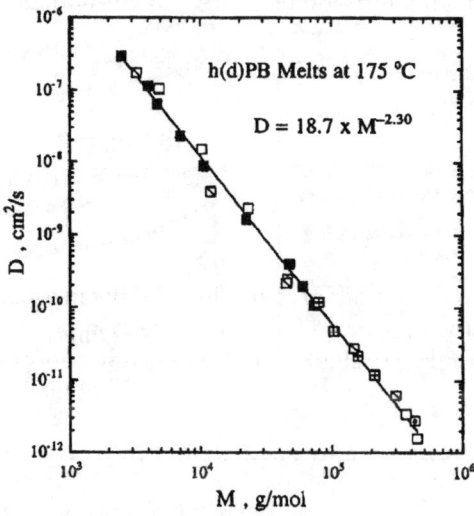

Fig. 3.28 Melt self-diffusion data for hydrogenated (or deuterated) polybutadiene samples adjusted to 175 °C as a function of molecular weight [82]: ■ [83], ⊠ [84], ▣[85], ◨ [86], □ [87]. (Reprinted with permission from [82]. Copyright 1999 The American Physical Society)

On the other hand, extensive viscosity measurements on PB by Colby et al. [88] show that above about 300 entanglement lengths N-dependence (the slope of η vs. N) turns from $N^{3.4}$ to N^3, indicating that the deviations from reptation are due to finite size effects (see also very recent results by Pyckhout et al. [89]).

Recently, NSE experiments have also been reported which were carried out to explore the effect of chain length on chain confinement [71]. The experiments were performed on PE melts spanning a molecular weight range from 12.4 kg/mol up to 190 kg/mol. Figure 3.29 displays the results. We first qualitatively discuss the data and compare the results at $Q=0.115$ Å^{-1}. For the highest molecular weights ($M_w=190$ kg/mol and $M_w=36$ kg/mol) the spectra are characterized by an initial fast decay, reflecting the unconstrained dynamics at early times, and very pronounced plateaus of $S(Q,t)$ at later times, signifying the tube constraints. The plateau values for the two samples are practically identical. Inspecting the results for smaller M_w, we realize that (i) the dynamic structure factor decays to lower values (0.5 at $M_w=190$ kg/mol, 0.4 at $M_w=24.7$ kg/mol, and nearly 0.2 at $M_w=12.4$ kg/mol); and (ii) the long-term plateaus start to slope the more the smaller M_w becomes, with $M_w=12.4$ kg/mol nearly losing the two-

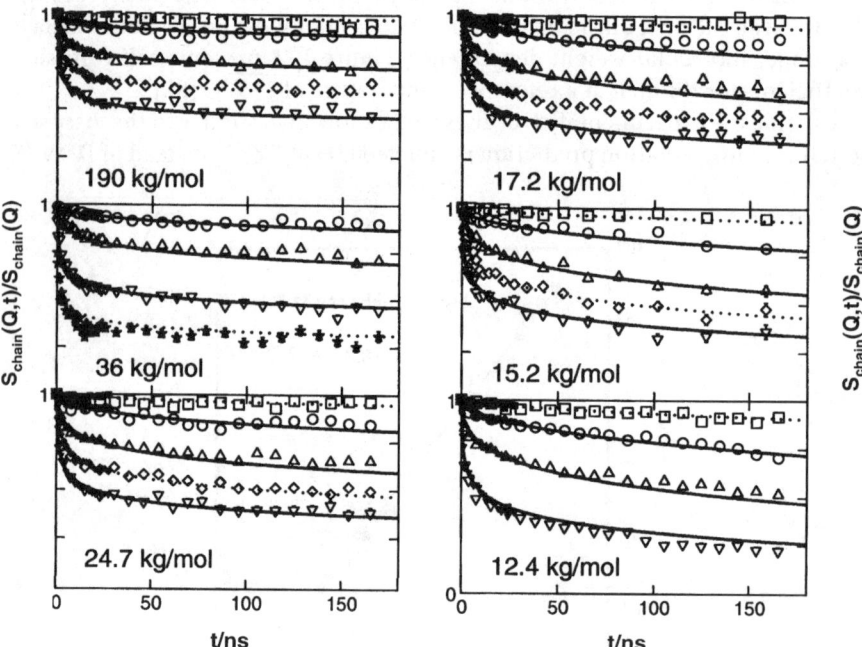

Fig. 3.29 Neutron-spin echo spectra from polyethylene melts of various molecular weights. The Q values correspond to: *squares* $Q=0.03$ Å^{-1}, *circles* $Q=0.05$ Å^{-1}, *triangles* (up) $Q=0.077$ Å^{-1}, *diamonds* $Q=0.096$ Å^{-1}, *triangles* (down) $Q=0.115$ Å^{-1}, *crosses* $Q=0.15$ Å^{-1}. *Filled symbols* refer to a wavelength of the incoming neutrons $\lambda=8$ Å and *open symbols* refer to $\lambda=15$ Å. For *lines*, see explanation in text (Reprinted with permission from [71]. Copyright 2002 The American Physical Society)

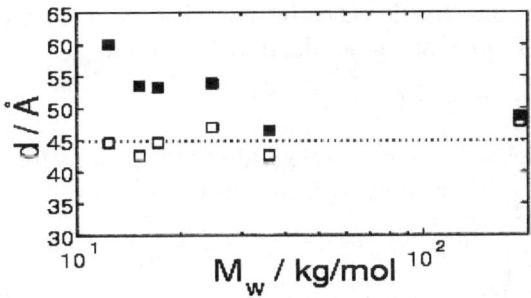

Fig. 3.30 Apparent tube diameters from model fits with pure reptation (*filled squares*) and reptation and contour-length fluctuations (*open squares*) as a function of molecular weight. The *dotted line* is a guide for the eye (Reprinted with permission from [71]. Copyright 2002 The American Physical Society)

relaxation step character of $S(Q,t)$. Apparently, the chain is disentangling from the tube and constraints are successively removed. If the data are analysed using the pure reptation model of Eq. 3.39, a tube diameter d is obtained, which increases with decreasing molecular weight. This is displayed in Fig. 3.30, where the tube diameter is plotted as a function of molecular weight. We observe an increase from about 45 to 60 Å. Secondly, it is found (not shown here) that the quality of the fit decreases with decreasing molecular weight [71].

As known from the broad crossover phenomena observed in the macroscopic chain dynamics (such as the molecular weight dependence of the melt viscosity) very important limiting mechanisms must exist that affect the confinement limit of the reptation process. These processes increase in importance as the chain length decreases. The two main mechanisms are [62]:

- Constraint release (CR). This takes place if a confining chain moves out of the way of a given chain and thus opens some freedom for lateral motion. This phenomenon is an intrinsic many-body effect and for monodisperse polymer melts becomes significant mainly in the creep regime.
- Contour Length Fluctuations (CLF). Fluctuations of the primitive path length of a confined chain are important for the removal of confinement in the local reptation regime, i.e. for times $t<\tau_R$. If a chain contracts in its tube, then after subsequent expansion it loses memory of the original tube. In this way, the effect of tube confinement is gradually diminished with increasing time and the tube confinement will be removed from the ends. In contrast to constraint release, CLF is a single body effect and more easily treatable [90].

In order to calculate the effects of CLF we have to ask how the fraction of monomers that is released through CLF at the chain ends grows with time. It has been recently shown that for $t<\tau_R$ the effect of reptation on escaping from the tube is negligible in comparison to CLF [90]. It is the first passage of a chain end that is assumed to relax the constraint of a tube segment on a chain. From the scale invariance of the Rouse equation (Eq. 3.7) an exact asymptotic result for the surviving tube fraction at early times may be obtained. The Rouse equation of motion is invariant under the transformation:

$$t = \lambda t'; R = \lambda^{1/4} R'; n = \lambda^{1/2} n' \tag{3.45}$$

With this, all relevant lengths scale with $t^{1/4}$. Therefore, for early times the fraction of still-confined tube segments must behave as:

$$\mu(t) = 1 - \frac{C_\mu}{Z}\left(\frac{t}{\tau_e}\right)^{1/4} \tag{3.46}$$

where $Z=N/N_e$ is the number of entanglements per chain. One-dimensional stochastic simulation led to $C_\mu=1.5\pm0.02$. The released segment fraction is then $\psi(t) = 1 - \mu(t)$.

Following an approach of Clark and McLeish [91], this result may be incorporated into the structure factor of Eq. 3.41. We assume that after time t all monomers between 0 and $s(t)$ and between $1 - s(t)$ and 1 have escaped from the tube. Here $s(t)= \Psi(t)/2$. This statement contains two approximations, first that the fraction of the chain escaped from the tube is the same for each end of the chain, and, second, it ignores the distribution of $s(t)$ by replacing it with a single average value. If we make these two assumptions the rest of the calculation is straightforward. We first use the fact that $r(s,t)$ is a Gaussian variable and therefore:

$$\langle \exp\{iQ[\underline{r}(s,t) - \underline{r}(s',0)]\}\rangle = \exp\left(-\frac{1}{2}\sum_{\alpha=x,y,z} Q_\alpha^2 \langle [r_\alpha(s,t) - r_\alpha(s',0)]^2\rangle\right) \tag{3.47}$$

Then we note:

$$\langle [r_\alpha(s,t) - r_\alpha(s',0)]^2\rangle =$$

$$\frac{dL}{3}\begin{cases} |s - s'| & s(t) < s < 1 - s(t) \quad \text{or} \quad s(t) < s' < 1 - s(t) \\ |2s(t) - s - s'| & s < s(t) \quad \text{and} \quad s' < s(t) \\ |2 - s - s' - 2s(t)| & s > 1 - s(t) \quad \text{and} \quad s' > 1 - s(t) \end{cases} \tag{3.48}$$

where $L=Zd$ is the contour length. The summation in Eq. 3.48 is then replaced by integration as:

$$S^{esc}(Q,t) = \int_0^1 ds \int_0^1 ds' \exp(-2\mu|s - s'|)$$

$$+ 2\int_0^{s(t)} ds \int_0^{s(t)} ds' \left(\exp\{-2\mu[2s(t) - s - s']\} - \exp(-2\mu|s - s'|)\right) \tag{3.49}$$

with μ as defined in Eq. 3.41.

The integrals can easily be evaluated and we obtain:

$$S^{esc}(Q,t) = \frac{N}{2\mu^2}[2\mu + e^{-2\mu} + 2 - 4\mu s(t) - 4e^{-2\mu s(t)} + e^{-4\mu s(t)}] \quad (3.50)$$

The lines in Fig. 3.29 display the result of a fit with Eq. 3.50. A group of three common Q values for all molecular weights is displayed by thick solid lines to facilitate an easy comparison of the molecular weight dependence of the curves. Additional Q values only available for some M_w's are represented by dotted lines.

The open squares in Fig. 3.30 show the resulting tube diameters. Aside from some small fluctuations they now stay constant, independent of M_w. We further note that particularly for smaller M_w the weighted error between fit and data is significantly smaller for this second approach. If we compare the experimental spectra with the model prediction (Eq. 3.39 and Eq. 3.50), we generally find good agreement. The gradually increasing decay of $S_{chain}(Q,t)$ with decreasing M_w is depicted very well both with respect to the magnitude of the effect as well as to the shape of $S_{chain}(Q,t)$. The largest disagreement is seen in the M_w=17.2 kg/mol data, where at lower Q the data appear to be somewhat irregular. At Q =0.05 Å^{-1}, $S(Q,t)$ tends to increase at longer times, possibly reflecting some systematic problems with measurement.

We now consider the tube diameters resulting from the two approaches (Fig. 3.30). At the highest molecular weight, contour length fluctuations are insignificant and both lines of fitting yield the same d. At M_w=36 kg/mol a slight difference appears, which increases strongly with decreasing length. At M_w= 12.4 kg/mol the difference in the fitted tube diameters between both approaches rises to nearly 50%, emphasizing the strong effect of the contour length fluctuation in loosening the grip of the entanglements on a given chain.

Thus, the comparison between the experimental chain length-dependent dynamic structure factor and theoretical predictions clearly shows that in the time regime $t \leq \tau_R$, contour length fluctuation is the leading mechanism that limits the chain confinement inherent to the reptation picture. Without any further assumption or fitting parameter – the tube diameter d stays constant with M_w – it is possible to describe the full M_w-dependence of $S_{chain}(Q,t)$ in terms of local reptation and the contour length fluctuation mechanism. Even for chain lengths corresponding to only 6–7 entanglements, the tube diameter appears to be a well-defined quantity, assuming the same value as for asymptotically long chains. The confinement is lifted from the chain ends inwards, while the chain centre remains confined in the original tube.

Concluding this paragraph we emphasize that CLF also affect the macroscopic melt properties in a significant way:

i. From the relation between $G(\omega)$ and the total tube survival probability it is immediately clear that the modifications of $S^{esc}(Q,t)$ should also be mirrored by $G(\omega)$. It has been shown that CLF introduce an $\omega^{-1/4}$ regime into the spectrum of $G''(\omega)$. However, the other limiting mechanism (CR) adds further significant modifications [62].

ii. The translational chain diffusion is necessarily also affected. First, the terminal time defining the diffusion "step" is reduced since reptation has only to relax the not-yet-released central parts of the tube and, secondly, the diffusive length is reduced since only the displacement of the central part counts. Both effects do not cancel and the net effect gives $D \sim Z^{-2.4}$ up to $Z \cong 300$, in very good agreement with experiment.
iii. Finally, the anomalous power law exponent of 3.4 for the viscosity mass relation has been attributed to CLF. The treatment of Millner and McLeish even predicts the crossover to pure reptation within the proper range of Z [92]. For experimental observations see also [88, 89, 93].

It is the virtue of the NSE experiments that they confirmed the CLF mechanism quantitatively on a molecular level in space and time. On the other hand, the constraint release process still remains to be probed on a molecular scale and it is expected that future NSE experiments on bimodal melts will clarify these processes on a molecular scale also.

4
Local Dynamics and the Glass Transition

Polymeric solids can be found in the crystalline and/or in the amorphous state, but most of them are either totally amorphous or semicrystalline. Non-crystalline solids are characterized by long range structural disorder, which is responsible for some very interesting basic properties, as well as for attractive and new technological applications (see e.g. [94]). Among the different kinds of non-crystalline materials, glass-forming systems (including polymers, metallic glasses, amorphous semiconductors...) are of utmost interest. A glass is a non-crystalline solid obtained by cooling the material from a temperature above the melting point T_M. While cooling, a system can either crystallize or stay in a liquid-like state. This supercooled metastable state is visco-elastic in the case of polymers. When the structural rearrangements characteristic for the supercooled state cannot follow the cooling rate, the system falls out of equilibrium and transforms into a glassy solid – a "frozen" liquid – within the experimental observation time. This phenomenon, known as the glass transition (see e.g. [95–103]), takes place in a temperature range that depends in general on the experimental cooling/heating conditions. Usually this temperature range is represented by only one temperature, the glass transition temperature T_g. Below T_g, the glassy state is unstable with respect not only to the supercooled liquid, but also to the crystalline solid. However, some degrees below T_g polymer glasses can be considered as "stable" from a practical point of view, i.e. their physical properties do not change over time scales of several years. Therefore, the study of the molecular dynamics at temperatures close to and above T_g is essential for understanding the phenomenon of the glass transition – "probably the deepest and most interesting unsolved problem in solid state theory", in the words of the Nobel Prize winner P.W. Anderson [104] – as well as for the control of many of the aspects of industrial processing of polymer systems.

Glasses show a series of "universal" features. For instance, they present some average short-range order, which extends to some neighbour shells of a given atom or molecule; their static structure factors $S_{pair}(Q)$ look similar independently of the kind of system. $S_{pair}(Q)$ shows broad diffraction peaks revealing inter- and intra-molecular correlations. Another general finding is the occurrence of a series of dynamic processes – in particular the so called α- and β-relaxations. The investigation of such common features at the length scales where glasses show similar structural properties, i.e. local scales, is highly important in order to scrutinize the problem of the glass transition.

What is the main feature distinguishing the dynamic behaviour of a glass-forming system and a simple liquid? Let us follow the time evolution of a given correlation function $\Phi(t)$ of a glass former for different temperatures, as schematically shown in Fig. 4.1a. At a high temperature (e.g. above the T_M), like T_1 in the figure, $\Phi(t)$ decays in a single step at times of the order of picoseconds – a behaviour expected for a simple liquid. If the system is cooled

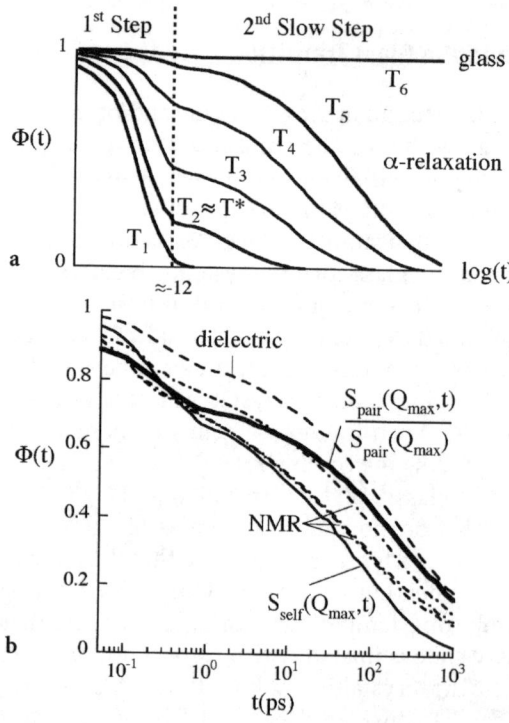

Fig. 4.1 a Typical time evolution of a given correlation function in a glass-forming system for different temperatures ($T_1>T_2>...>T_6$). **b** Molecular dynamics simulation results [105] for the time decay of different correlation functions in polyisoprene at 363 K: normalized dynamic structure factor at the first static structure factor maximum (*solid thick line*), intermediate incoherent scattering function of the hydrogens (*solid thin line*), dipole-dipole correlation function (*dashed line*) and second order orientational correlation function of three different C–H bonds measurable by NMR (*dashed-dotted lines*)

down, there is a temperature range, which we will call T^*, where a second step in $\Phi(t)$ develops, slowing down the decay of the correlations at longer times. This second step becomes more and more important when the temperature of the system decreases. The corresponding characteristic time shows a very strong T-dependence. The state of the system in this temperature region $T<T^*$ is what is known as supercooled liquid state. In the neighbourhood of T_g the correlations of the system are finally frozen and the obtained state is glassy (T_6 in Fig. 4.1a). The characteristic feature of the supercooled liquid in contraposition to the simple liquid state is thus the presence of the second step in $\Phi(t)$, which always displays a typical non-exponential time decay. This second step is called the α-relaxation, independently of the correlator observed or the experimental technique used. Since the α-process is a universal feature of the dynamics of supercooled liquids, it is nowadays generally accepted that the α-relaxation origins from a structural relaxation at the intermolecular level.

4 Local Dynamics and the Glass Transition

Thereby, the features of the α-relaxation observed by different techniques are different projections of the actual structural α-relaxation. Since the glass transition occurs when this relaxation freezes, the investigation of the dynamics of this process is of crucial interest in order to understand the intriguing phenomenon of the glass transition. The only microscopic theory available to date dealing with this transition is the so-called mode coupling theory (MCT) (see, e.g. [95, 96, 106] and references therein); recently, landscape models (see, e.g. [107–110]) have also been proposed to account for some of its features.

In the case of polymers, the α-relaxation has been well characterized for many years, e.g. by dielectric spectroscopy and mechanical relaxation (see, e.g. [34, 111]). The main experimental features extracted from relaxation spectroscopies are:

i. The non-exponential decay of the correlation function, known as stretching. The decay of a given correlator $\Phi(t)$ through the α-relaxation can be well described by a stretched exponential or Kohlrausch-Williams-Watts (KWW) function [112, 113]:

$$\Phi(t) \propto \exp\left[-\left(\frac{t}{\tau_\alpha(T)}\right)^\beta\right] \quad (4.1)$$

$\tau_\alpha(T)$ is the associated characteristic relaxation time and β is the stretching parameter ($0<\beta\leq1$).

ii. The time–temperature superposition, implying that the functional form does not appreciably depend on temperature (see e.g. [34, 111]). For instance, mechanical or rheological data corresponding to different temperatures can usually be superimposed if their time/frequency scales are shifted properly taking a given temperature T_R as reference.

iii. The non-Arrhenius temperature-dependence of the relaxation time. It shows a dramatic increase when the glass transition temperature region is approached. This temperature dependence is usually well described in terms of the so called Vogel-Fulcher temperature dependence [114, 115]:

$$\tau_\alpha(T) = \tau_{\alpha_0} \exp\left[\left(\frac{B}{T-T_0}\right)\right] \quad (4.2)$$

T_0 is known as the Vogel-Fulcher temperature and is located about 30 K below T_g. τ_{α_0} is the asymptotic value of the relaxation time of the correlator Φ for $T\to\infty$. Also the rheological shift factors $a_R(T)$ mentioned above approximately follow such temperature dependences [34]:

$$a_R(T) = a_{R_0}(T_R) \exp\left(\frac{B_R}{T-T_{0R}}\right) \quad (4.3)$$

The prefactor $a_{R_0}(T_R)$ is defined such that at the chosen reference temperature T_R, $a_R(T_R)=1$.

However, though the establishment of such general features is very important, the relaxation techniques employed provide only indirect observations of the structural relaxation. In fact, the decay of different correlators in the α-relaxation regime may differ, as shown in Fig. 4.1b.

At this point, we realize the advantages offered by NSE for the investigation of the dynamics in the α-relaxation regime. First of all, in providing space-time information NSE may focus directly on the structural relaxation and observe the relaxation of the interchain structure factor peak. Secondly, microscopic and mesoscopic regions can be explored that are not easily accessed by relaxation techniques. On the other hand, NSE allows to follow different correlators: the pair correlation function (collective dynamics) from the study of coherent scattering in fully deuterated samples or the self-motions of hydrogens in protonated samples. Compared to neutron backscattering, NSE has the advantage of allowing a precise determination of the spectral shape since it directly accesses the correlations in the time domain so that deconvolution from the instrumental resolution is possible. Moreover, NSE offers the best energy resolution and the widest dynamic window among neutron scattering techniques. Since the fastest times accessible by NSE are of the order of picoseconds, the microscopic dynamics – giving rise to the first fast decay of the correlations (see Fig. 4.1) – takes place out of its window. This dynamics, which comprises the Boson peak and the so-called fast dynamics (see e.g. [97–102]), can be investigated by time-of-flight instruments and is usually interpreted in terms of the MCT.

Below T_g, in the glassy state the main dynamic process is the secondary relaxation or the β-process, also called Johari-Goldstein relaxation [116]. Again, this process has been well known for many years in polymer physics [111], and its features have been established from studies using relaxation techniques. This relaxation occurs independently of the existence of side groups in the polymer. It has traditionally been attributed to local relaxation of flexible parts (e.g. side groups) and, in main chain polymers, to twisting or crankshaft motion in the main chain [116]. Two well-established features characterize the secondary relaxation.

It can not be described by means of a single Debye process, but more complicated relaxation functions involving distributions of relaxation times (like the Cole-Cole function [117]) or distributions of energy barriers (like log-normal functions [118]) have to be used for its description. Usually a narrowing of the relaxation function with increasing temperature is observed. The Arrhenius temperature dependence of the associated characteristic time is:

$$\tau_\beta(T) = \tau_{\beta_0} \exp\left(\frac{E_0}{k_B T}\right) \quad (4.4)$$

$\tau_\beta(T)$ is defined for instance as the inverse of the position of the maximum of the relaxation peak $\tau_\beta(T) = \frac{1}{\omega_{\beta\max}}$.

These results are commonly interpreted in terms of distributions of elemental local jump processes [116]. For instance, a superposition of Debye elemental processes with a Gaussian distribution of energy barriers $g(E)$:

$$g(E) \sim \frac{1}{\sqrt{\pi}\sigma} \exp\left[-\left(\frac{E-E_0}{\sigma}\right)^2\right] \tag{4.5}$$

is often invoked for describing this relaxation. Here σ is the width of the distribution and E_0 is the mean activation energy. The resulting relaxation function in either the time or frequency domain reads as:

$$\Phi_\beta(t) = \int_{-\infty}^{+\infty} g(E) \exp\left[-\frac{t}{\tau_{\beta_0} \exp\left(\frac{E}{k_B T}\right)}\right] dE \tag{4.6}$$

$$\Phi_\beta^*(\omega) = \int_{-\infty}^{+\infty} g(E) \frac{1}{1 + i\omega\tau_{\beta_0} \exp\left(\frac{E}{k_B T}\right)} dE \tag{4.7}$$

The origin of the distribution should be related to the inherent disorder in the glass. However, the microscopic nature of the β-relaxation still remains unclear.

Presently NSE cannot access this process within the glassy state since the associated characteristic times below T_g are too long. Nevertheless, the secondary process is also active above T_g and in favourable cases this feature allows its observation in the mesoscopic time window. The α- and β-relaxations, as observed by spectroscopic techniques, coalesce in what is usually called an $\alpha\beta$-process in a temperature range 10%–20% above T_g, which we will refer to as merging temperature T_m. Though T_m depends on the particular system (e.g. on the microstructure), the merging process is usually observed at about 10^{-7} s by relaxation techniques. This fact naturally poses serious difficulties for an isolated observation of the β-process by NSE, and generally the global $\alpha\beta$-process is accessed in the NSE time window.

Finally we would like to mention other dynamic processes in the glassy state, e.g. vibrations including the Boson peak, tunnelling processes, methyl group rotations etc. However, these processes either cannot be accessed by NSE (e.g. vibrations can be studied by time-of-flight spectrometers in the meV region, see [97–102, 119]) or are better investigated by other techniques. This would be the case of methyl group dynamics including tunnelling and stochastic motions [10, 120, 121].

4.1
α-Relaxation

4.1.1
Structural α-Relaxation: Dynamic Structure Factor

The short range order in glasses leads to the presence of broad peaks in the static structure factor $S_{pair}(Q)$. $S_{pair}(Q)$ can be measured by neutron diffraction on fully deuterated samples, reflecting the pair correlation function of deuterons and carbons in an indistinguishable way. A precise assignment of particular atom pair correlations to the different peaks present in $S_{pair}(Q)$ is not evident and requires additional experimental information on different partial structure factors and, usually, comparison with fully atomistic molecular dynamics simulations (see, e.g. [122]). However, phenomenological studies based on the temperature dependence of the different peaks of $S_{pair}(Q)$ allow determination of the nature of the main contributions to these peaks. As an example, Fig. 4.2 shows $S_{pair}(Q)$ for the archetypical polymer 1,4-polybutadiene (PB) [123]. The mean position of its first peak (centred at $Q_{max} \approx 1.5$ Å^{-1}) moves strongly with temperature. This observation implies that this peak mainly originates from interchain correlations, where weak van der Waals interactions give rise to thermal expansion. The second peak (around 3 Å^{-1}) relates mainly to intrachain correlations, as may be deduced from the T-independence of its position indicative for covalent bonds. Similar properties are generally found in all glass-forming polymers. Thus, the predominantly interchain nature of the correlations contributing to the first peak of $S_{pair}(Q)$ seems to be a common feature in polymers, similar to the intermolecular origin of the first peak in low molecular weight glasses.

The dynamic process driving the decay of the interchain correlations in glass-forming polymers is the structural α-relaxation. The direct microscopic obser-

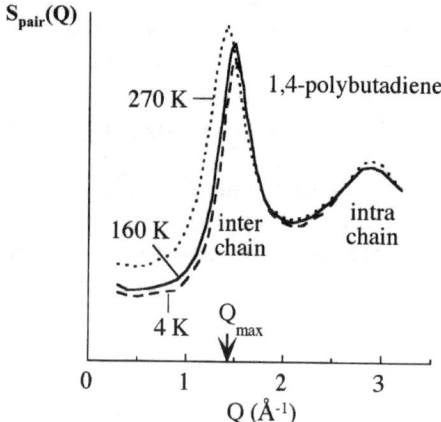

Fig. 4.2 Static structure factor measured on fully deuterated 1,4-polybutadiene [123] at the temperatures indicated

vation of the structural relaxation is realized by looking at the temporal evolution of such correlations, i.e. through the study of the dynamic structure factor at the Q-value corresponding to the first structure factor maximum, $S_{\text{pair}}(Q_{\max},t)$ [let us remember that $S_{\text{pair}}(Q)=S_{\text{pair}}(Q,t=0)$]. NSE studies of $S_{\text{pair}}(Q_{\max},t)$ on a variety of polymers (see, e.g. [8, 124–129]) have established the two main features of the α-process also for the true structural relaxation, namely:

i. The stretching of the relaxation: the spectral shape deviates from the single exponential form and, as previously mentioned for spectroscopic results, can be well described by a Kohlrausch-Williams-Watts (KWW) or stretched exponential function:

$$\frac{S_{\text{pair}}(Q_{\max},t)}{S_{\text{pair}}(Q_{\max},0)} = f_{Q\max}(T) \exp\left[-\left(\frac{t}{\tau_s(T)}\right)^\beta\right] \quad (4.8)$$

Here, $f_{Q\max}(T)$ is a generalized Debye-Waller factor giving account for the decay of the correlations at faster times. In the framework of the MCT it is also called non-ergodicity factor. The characteristic timescale $\tau_s(T)$ is the structural relaxation time and β is the stretching parameter ($0<\beta\leq 1$).

ii. The scaling: the functional shape hardly depends on temperature. Curves corresponding to different temperatures superimpose in a single master curve when they are represented against a reduced time variable that includes a T-dependent shift factor.

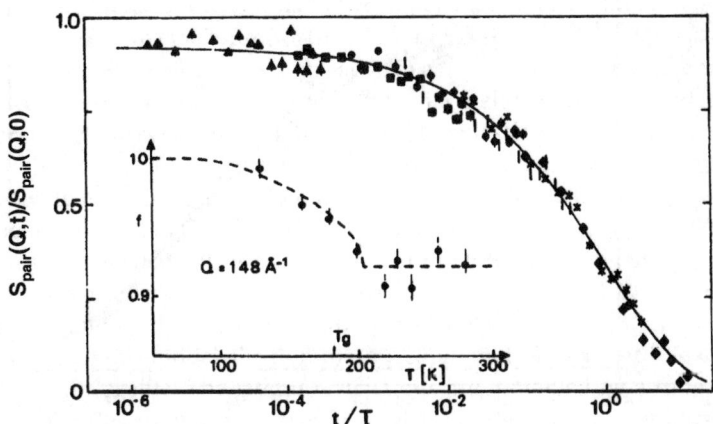

Fig. 4.3 Scaling representation of the spin-echo data at the first static structure factor peak $Q_{\max}$. Different symbols correspond to different temperatures. *Solid line* is a KWW description (Eq. 4.8) of the master curve for 1,4-polybutadiene at $Q_{\max}=1.48$ Å^{-1}. The scale $\tau(T)$ is taken from a macroscopic viscosity measurement [130]. *Inset*: Temperature dependence of the non-ergodicity parameter $f(Q)$ near $Q_{\max}$; the lines through the points correspond to the MCT predictions (Eq. 4.37) (Reprinted with permission from [124]. Copyright 1988 The American Physical Society)

These two features are demonstrated in Fig. 4.3 for 1,4-polybutadiene (PB), in Fig. 4.4 for polyisobutylene (PIB) and atactic polypropylene (aPP), in Fig. 4.5 for polyurethane (PU) and poly(vinyl chloride) (PVC) and in Fig. 4.6 for poly(vinyl ethylene) (1,2-polybutadiene, PVE) (see Table 1.1 for microstructure). In each case $S_{pair}(Q_{max},t)/S_{pair}(Q_{max},0)$ is very well described by the KWW laws.

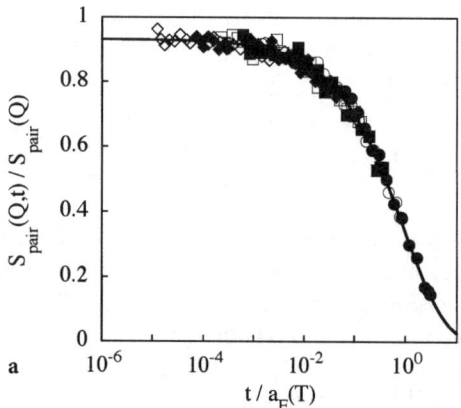

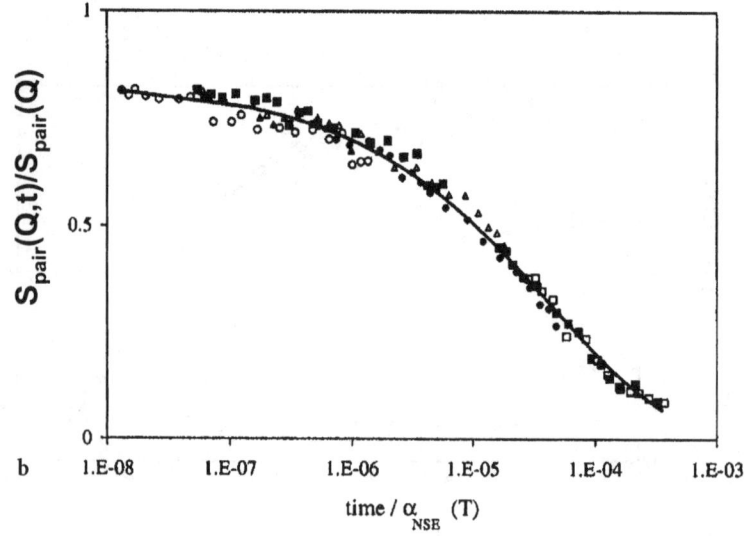

Fig. 4.4 Scaling representation of the spin-echo data at Q_{max}. Different symbols correspond to different temperatures. *Solid line* is a KWW description (Eq. 4.8) of the master curve. The shift factors are taken from macroscopic viscosity measurements. **a** Polyisobutylene at $Q_{max}=1.0$ Å^{-1} [125] measured on IN11 (viscosity data from [34]). **b** Atactic polypropylene at $Q_{max}=1.11$ Å^{-1} (viscosity data from [131]). (**b** Reprinted with permission from [126]. Copyright 2001 Elsevier)

4 Local Dynamics and the Glass Transition

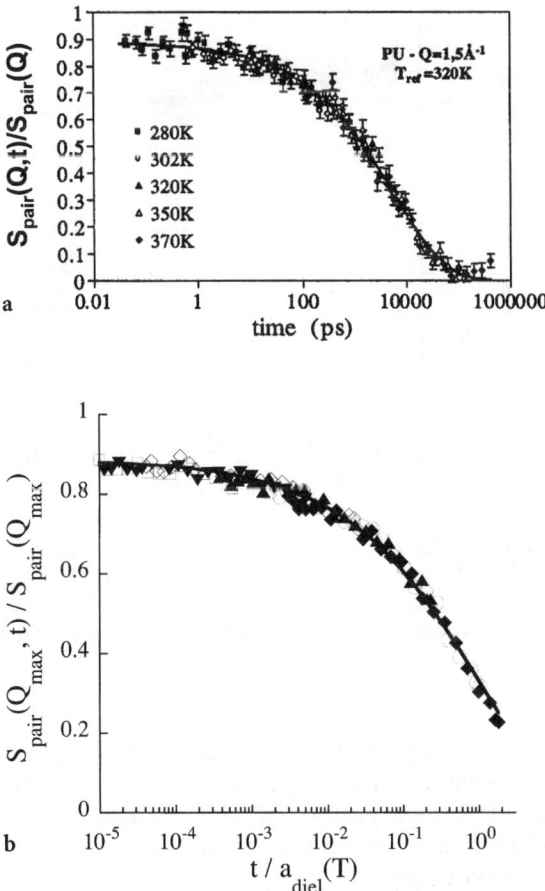

Fig. 4.5 Scaling representation of the spin-echo data at Q_{max}. Different symbols correspond to different temperatures. *Solid line* is a KWW description (Eq. 4.8) of the master curve. **a** Polyurethane at $Q_{max}=1.5$ Å^{-1}. The shift factors have been obtained from the superposition of the NSE spectra. (Reprinted with permission from [127]. Copyright 2002 Elsevier). **b** Poly(vinyl chloride) at $Q_{max}=1.2$ Å^{-1}. The shift factors have been obtained from dielectric spectroscopy. (Reprinted with permission from [129]. Copyright 2003 Springer, Berlin Heidelberg New York)

The values of the stretching parameter for all the polymers investigated so far are listed in Table 4.1. The β-values are found to be in the range 0.4–0.6. The superposition of the different spectra into a single master curve indicates that the time–temperature superposition principle is fulfilled to a large extent. Individual fits of Eq. 4.8 to the decaying curves may, however, lead to slightly increasing values of β with increasing temperature, as reported in [127] for PU. The uncertainties in the determination of the β-value were discussed in [129] for the case of PB.

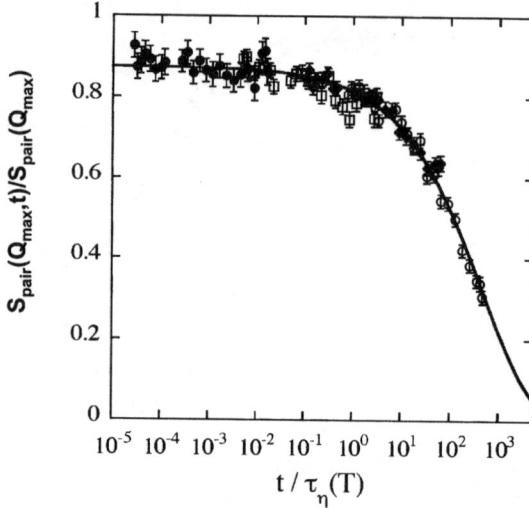

Fig. 4.6 Scaling representation of the spin-echo data corresponding to poly(vinyl ethylene) at $Q_{max}=1.0$ Å^{-1} and 340 K (*empty circle*), 320 K (*filled diamond*), 300 K (*empty square*, taken as reference temperature), 280 K (*filled circle*). *Solid line* is a KWW description (Eq. 4.8) of the master curve with $\beta=0.43$. The scale $\tau(T)$ is taken from a macroscopic viscosity measurement [132]

Concerning this scaling universality we have to add the observation that in all the reported studies the shift factors based on viscosity measurements allow a fairly accurate superposition of the microscopic data [8, 124–126] (Figs. 4.3a and 4.4). The temperature dependence of the characteristic time associated with the viscosity $\tau_\eta(T)$ is related to the temperature-dependent rheological shift factors (Eq. 4.3) through $\tau_\eta(T) \sim a_R(T)/T$ [34]. Considering the steep temperature dependence of $a_R(T)$ and the narrow temperature range usually investigated by NSE, the $1/T$ correction factor is sometimes neglected. By shifting the NSE timescale with $\tau_\eta(T)$ well-defined master curves are obtained. This also holds for the case of PVE data, shown here for the first time (Fig. 4.6). This implies that the macroscopic viscosity is directly related to the structural relaxation leading to the decay of the interchain correlations at a microscopic level. It is noteworthy that the relation between the structural relaxation observed with neutrons at the structure factor maximum and viscosity flow also holds for molecular glass formers such as ortho-terphenyl [134]. For different polymers investigated to date, Table 4.1 compiles the laws describing the temperature dependence of the structural relaxation time. In each case, the Vogel-Fulcher parameters have been fixed from the macroscopic measurements used to scale the NSE data.

At this point the question arises whether the feature usually termed as "universality" is also fulfilled: do the timescales deduced from the investigation of different correlators in the α-relaxation regime show the same temperature

4 Local Dynamics and the Glass Transition

Table 4.1 Parameters related to the structural relaxation for the polymers investigated by NSE: glass transition temperature T_g, position of the first static structure factor peak Q_{max}, shape parameter β, magnitude considered to perform the scaling of the NSE data, and temperature dependence of the structural relaxation time τ_s

Polymer	T_g (K)	Q_{max} (Å^{-1})	β	Scaled with	τ_s (ns)	Ref.
PB	181	1.48	0.45	a_R/T	$\dfrac{5.9e-3}{T}\exp\left(\dfrac{1404.5}{T-128}\right)$	[124]
PI	205–213	1.44[a]	0.38	τ_{DS}		[8]
PIB	205	1.0	0.55	a_R	$3.9e-7\exp\left(\dfrac{4369.7}{T-89.2}\right)$	[125]
aPP-I[b]	210	1.1	0.59	a_R	$5.0e-5\exp\left(\dfrac{1284.8}{T-168}\right)$	[126]
aPP-II[b]	237		0.52	a_R	$5.0e-5\exp\left(\dfrac{1284.8}{T-200}\right)$	
PU	250	1.5	0.43[c]	τ_{DS}	$2.5e-6\exp\left(\dfrac{1789}{T-194}\right)$	[127]
PVC	358	1.2	0.50	τ_{DS}	$9.6e-4\exp\left(\dfrac{1005.6}{T-317.24}\right)$	[128, 129]
PVE	272	1.0	0.43	a_R/T	$\dfrac{8.4e-2}{T}\exp\left(\dfrac{1032.8}{T-226}\right)$	This work

[a] Estimated from diffraction measurements at low temperature. In the temperature range investigated Q_{max} is close to 1.3 Å^{-1} [122].
[b] Two samples with different molecular weights were studied I: M_n=1270 g/mol, II: M_n= 2570 g/mol (M_n number-average molecular weight).
[c] If the non-ergodicity parameter is fixed to a T-independent value, an increase of β from 0.36 to 0.45 with increasing T in the T-range studied is reported.

dependence as the structural relaxation directly observed from NSE at Q_{max}? The available results concerning this question are summarized in the following section.

4.1.1.1
Dielectric Spectroscopy (DS)

In most cases where the comparison has been made it seems that the temperature dependencies obtained are very much compatible. This is clearly demonstrated for PI [8, 9] and for PVC (Fig. 4.5b) [128, 129], where the shift factors

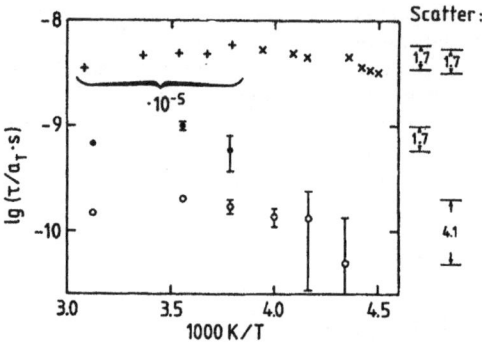

Fig. 4.7 Temperature dependence of the mean relaxation time $\langle\tau\rangle$ divided by the rheological shift factor for the dielectric normal mode (*plus*) the dielectric segmental mode (*cross*) and NSE at Q_{max}=1.44 Å^{-1} (*empty circle*) and Q=1.92 Å^{-1} (*empty square*) [7] (Reprinted with permission from [8]. Copyright 1992 Elsevier)

used for building the NSE master curves are indeed those obtained from DS. Figure 4.7 [8] shows that for PI the thermal behaviour of both the segmental mode as well as the normal mode (reflecting the end-to-end vector dynamics) investigated by DS are the same as those of the viscosity and the structural relaxation time. Also for PB a very nice agreement is found between the DS results on the α-relaxation and the viscosity results – and hence the structural relaxation results from NSE – as can be appreciated in Fig. 4.8 [133]. However, in this case, it is important to emphasize that this coincidence is only achieved when the influence of the secondary β-relaxation is removed from the DS spectra [133]. If the maxima of the global DS spectra are considered as representative for the α-process (dotted line in Fig. 4.8), a clear disagreement is found with the structural relaxation data. The separation of the contributions of the two processes to the DS spectra could be realized by applying a deconvolution procedure based on the statistical independence of the structural and secondary relaxations. This hypothesis was supported by a parallel study of the dynamic structure factor by NSE in a wide Q- and T-range (see below). For PIB the dielectric strength is very weak, preventing a good characterization of the temperature dependence of the dielectric relaxation time at high temperatures, i.e. in the overlapping region with the NSE studies on the dynamic structure factor. Figure 4.9 shows that the thermal behaviour of both sets of data could be compatible [125]. Finally, the comparison of DS results and the structural relaxation times in PU (Fig. 4.10) shows a very good agreement with respect to their temperature dependencies [127].

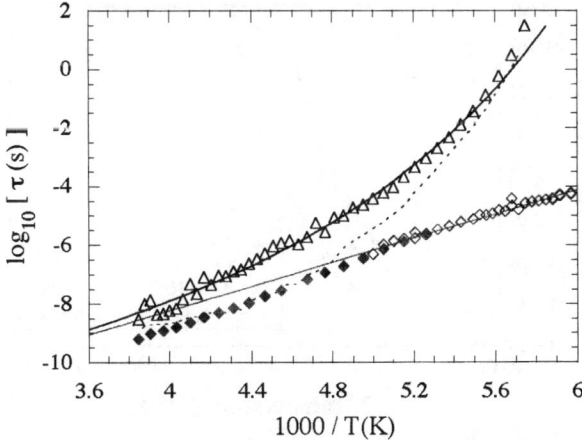

Fig. 4.8 Temperature dependence of the dielectric characteristic times obtained for PB: τ_α for the α-relaxation (*empty triangle*) for the τ^β_{max}-relaxation (*empty diamond*), and $\tau^{\beta eff}_{max}$ for the contribution of the β-relaxation modified by the presence of the α-relaxation (*filled diamond*). They have been obtained assuming the α- and β-processes as statistically independent. The Arrhenius law shows the extrapolation of the temperature behaviour of the β-relaxation. The *solid line* through τ_α points shows the temperature behaviour of the time-scale associated to the viscosity. The *dotted line* corresponds to the temperature dependence of the characteristic timescale for the main peak. (Reprinted with permission from [133]. Copyright 1996 The American Physical Society)

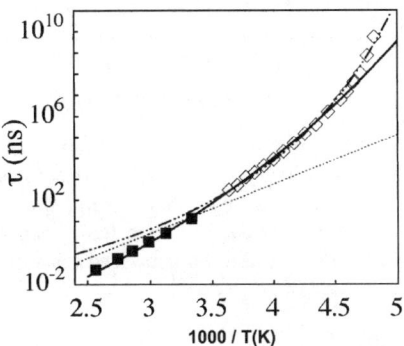

Fig. 4.9 Temperature dependence of the characteristic time of the α-relaxation in PIB as measured by dielectric spectroscopy (defined as $(2\pi f_{max})^{-1}$) (*empty diamond*) and of the shift factor obtained from the NSE spectra at $Q_{max}=1.0$ Å^{-1} (*filled square*). The different lines show the temperature laws proposed by Törmälä [135] from spectroscopic data (*dashed-dotted*), by Ferry [34] from compliance data (*solid*) and by Dejean de la Batie et al. from NMR data (*dotted*) [136]. (Reprinted with permission from [125]. Copyright 1998 American Chemical Society)

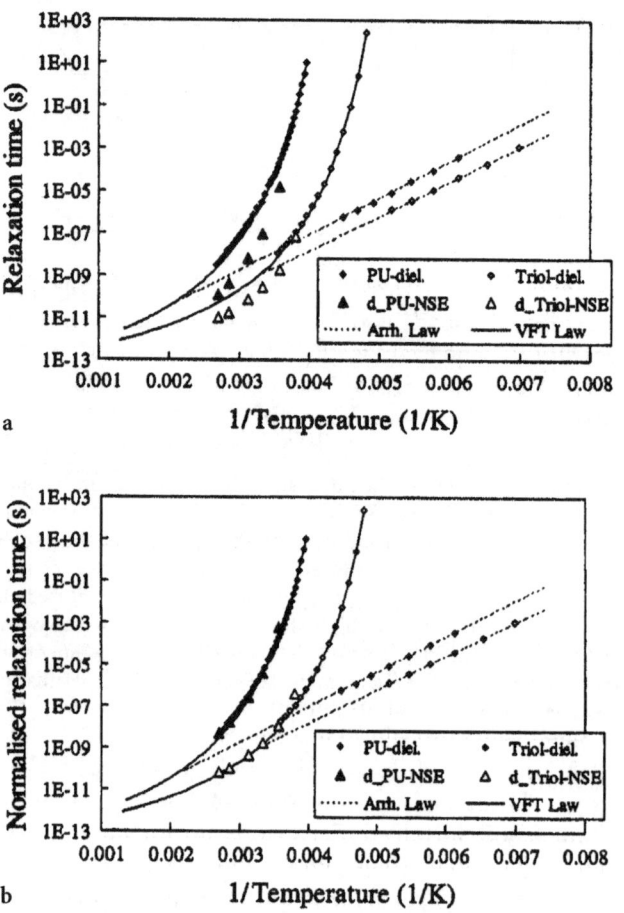

Fig. 4.10 a Characteristic relaxation times determined from dielectric measurements [137] (*diamonds*), and from NSE spectra at Q_{max} (*triangles*) for triol (*open symbols*) and PU (*solid symbols*). The *full lines* correspond to Vogel-Fulcher and the *dotted lines* to Arrhenius descriptions. **b** Relaxation times from NSE spectra have been arbitrarily multiplied by a factor 6 for triol and 40 for PU to build a normalized relaxation map. (Reprinted with permission from [127]. Copyright 2002 Elsevier)

4.1.1.2
Nuclear Magnetic Resonance (NMR)

For PIB the apparent activation energy found for the structural relaxation time in the NSE window is almost twice that determined by ^{13}C NMR [136] (see Fig. 4.9 [125]). For aPP, the temperature dependence of ^{13}C NMR results [138] seems, however, to be quite compatible with that of the NSE data; nevertheless, 2D exchange NMR studies on this polymer [139] reveal a steeper dependence. This can be seen in Fig. 4.11 [126].

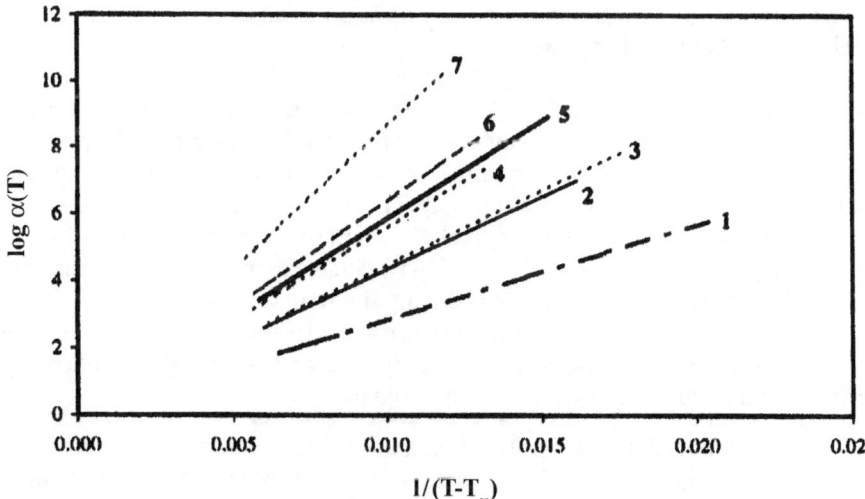

Fig. 4.11 Temperature dependence of the shift factors as reported in the literature for atactic polypropylene: *1* dynamic mechanical measurements [140], *2* NMR data of Pschorn et al. [141], *3* photon correlation spectroscopy [142], *4* from ^{13}C NMR measurements of Moe et al. [138], *5* terminal relaxation measured from dynamic mechanical analysis [143], *6* from viscosity measurements [131] and *7* from 2D exchange NMR and spin relaxation times [139]. (Reprinted with permission from [126]. Copyright 2001 Elsevier)

4.1.1.3
Results from Other Spectroscopic Techniques and Photon Correlation Spectroscopy

Results from other spectroscopic techniques and photon correlation spectroscopy have been compared for aPP in [126] (see Fig. 4.11). A scaling of the dynamic structure factor at Q_{max} could not be achieved on the basis of the dynamic data reported in [140]. The other temperature dependencies obtained seem to be compatible with the neutron data. Finally, the temperature dependence deduced by Törmälä for PIB from the compilation of different spectroscopic data does not agree with the result of the microscopic observation of the structural relaxation (see Fig. 4.9 [125]).

From this comparison it follows that the observation of the structural relaxation by standard relaxation techniques in general might be hampered by contributions of other dynamic processes. It is also noteworthy that the structural relaxation time at a given temperature is slower than the characteristic time determined for the α-relaxation by spectroscopic techniques [105]. An isolation of the structural relaxation and its direct microscopic study is only possible through investigation of the dynamic structure factor at the interchain peak – and NSE is essential for this purpose.

4.1.2
Self-Atomic Motions of Protons

An additional microscopic insight into the structural relaxation can be obtained by means of NS through an *indirect* probe as it is the scattered intensity from protonated samples. As explained in Sect. 2, this results in the intermediate incoherent scattering function $S_{self}(Q,t)$, which is the Fourier transform of the self part of the van Hove correlation function $G_s(r,t)$ corresponding to the hydrogens in the system (see, e.g. [144]). Such experiments investigate the individual motion of the hydrogens during the structural relaxation.

To date, incoherent quasi-elastic neutron scattering experiments on the α-relaxation regime of glass-forming polymers have revealed the following main features for the self-motion of hydrogens:

i. The stretched time behaviour of $S_{self}(Q,t)$. This is easily recognized from Fig. 4.12a. There $S_{self}(Q,t)$ measured by NSE on protonated PI with deuterated methyl groups PId3, relating thus to the main chain hydrogens of the polymer, is shown for different Q-values at the same temperature. The dotted line shows a single exponential decay. Evidently, the experimental data decay in a markedly slower way.
ii. A strongly Q-dependent characteristic time indicating a diffusive-like character (see Fig. 4.13). Without any kind of data evaluation, it is clear that the decays in the time domain (Fig. 4.12a) take place in a time range that clearly becomes faster with increasing Q-value. In a similar way, the spectra measured by backscattering techniques on fully protonated PB 110 K above T_g (Fig. 4.12b) present an increasing line broadening when the Q-value increases, implying a decrease of the associated timescale with increasing Q. Dispersion is a signature of diffusion [150]. Such a diffusive character of the hydrogen motion in the α-relaxation regime might be somehow striking since most of the investigations on the α-relaxation have been performed by relaxation techniques. As in most of these techniques reorientational motions are probed, traditionally a rotational character has been implicitly assumed for this process.

To date most of the studies performed on the incoherent scattering of protonated polymers in the α-relaxation regime have been realized by using BS techniques [146, 148, 151]. These investigations, that in general are restricted to the Q-range $0.2 \leq Q \leq 1.5$ Å^{-1}, have established the two features discussed above. In fact, following the common wisdom and assuming again a KWW function for $S_{self}(Q,t)$:

$$S_{self}(Q,t) = A(Q,T) \exp\left[-\left(\frac{t}{\tau_{self}(Q,T)}\right)^{\beta}\right] \quad (4.9)$$

where τ_{self} is the Q- and T-dependent KWW-relaxation time and $\beta<1$ the stretching exponent. Good representations of the experimental results were ob-

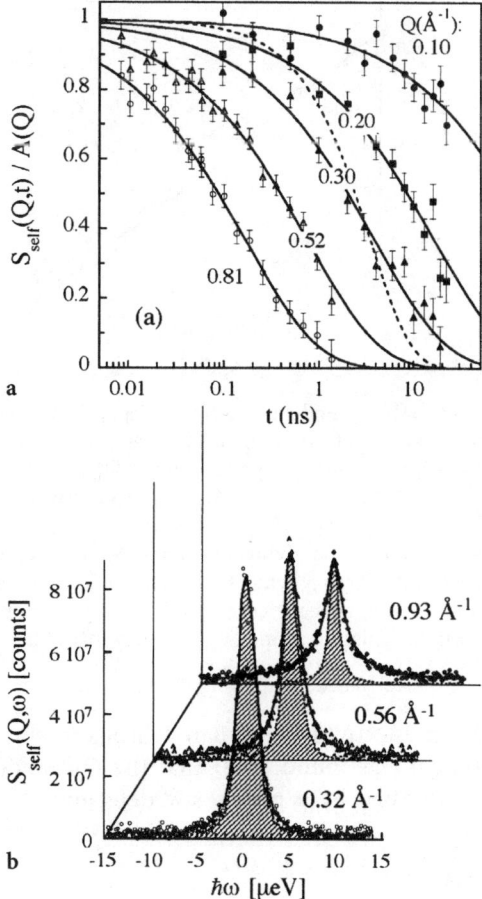

Fig. 4.12 a Incoherent intermediate scattering function normalized to the value of the LMF obtained from NSE (*full symbols* FZ-Jülich, *empty* ILL Grenoble) on PId3 at 340 K and the Q-values indicated [9]. *Solid lines* are KWW descriptions with β=0.57. The *dotted line* shows a single exponential decay for comparison. b Incoherent scattering function measured by the BSS Jülich for PB at 280 K and the Q-values indicated. The *hatched areas* under the dotted lines show the instrumental resolution functions and the *solid lines* are fitting curves based on KWW descriptions. (b Reprinted with permission from [145]. Copyright 2002 Springer, Berlin)

tained. In this comparison the Fourier transform of Eq. 4.9 has to be convoluted with the instrumental resolution function and fitted to the experimental data. An example is shown in Fig. 4.12b. The prefactor $A(Q,T)$ in Eq. 4.9 is a Lamb-Mössbauer factor (LMF) or Debye-Waller factor (DWF) accounting for faster processes:

$$A(Q,T) = \exp\left(-\frac{\langle u^2 \rangle}{3} Q^2\right) \tag{4.10}$$

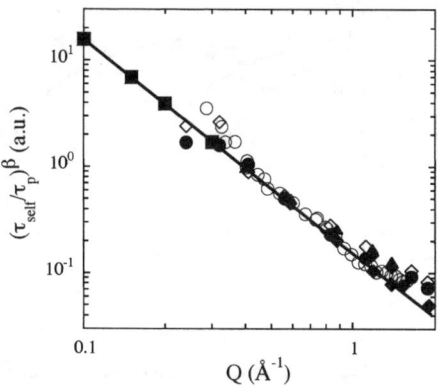

Fig. 4.13 Momentum transfer dependence of the characteristic time associated to the self-motion of protons in the α-relaxation regime: Master curve (time exponentiated to β) constructed with results from six polymers: polyisoprene (340 K, β=0.57) (*filled square*) [9]; polybutadiene (280 K, β=0.41) (*filled circle*) [146]; polyisobutylene (390 K, β=0.55) (*empty circle*) [147]; poly(vinyl methyl ether) (375 K, β=0.44) (*filled triangle*) [148]; phenoxy (480 K, β=0.40) (*filled diamond*) [148] and poly(vinyl ethylene) (340 K, β=0.43) (*empty diamond*) [146]. The data have been shifted by a polymer dependent factor τ_p to obtain superposition. The *solid line* displays a Q^{-2}-dependence corresponding to the Gaussian approximation (Eq. 4.11). (Reprinted with permission from [149]. Copyright 2003 Institute of Physics)

It is characterized by an effective mean squared displacement of the proton $\langle u^2 \rangle$.

From such studies it was found [148] that the Q-dependence of τ_{self} can approximately be described by a power law determined by the stretching exponent β:

$$\tau_{\text{self}}(Q,T) = a_0(T)Q^{-2/\beta} \tag{4.11}$$

where $a_0(T)$ does not depend on Q. As polymers show typical values for β of around 0.4–0.6, Eq. 4.11 predicts much stronger Q-dependencies than that characteristic for simple diffusion (Q^{-2}). Such dispersions have been found in a large number of polymer systems [146, 147, 148, 152, 153] in the low Q-regime ($Q \leq 1$ Å^{-1} approximately). As a demonstration, a master curve considering data corresponding to six different polymer systems is depicted in Fig. 4.13. Since the β-value depends on the system investigated, in order to check the generality of Eq. 4.11 the representation of Fig. 4.13 shows the results for τ_{self} exponentiated to β. Polymer-dependent shift factors have then been applied to superimpose the data in a single curve. In such a plot, the Q-dependence expressed by Eq. 4.11 translates in a Q^{-2}-law. It is clear that the power law proposed (solid line) accounts for the Q-dependence experimentally found in all these systems for $Q \leq 1$ Å^{-1}. The good agreement between different data sets implies universal behaviour.

Though the BS experiments have been very useful for gaining insight into the hydrogen motions taking place during α-relaxation, they are affected by a number of shortcomings:

i. Due to the convolution with the instrumental resolution, an accurate determination of the spectral shape is not possible. Therefore, the value of β has to be determined by other methods or from coherent scattering measurements.
ii. The dynamic range is quite narrow.
iii. The energy or time resolution is worse than that of NSE.

NSE overcomes these problems. Though NSE studies on incoherently scattering samples are more difficult to perform than on coherent scatterers, careful measurements on PIB and later on PI have led to results as nice as those depicted in Fig. 4.14a and Fig. 4.12a, respectively. As shown in these figures, the assumption of a KWW functional form is indeed very adequate for the description of $S_{\text{self}}(Q,t)$. Moreover, precise values for the stretching exponent can be obtained from such experimental curves. For PIB a β-value of 0.55 delivers good quality fits in the whole T- and Q-range investigated [147]. This is shown in Fig. 4.14a for several Q-values at the highest temperature investigated, 390 K. The value obtained for β in the incoherent case denotes the same stretching as that observed for the structural relaxation in this polymer (see Table 4.1). While in the case of PIB the time–temperature superposition principle seems to be fulfilled for $S_{\text{self}}(Q,t)$, the results on PI show a tendency of the shape parameter β to increase with increasing temperature (from about 0.40 to 0.57 in 280 K$\leq T \leq$340 K). The reported value for β obtained from the dynamic structure factor at 1.44 Å^{-1}, i.e. close to Q_{max}, is 0.38 [8]. A similar value has also been found from DS measurements [8]. Regarding a possible Q-dependence of the stretching, from the experimental spectra it is not possible to resolve a clear variation. The data are compatible with a description in terms of a Q-independent β-value.

In addition, NSE measurements have allowed confirmation of the relation between non-Debye behaviour and Q-dispersion of τ_{self} (Eq. 4.11) proposed from the early BS studies of $S_{\text{self}}(Q,t)$ [148]. Figure 4.14b displays for PIB the characteristic times determined from the NSE experiments. The law $Q^{-2/\beta}$ provides a nearly perfect description of the Q-dependence of these times, at least in the Q-range below 1 Å^{-1}. Above this value, some deviations towards a weaker dependence could be envisaged. But, within the uncertainties and below 1 Å^{-1}, it is clear that (i) the Q-dependence is well captured by the proposed law $\tau_{\text{self}} \propto Q^{-2/\beta}$ and (ii) it is fulfilled for all temperatures investigated. This is evidenced by the factorization of the Q- and T-dependencies of τ_{self}. Analogous results could be deduced from the PI study [9, 154] in a similar Q-range. Thus, the NSE studies on these two polymers provide strong support to the previous BS results on the Q-dependence of τ_{self}. This can also be realized from Fig. 4.13, which includes the times determined from NSE for PI and PIB. Very recent NSE experiments on a third polymer, PVE [155] already show some slight deviations from Gaussian behaviour at $Q \approx 0.5$ Å^{-1}, which will be discussed later.

We may now discuss the implications of the results found for the self-motion of hydrogens in the α-relaxation regime by neutron scattering. It is well known that for some simple cases – free nuclei in a gas, harmonic crystals,

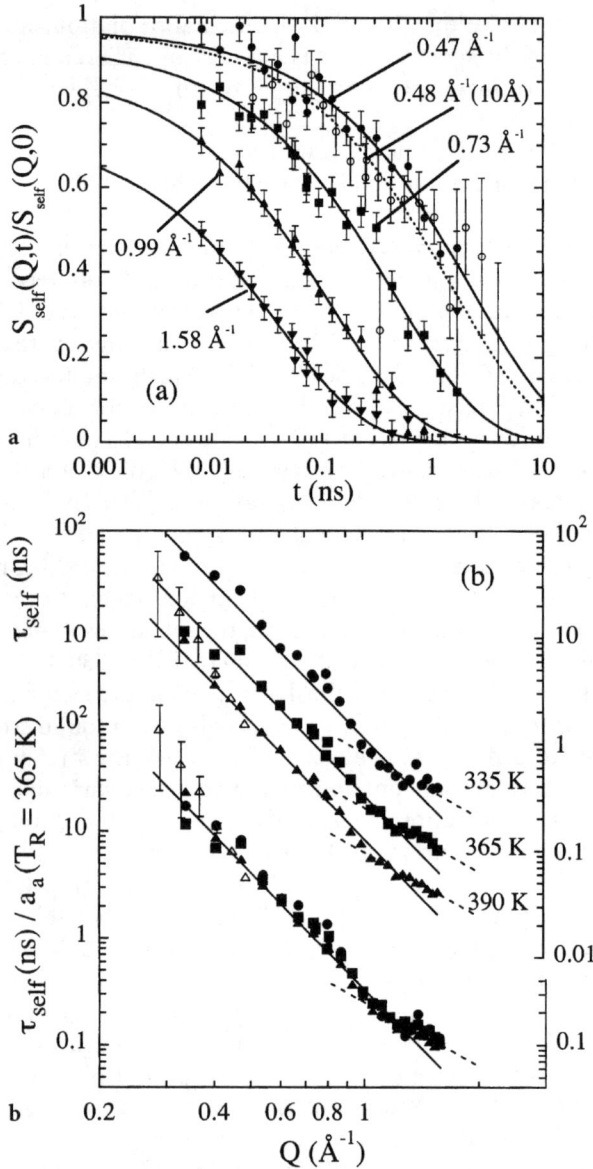

Fig. 4.14 Results on fully protonated PIB by means of NSE [147]. **a** Time evolution of the self-correlation function at the Q-values indicated and 390 K. *Lines* are the resulting KWW fit curves (Eq. 4.9). **b** Momentum transfer dependence of the characteristic time of the KWW functions describing $S_{self}(Q,t)$ at 335 K (*circles*), 365 K (*squares*) and 390 K (*triangles*). In the scaling representation (*lower part*) the 335 K and 390 K data have been shifted to the reference temperature 365 K applying a shift factor corresponding to an activation energy of 0.43 eV. *Solid (dotted) lines* through the points represent $Q^{-2/0.55}$ (Q^{-2}) power laws. *Full symbol*: λ=6 Å, *empty symbols* λ=10 Å (Reprinted with permission from [147]. Copyright 2002 The American Physical Society)

simple diffusion at long times – $G_s(r,t)$ is a Gaussian function [150, 156]; in an isotropic system this implies:

$$G_s^{\text{gauss}}(r,t) = \left[\frac{\alpha(t)}{\pi}\right]^{3/2} \exp[-\alpha(t)\,r^2] \tag{4.12}$$

In such a Gaussian case the intermediate scattering function is entirely determined by the mean squared displacement of the atom $\langle r^2(t)\rangle$:

$$S_{\text{self}}^{\text{gauss}}(Q,t) = \exp\left[-\frac{\langle r^2(t)\rangle}{6}Q^2\right] \tag{4.13}$$

In the light of the results described above (Eq. 4.9, Eq. 4.10 and Eq. 4.11), for the polymer hydrogens in the α-regime the full expression for $S_{\text{self}}(Q,t)$ reads:

$$\begin{aligned}S_{\text{self}}(Q,t) &= \exp\left(-\frac{\langle u^2\rangle}{3}Q^2\right)\exp\left[-\left(\frac{t}{a_0(T)\,Q^{-2/\beta}}\right)^{\beta}\right]\\ &= \exp\left(-\frac{\left\{2\langle u^2\rangle + 6\left[\dfrac{t}{a_0(T)}\right]^{\beta}\right\}}{6}Q^2\right)\end{aligned} \tag{4.14}$$

We immediately realize that this function has exactly the Q-dependence predicted by the Gaussian approximation (Eq. 4.13). From the comparison between these two expressions (Eq. 4.14 and Eq. 4.13), the time-dependent mean square displacement of the hydrogens can easily be extracted:

$$\langle r^2(t)\rangle = 2\langle u^2\rangle + 6\left[\frac{t}{a_0(T)}\right]^{\beta} \tag{4.15}$$

This finding implies a sublinear increase of $\langle r^2(t)\rangle$ with time. Thus, the incoherent neutron scattering studies qualify the motion of the hydrogens as an anomalous diffusion involving a sublinearly increasing $\langle r^2(t)\rangle$.

The fact that the Gaussian approximation is fulfilled in the Q-range $Q \leq 1$ Å^{-1} has important implications concerning the origin of the non-exponential nature of the α-process [146, 153]. The understanding of the microscopic mechanism behind this peculiar behaviour is currently a topic of strong debate (see, e.g. [157] and references therein). Two limiting scenarios are invoked to explain the KWW functional form. Since a monotonous function can always be written as Laplace transform of a non-negative function, the KWW function can easily be interpreted as arising from the superposition of different simple exponential relaxations weighted by a broad distribution of relaxation times $g(\ln \tau)$:

$$\Phi(t) = \exp\left[-\left(\frac{t}{\tau_w}\right)^{\beta}\right] = \int_{-\infty}^{+\infty} g(\ln \tau)\exp\left(-\frac{t}{\tau}\right)d(\ln \tau) \tag{4.16}$$

This picture is usually known as the "heterogeneous" scenario. The distribution of relaxation times g (ln τ) can be obtained from $\Phi(t)$ by means of inverse Laplace transformation methods (see, e.g. [158] and references therein) and for β=0.5 it has an exact analytical form. It is noteworthy that if this scenario is not correct, i.e. if the integral kernel, exp($-t/\tau$), is conceptually inappropriate, g(ln τ) becomes physically meaningless. The other extreme picture, the "homogeneous" scenario, considers that all the particles in the system relax identically but by an intrinsically non-exponential process.

The "heterogeneous" picture was involved in some of the first theoretical approaches to the dynamics of supercooled liquids [159–161]. This scenario has also recently been invoked in connection with the question of cooperative rearranging regions in glass-forming systems [162]. On the other hand, most of the experimental work accumulated over the last few years was analysed in the framework of the MCT [95, 96, 106], which does not address this question. Recent experimental results from different spectroscopic techniques (see, e.g. [163–165]) and computer simulations in Lennard-Jones systems [166, 167] have stimulated a new revival of the "heterogeneous" picture, which is usually related to some kind of spatial heterogeneity. It has been suggested [164] that the size of the spatial heterogeneities close to the glass transition temperature T_g should range in the nanometer scale (20–50 Å for low molecular weight glass-forming systems and up to about 100 Å for polymers). The above-mentioned results, in particular those related to the so called rotation-translation paradox [168], seem also to suggest that the spatial heterogeneities mainly develop in the temperature range below about 1.2 T_g, while at higher temperatures the glass-forming matrix appears as essentially homogeneous. For glass-forming polymers, on the other hand, it has also been suggested [169] that a heterogeneous structure should be present even at $T>1.2\ T_g$.

The above experimental results largely relate to spectroscopic techniques, which do not give direct information about the spatial scale of the molecular motions. The size of the spatial heterogeneities is estimated by indirect methods such as sensitivity of the dynamics to the probe size or from the differences between translational and rotational diffusion coefficients (rotation-translation paradox). It might be expected that the additional spatial information provided by neutron scattering could help to discriminate between the two scenarios proposed.

In the "heterogeneous" scenario, and in parallel with the procedure followed for developing Eq. 4.16, we may consider a distribution of local diffusivities associated with the different regions of the sample. The resulting intermediate scattering function can be expressed as:

$$S_{\text{self}}(Q,t) = \exp\left[-\left(\frac{t}{\tau_w}\right)^\beta\right] = \int_{-\infty}^{+\infty} g(\ln D^{-1}) \exp\left(-\frac{Q^2 t}{D^{-1}}\right) d(\ln D^{-1}) \qquad (4.17)$$

where exp($-Q^2Dt$) is the intermediate scattering function corresponding to simple diffusion in the Gaussian approximation and D the corresponding diffusion coefficient. As in the case of Eq. 4.16, by properly choosing the distribution $g(\ln D^{-1})$ a KWW time dependence of the resulting $S_{\text{self}}(Q,t)$ can be easily reproduced. However, now the "stretched variable" is $X=Q^2 t$, i.e. the "conjugated" variable (showing the same dimension) of the distributed magnitude D^{-1}. As consequence, the resulting $S_{\text{self}}(Q,t)$ reads as:

$$S_{\text{self}}(Q,t) = \exp\left[-\left(\frac{Q^2 t}{D_w^{-1}}\right)^\beta\right] = \exp\left[-\left(\frac{t}{D_w^{-1} Q^{-2}}\right)^\beta\right] \qquad (4.18)$$

which can also be written in the phenomenological KWW form of Eq. 4.9 with $\tau_w = Q^{-2} D_w^{-1}$. Therefore, in the "heterogeneous" scenario the Q-dependence of the phenomenological KWW relaxation time results to be the same as the Q-dependence of each of the elementary diffusion times associated to each spatial region: $\tau=Q^{-2}D^{-1}$. This is in clear contradiction with the experimental results collected so far for the Q-dependence of the characteristic time of the hydrogen self-motions in the α-relaxation regime (Eq. 4.11). We conclude that neutron scattering studies on a microscopic level rule out the origin of the non-exponential character of the α-relaxation as a superposition of single exponential diffusive processes.

An anomalous sublinear time-dependence of the mean square displacement is obtained in the framework of many different theoretical approaches of transport in disordered systems in general (see for example [170] as a comprehensive review). Most of these approaches are based on random walks in fractal structures, which are considered as good models for the geometrical structure of most disordered materials. However, it is worth emphasizing that, independently of the geometrical structure considered, these approaches should be classified as homogeneous, in the meaning that the sub-linear diffusion does not emerge as a consequence of a superposition of regular diffusion processes weighted by a distribution of local diffusivities.

In general, deviations of $G_s(r,t)$ from the Gaussian form (Eq. 4.12) may be expected in certain space-time domains. These can be quantified in lowest order by the so-called second order non-Gaussian parameter α_2 defined as [171]:

$$\alpha_2(t) = \frac{3}{5} \frac{\langle r^4(t) \rangle}{\langle r^2(t) \rangle^2} - 1 \qquad (4.19)$$

where the moments are defined in the usual way:

$$\langle r^{2n} \rangle = \int_0^\infty r^{2n}\, G_s(r,t)\, 4\pi r^2\, dr \qquad (4.20)$$

In the Gaussian approximation (Eq. 4.12) the mean squared displacement is given by $\langle r^2(t) \rangle = 3/[2\alpha(t)]$, and $\alpha_2(t)$ is zero of course. In the light of the above results obtained by neutron scattering (summarized in Eq. 4.14), the values of the non-Gaussian parameter for this process should be very small. However, this result is in apparent contradiction to recent molecular dynamics (MD) simulation results on glass forming systems of different nature like water [172], Lennard-Jones liquids [166], selenium [173] and orthoterphenyl [174]. In all these cases, an almost universal behaviour for this parameter is found (see e.g. [173]): it takes positive values showing a maximum that shifts towards longer times with decreasing temperature. Experimentally, the validity of Eq. 4.11 had been mainly checked in the restricted range $Q \leq 1$ Å^{-1}. We note, however, that in the case of poly(vinyl methyl ether) (PVME), where the study was extended up to $Q \approx 5$ Å^{-1} by the thermal BS instrument IN13 at the Institut Laue Langevin (ILL, Grenoble, France), indications of deviations from Eq. 4.11 appear at high Q [153].

These two facts motivated a critical check of the validity of Eq. 4.11 in a wide Q-range [9, 105, 154, 155]. For this purpose the information obtainable from fully atomistic MD simulations was essential. The advantage of MD simulations is that, once they are validated by comparison with results on the real system, magnitudes that cannot be accessed by experiments can be calculated, as for example the time dependence of the non-Gaussian parameter. The first system chosen for this goal was the archetypal polymer PI. The analysis of the MD simulations results [105] on the self-motion of the main chain hydrogens was performed in a similar way to that followed with experimental data. This led to a confirmation of Eq. 4.11 beyond the uncertainties for $Q \leq 1.3$ Å^{-1} (see Fig. 4.15). However, clear deviations from the Q-dependence of the Gaussian behaviour

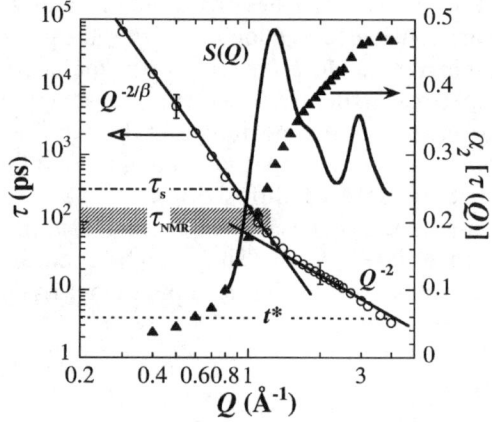

Fig. 4.15 Momentum transfer (Q)-dependence of the characteristic time $\tau(Q)$ of the α-relaxation obtained from the slow decay of the incoherent intermediate scattering function of the main chain protons in PI (O) (MD-simulations). The *solid lines* through the points show the Q-dependencies of $\tau(Q)$ indicated. The estimated error bars are shown for two Q-values. The Q-dependence of the value of the non-Gaussian parameter at $\tau(Q)$ is also included (*filled triangle*) as well as the static structure factor $S(Q)$ on the linear scale in arbitrary units. The horizontal *shadowed area* marks the range of the characteristic times τ_{NMR}. The values of the structural relaxation time τ_s and t^* are indicated by the *dashed-dotted* and *dotted lines*, respectively (see the text for the definitions of the timescales). The temperature is 363 K in all cases. (Reprinted with permission from [105]. Copyright 2002 The American Physical Society)

were found at higher Q-values (see Fig. 4.15). The calculated non-Gaussian parameter showed a double peak structure (see Fig. 4.16) where the short-time maximum is related to the fast librational motions of C–H bonds and the other is centred at a time $t^*\approx 4$ ps, i.e. in the so-called decaging regime of $\langle r^2(t)\rangle$ (terminology of mode coupling theory [95, 96, 106]). This second peak of $\alpha_2(t)$ shows a similar behaviour to that observed in the computer simulations previously mentioned. Thus, glass-forming polymers are not an exception regarding this question. Once the sublinear behaviour of $\langle r^2(t)\rangle$ is well established (the α-relaxation governs the dynamics), $\alpha_2(t)$ decreases to its long-term limit of zero. The deviations from Gaussianity evidenced by the second peak of $\alpha_2(t)$ manifest in the weakening of the Q-dependence of $\tau_{\text{self}}(Q)$ at high Q-values.

In a feedback-based procedure, careful neutron scattering measurements on the real sample (PI with deuterated methyl groups) were carried out. The aim was to establish the experimental evidence of the crossover from Gaussian to non-Gaussian behaviour found from the MD-simulations for self-motion of PI main chain protons in the α-relaxation regime. By combining three spectrometers, the NSE spectrometers IN11c and that in Jülich and the BS instrument IN13, the widest Q-range available ($0.1\leq Q\leq 5$ Å^{-1}) was covered [9, 154]. Some examples of the data acquired by NSE are shown in Fig. 4.12a. Due to the strong Q-dependencies of the spectra and the limited dynamic window of the spectrometers used, measurements at different temperatures had to be combined in order to cover the whole Q-range under investigation. A slight increase was

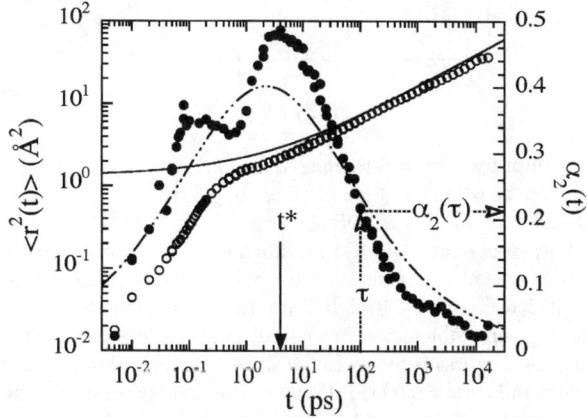

Fig. 4.16 Time evolution of the mean squared displacement $\langle r^2\rangle$ (*empty circle*) at 363 K and the non-Gaussian parameter α_2 obtained from the simulations at 363 K (*filled circle*) for the main chain protons of PI. The *solid vertical arrow* indicates the position of the maximum of α_2, t^*. At times $\tau>\tau(Q_{\text{max}})$, the crossover time, α_2 assumes small values, as in the example shown by the *dotted arrows*. The corresponding functions $\langle r^2\rangle$ and α_2 are deduced from the analysis of the experimental data at 320 K in terms of the jump anomalous diffusion model and are displayed as *solid lines* for $\langle r^2\rangle$ and *dashed-dotted lines* for α_2. (Reprinted with permission from [9]. Copyright 2003 The American Physical Society)

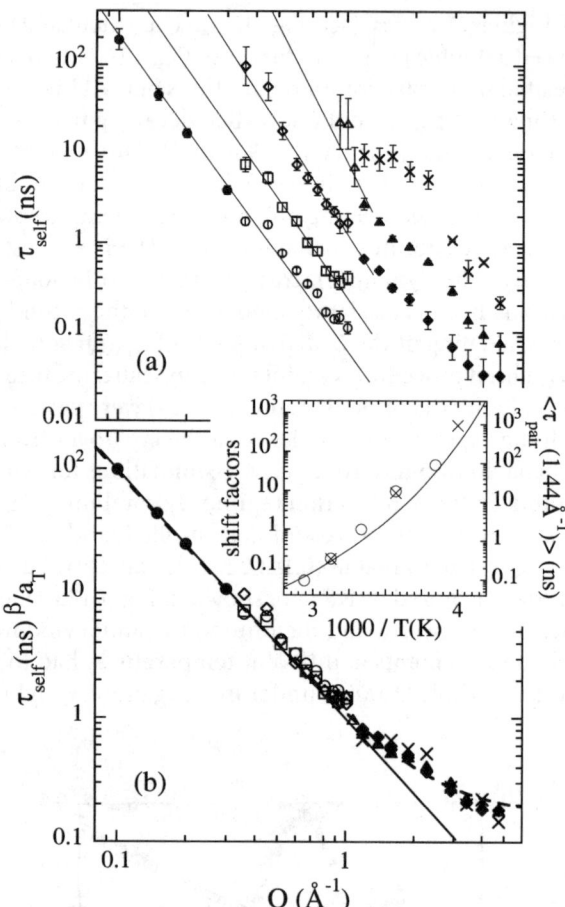

Fig. 4.17 a Momentum transfer dependence of the characteristic times obtained from the KWW description of the results on PId3 obtained by BS (ILL Grenoble) (*cross* 260 K, *filled triangle* 280 K, *filled diamond* 300 K), IN11C (*empty triangle* 280 K, *empty diamond* 300 K, *empty square* 320 K, *empty circle* 340 K) and Jülich NSE spectrometer (*filled circle* 340 K). **b** Master curve constructed exponentiating the points in **a** to the power of β and shifting them by the appropriate factors a_T. Straight lines show Gaussian behaviour. The *dashed line* in **b** shows the description of the master in terms of an anomalous jump diffusion model (see text) with $\ell_{\alpha 0}=0.42$ Å. The *inset* compares the shift factors obtained from dielectric spectroscopy (*solid line*) and from $S_{self}(Q,t)$ [9] (*cross*). The average times from the pair correlation function measured by IN11 at 1.44 Å^{-1} on PId8 are also displayed (×) (Reprinted with permission from [9]. Copyright 2003 The American Physical Society)

found for the value of the shape parameter β with increasing temperature ($\beta \approx 0.4$ at 260 K and $\beta = 0.57$ at 340 K), leading to a variation of the Q-dependence of the characteristic time with temperature (see Fig. 4.17a). Therefore, the best way to bring together all the experimental data at different temperatures was to use a representation analogous to that shown in Fig. 4.13. Exponentiating the timescales to the corresponding β value, the Gaussian behaviour translates in a Q^{-2} power law. Then the application of the temperature-dependent shift factors shown in the inset allowed the building of the master curve presented in Fig. 4.17b. This curve unequivocally demonstrates the existence of the crossover in the experimental data also.

The deviations from Gaussian behaviour were successfully interpreted as due to the existence of a distribution of finite jump lengths ℓ_α underlying the sublinear diffusion of the proton motion [9, 149, 154]. A most probable jump distance of $\ell_{\alpha_0} = 0.42$ Å was found for PI main-chain hydrogens. With the model proposed not only a good description of the Q-dependence of the characteristic time is achieved (Fig. 4.17b), but it also accounts for the $\alpha_2(t)$ time dependence in an approximate way (Fig. 4.16).

A similar combined study involving NSE and MD simulations has recently been performed on the hydrogen motion of PVE [155]. The almost perfect agreement between experiment and simulation can be appreciated from Fig. 4.18. For PVE, the Q-range where the correlation expressed by Eq. 4.11 holds is restricted to $Q \leq 0.5$ Å^{-1} approximately (Fig. 4.19a). With increasing Q, $\tau_{self}(Q)$ bends systematically upwards, deviating from the Gaussian prediction (power law indicated by the dotted line). We note that the slight "bump" in the experimental data

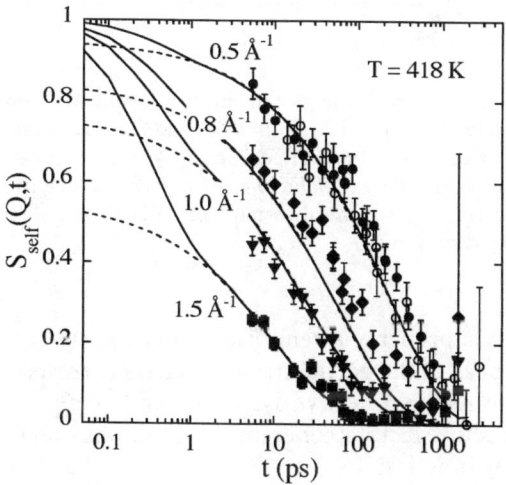

Fig. 4.18 $S_{self}(Q,t)$ for all protons in PVE at 418 K: MD simulations (*lines*) and NSE results (*empty symbols* $\lambda = 8$ Å, *full symbols* $\lambda = 5.6$ Å) at the Q-values indicated. The *dotted lines* are KWW descriptions of the second decay with $\beta = 0.55$. For the purpose of comparison with the MD-simulation the NSE-data were corrected for "band pass" effects [155]

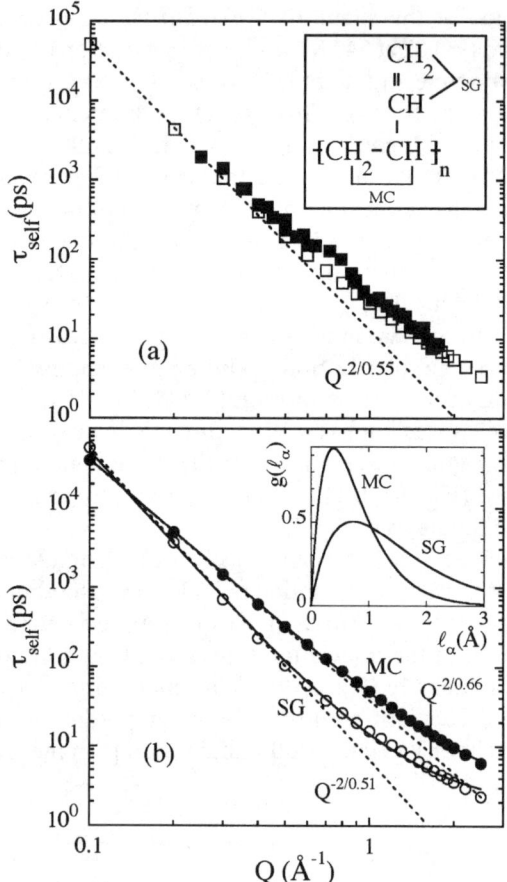

Fig. 4.19 $\tau_{self}(Q)$ obtained for **a** all the protons in PVE (*empty* MD simulations, *full* NSE, β=0.55) and **b** the main chain (*filled circle*, β=0.66) and the side group hydrogens (*empty circle*, β=0.51), both from the MDS. *Dotted lines* are expected Q-dependence from the Gaussian approximation in each case. *Solid lines* are description in terms of the anomalous jump diffusion model. *Insets*: Chemical formula of PVE (**a**) and distribution functions obtained for the jump distances (**b**)

around Q=0.7 Å^{-1} is probably a reminiscence of the coherent contributions close to the main peak of the partial structure factor corresponding to the fully protonated sample. As the collective dynamics tends to slow down close to the maxima of the structure factors, coherent contributions lead to effective longer times for the decay function.

The stronger deviations from Gaussian behaviour for PVE, a polymer with large side groups (see Table 1.1), could be caused by the different mobility of the hydrogens linked to the main chain carbons and those in the side groups. This hypothesis could be confirmed by the MD simulations. Distinguishing the main chain from the side group hydrogens, the respective $\tau_{self}(Q)$ showed the deviations at higher Q-values, as can be seen in Fig. 4.19b. For both subsystems of atoms, the anomalous jump diffusion model proposed in [154] could account for the Q-dependence of $\tau_{self}(Q)$, obtaining different values for the mean jump distances involved in each case (see inset of Fig. 4.19b).

Thus a possible interpretation of deviations from Gaussianity – beyond those expected from the jump effect – can be based on the existence of distributions of atoms with different degrees of mobility in the sample, and hence particularly strong deviations from the Gaussian behaviour of the incoherent scattering function might be assigned to such dynamic heterogeneities in the sample. This phenomenon is well known in blends of polymers (see Sect. 6.1) [175]. There is also another homopolymer, PVC, where an extremely non-Gaussian behaviour has been found [128]. The nature of this polymer has turned out to be so particular [176] that books have even been devoted to it [177]. Strong controversies on its "crystalline" nature can be found in the literature [176–184], and even more than sixty years ago it was established that the α-relaxation in this polymer shows an anomalous rapidly increasing broadening with decreasing temperature approaching T_g [185]. In [128] the structural features of PVC were studied by SANS and the dynamic behaviour by a combination of DS, NSE on the dynamic structure factor at Q_{max} and BS on the self-motions of protons. This combined effort allowed the establishment of the existence of dynamic heterogeneities, which could be explained by the coexistence of regions with different dynamic properties leading to a distribution of characteristic relaxation times. For all regions, the same functional form for the α-relaxation was assumed, which could be unequivocally determined from the NSE data at Q_{max}. The distribution of relaxation times found was compatible with the distribution of only one variable, the glass transition temperature of the region. This could be understood by taking into account structural heterogeneities related to the presence of microcrystallites in this peculiar polymer. This example shows how the microscopic insight offered by neutron scattering can be of utmost relevance for unravelling long-standing problems in polymer physics.

Finally we compare the temperature dependencies reported for the structural relaxation and the self-motion of hydrogens studied by NSE. For PI, the shift factors used for the construction of the master curve on $\tau_{self}(Q,T)^{\beta}$ (Fig. 4.17) are identical to those observed for the structural relaxation time [8]. This temperature dependence also agrees with DS and rheological studies. The case of PIB is more complex [147]. The shift factors obtained from the study of $\tau_{self}(Q,T)$ (Fig. 4.14b) reveal an apparent activation energy close to that reported from NMR results (≈ 0.4 eV) [136]. This temperature dependence is substantially weaker than that observed for the structural relaxation time (≈ 0.7 eV, coinciding with rheological measurements) in the same temperature range (see Fig. 4.20).

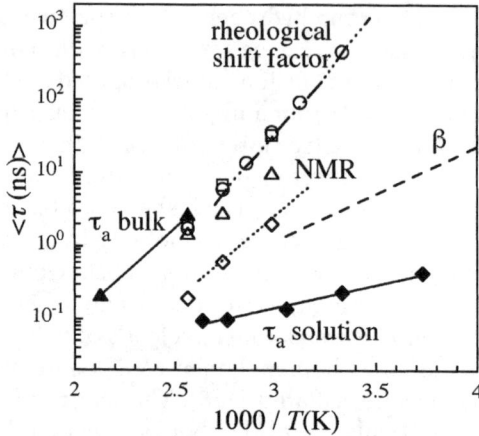

Fig. 4.20 Temperature dependence of the average relaxation times of PIB: results from rheological measurements [34] (*dashed-dotted line*), the structural relaxation as measured by NSE at Q_{max} (*empty circle* [125] and *empty square*), the collective time at 0.4 Å^{-1} (*empty triangle*), the time corresponding to the self-motion at Q_{max} (*empty diamond*), NMR (*dotted line* [136]), and the application of the Allegra and Ganazzoli model to the single chain dynamic structure factor in the bulk (*filled triangle*) and in solution (*filled diamond*) [186]. *Solid lines* show Arrhenius fitting curves. *Dashed line* is the extrapolation of the Arrhenius-like dependence of the β-relaxation as observed by dielectric spectroscopy [125]. (Reprinted with permission from [187]. Copyright 2003 Elsevier)

4.2
β-Relaxation

In the previous section it has been shown that the temporal evolution of the pair correlations at the interchain peak is governed by the structural relaxation. If we move now towards more local scales – i.e. higher Q-values – we see that the static correlations observed in $S_{pair}(Q)$ correspond to pair correlations along a given chain. It is then natural to think that their time dependence might relate to dynamic processes other than the structural relaxation.

The first experimental observation of such an effect in the decay of the dynamic structure factor was made on PB, when it was explored close to the first minimum of $S_{pair}(Q)$ (Q_{min}=1.88 Å^{-1}, see Fig. 4.2) [188]. Two salient features were reported in that work:

i. The spectral shape was different from that measured for the α-relaxation at the first peak Q_{max}
ii. At temperatures above T_g+40 K the temperature dependence of the microscopic relaxation agreed with the viscosity scale [124] and at lower temperatures strong decoupling effects were found

These are clearly evidenced in the scaling representation shown in Fig. 4.21. Thus, at high temperatures the decay of the correlations at Q_{min} seem to relate

4 Local Dynamics and the Glass Transition

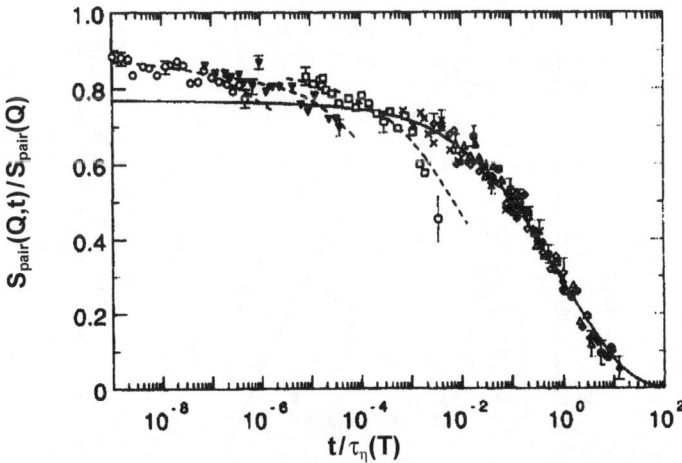

Fig. 4.21 Scaling representation of the NSE data on fully deuterated PB near the first valley of $S(Q)$ (Q=1.88 Å^{-1}). The scale $\tau_\eta(T)$ is taken from viscosity measurements. The *solid line* represents the master function obtained for the spectra at temperatures ≥220 K (KWW with β=0.37). The *dashed lines* are the result of fitting stretched exponentials to the different low-T spectra keeping β=0.37 fixed: (*filled diamond* 280 K, *filled circle* 260 K, *empty triangle* 250 K, *empty diamond* 240 K, *cross* 230 K, *empty square* 205 K, *filled square* 190 K, *empty circle* 180 K) (Reprinted with permission from [188]. Copyright 1992 The American Physical Society)

to the structural relaxation (the temperature dependence there coincides with the structural time). However, at lower temperatures a different relaxation mechanism taking place at these more local scales is active in the NSE window. An approximate Arrhenius-like behaviour was found for this additional process (see Fig. 4.22). This observation, together with later dielectric measurements on the same sample, brought evidence that below the decoupling temperature, which coincides with the temperature where the dielectric α- and β-relaxations merge (T_m, see Fig. 4.8), the NSE relaxations appeared to be related to the β-process observed by relaxation techniques.

A later systematic study carried out on this polymer in a wide Q-range extending from the first to the second maxima of $S_{\text{pair}}(Q)$ and exploring lower temperatures established the existence of a microscopic secondary process, which can be related to the dielectric β-relaxation [133, 189]. In particular, the measurements of $S_{\text{pair}}(Q,t)$ at Q=2.71 Å^{-1}, i.e. close to the second $S_{\text{pair}}(Q)$ maximum, revealed severe deviations from scaling with the structural relaxation (see Fig. 4.23).

A phenomenological description of the dynamic structure factor at this Q-value by KWW functions:

$$\frac{S_{\text{pair}}(Q,t)}{S_{\text{pair}}(Q,0)} = f_Q(T) \exp\left\{-\left[\frac{t}{\tau_{\text{pair}}(Q,T)}\right]^\beta\right\} \qquad (4.21)$$

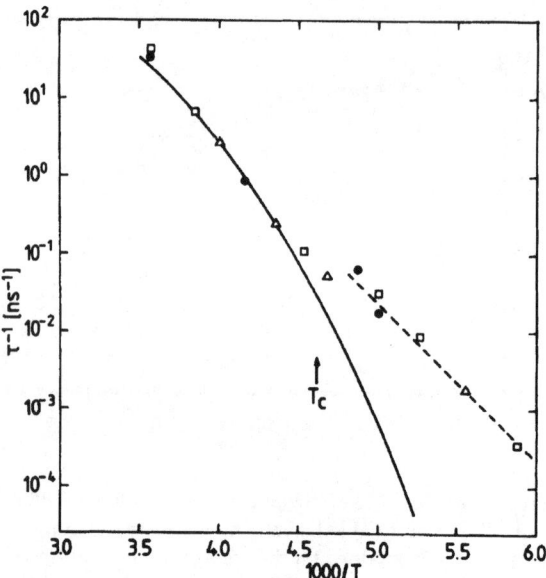

Fig. 4.22 Arrhenius representation of the relaxation rates obtained from fitting stretched exponentials to the spectra of PB at $Q=1.88$ Å^{-1} at different temperatures. The three symbols represent three different sets of experiments carried out in separate experimental runs. The *solid line* displays the viscosity time scale. The *dashed line* indicates the Arrhenius behaviour of the low-temperature branch. (Reprinted with permission from [188]. Copyright 1992 The American Physical Society)

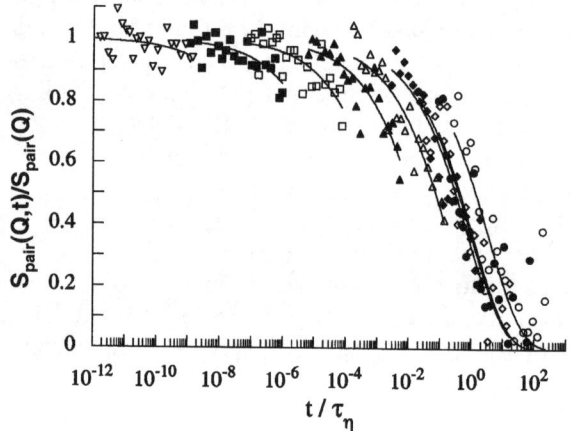

Fig. 4.23 Scaling representation of the dynamic structure factor of PB data at $Q=2.71$ Å^{-1} (*empty circle* 300 K, *filled circle* 280 K, *empty diamond* 260 K, *filled diamond* 240 K, *empty triangle (up)* 220 K, *filled triangle (up)* 205 K, *empty square* 190 K, *filled square* 180 K, *empty triangle (down)* 170 K). *Solid lines* correspond to KWW functions with $\beta=0.41$ [133] (Reprinted with permission from [133]. Copyright 1996 The American Physical Society)

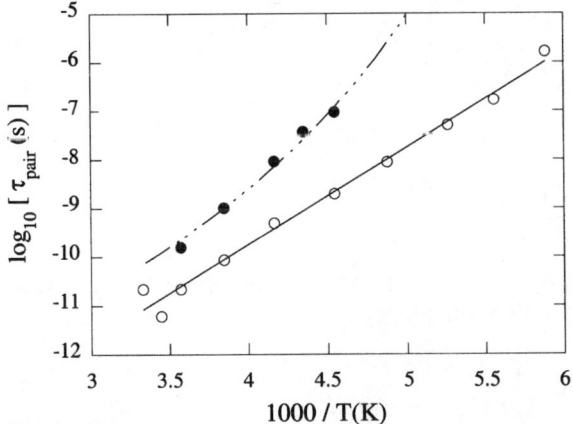

Fig. 4.24 Temperature dependence of the characteristic times obtained from the fits of $S_{pair}(Q,t)$ to stretched exponentials with $\beta=0.41$ at $Q_{max}=1.48$ Å^{-1} (*filled circle*) and 2.71 Å^{-1} (*empty circle*). *Dashed-dotted line* corresponds to the Vogel-Fulcher-like temperature dependence of the viscosity and the *solid line* to the Arrhenius-like temperature dependence of the dielectric β-relaxation. (Reprinted with permission from [189]. Copyright 1996 The American Physical Society)

with a fixed value of the shape parameter ($\beta=0.41$) led to an Arrhenius-like behaviour of the characteristic times $\tau_{pair}(Q=2.71$ Å$^{-1})$ as displayed in Fig. 4.24 (empty circles). Comparison with the temperature dependence of the dielectric β-process (solid line in the figure, corresponding to an activation energy of 0.41 eV) [189, 190] strongly suggests that the decay of the intrachain correlations at the second structure factor peak indeed relates to the secondary relaxation observed by DS. However, the absolute values of the characteristic times obtained from both techniques differed by about two orders of magnitude [133, 189]. Thus, these NSE experiments revealed the very astonishing result that in PB the density fluctuations at intrachain scales, directly seen by neutrons, seem to decay about 100 times faster than the dipole orientations observed by dielectric spectroscopy, but with the same activation energy.

4.2.1
Model for β-Relaxation

In [189] a simple two state model for the dynamic structure factor corresponding to the Johari-Goldstein β-process was proposed. In this model the β-relaxation is considered as a hopping process between two adjacent sites. For such a process the self-correlation function is given by a sum of two contributions:

$$S_{self}^{hop}(Q,t) = S_{self}^{el}(Q) + S_{self}^{inel}(Q) \exp\left[-\frac{2t}{\tau(E)}\right] \quad (4.22)$$

with

$$S_{\text{self}}^{\text{inel}}(Q) = \frac{1}{2}\left[1 - \frac{\sin(Qd_\beta)}{Qd_\beta}\right] \tag{4.23}$$

and

$$S_{\text{self}}^{\text{el}}(Q) = \frac{1}{2}\left[1 + \frac{\sin(Qd_\beta)}{Qd_\beta}\right] \tag{4.24}$$

Here d_β is the distance between the two sites and $\tau(E)$ is the jump time corresponding to an activation energy E:

$$\tau(E) = \tau_o \exp\left(\frac{E}{k_B T}\right) \tag{4.25}$$

The elastic contribution $S_{\text{self}}^{\text{el}}$ is also called elastic incoherent structure factor (EISF). It may be interpreted as the Fourier transformed of the asymptotic distribution of the hopping atom for infinite times. In an analogous way to the relaxation functions (Eq. 4.6 and Eq. 4.7), the complete scattering function is obtained by averaging Eq. 4.22 with the barrier distribution function $g(E)$ obtained, e.g. by dielectric spectroscopy (Eq. 4.5)

$$S_{\text{self}}^\beta(Q,t) = \int_0^{+\infty} g(E)\, S_{\text{self}}^{\text{hop}}(Q,t)\, dE \tag{4.26}$$

The Q-dependence of the two contributions to Eq. 4.22 (or equivalently, to Eq. 4.26) is displayed in Fig. 4.25 for the case $d_\beta=1.5$ Å. From the oscillation of both contributions with Q the jump distance may be obtained. The associated timescale may be found from the time decay of the inelastic part.

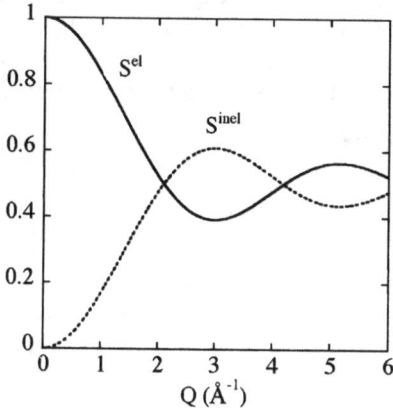

Fig. 4.25 Elastic and inelastic contribution to the incoherent scattering function for jump motion between two sites separated by 1.5 Å

4 Local Dynamics and the Glass Transition

The coherent counterpart is more difficult to obtain, since now we have to deal with a change of configurations of atoms rather with single atom jumps. The conceptual difference between the pair and self-correlation function for jump processes may be visualized most easily considering rotational jumps. Let us regard, for example, the 120° rotational jumps of a methyl group around its symmetry axis. The self-correlation reveals the atomic jumps of the associated hydrogens. The pair correlation reflects the change of atomic configurations before and after the jump. Since a 120° jump does not change the configuration, a coherent scattering experiment would not reveal any dynamic effect.

A reasonable approximation for the pair correlation function of the β-process may be obtained in the following way. We assume that the inelastic scattering is related to uncorrelated jumps of the different atoms. Then all interferences for the inelastic process are destructive and the inelastic form factor should be identical to that of the self-correlation function, $S_{\text{self}}^{\text{inel}}(Q)$ given by Eq. 4.24. On the other hand, for $t=0$ the pair correlation function is reflected by the static structure factor $S_{\text{pair}}(Q)$. As this condition has to be fulfilled also by $S_{\text{pair}}^{\beta}(Q,t)$, its elastic contribution has to be $S_{\text{pair}}(Q)-S_{\text{self}}^{\text{inel}}(Q)$. For the normalized dynamic structure factor, we arrive at:

$$\frac{S_{\text{pair}}^{\beta}(Q,t)}{S_{\text{pair}}(Q)} = \left\langle \frac{S_{\text{pair}}(Q) - S_{\text{self}}^{\text{inel}}(Q)}{S_{\text{pair}}(Q)} + \frac{S_{\text{self}}^{\text{inel}}}{S_{\text{pair}}(Q)} e^{-2t/\tau(E)} \right\rangle \qquad (4.27)$$

This incoherent approximation does not reveal symmetry e.g. related cancellations, but displays a major feature of the corresponding dynamic structure factor, namely the relative suppression of the inelastic contributions from local jump processes at the maximum of the static structure factor. The application of this model to the case of PB resulted in a very good description of the experimental data in the temperature range where the α-relaxation contribution is sufficiently slow (well below the dielectric merging T_{m}), (Fig. 4.26). The jump distance of the β-process was found to be 1.5 Å [133, 189]. Fig. 4.27 displays the corresponding inelastic dynamic structure factor (Eq. 4.27). It is strongly reduced at the position of the first peak, while it contributes significantly at higher Q. This picture suggests a Q-selectivity for the different relaxation processes: at Q_{max} local jump processes contribute only weakly and the relaxation due to interchain motion dominate. On the other hand at larger Q, in particular in the minimum of $S_{\text{pair}}(Q)$, the secondary relaxation reveals itself.

A more realistic model for the secondary relaxation needs to consider motions of a molecular group (considered as a rigid object) between two levels. The group may contain N atoms with the scattering length b_i at positions $\underline{r}_i$, ($i=1,N$). The associated motion may consist of a rotation around an arbitrary axis, e.g. through the centre of mass depicted by a rotational matrix Ω and a displacement by a translational vector $\underline{R}$. In order to evaluate the coherent dynamic structure factor, scattering amplitudes of the initial (1) and final (2) states have to be calculated:

$$A_1(Q) = \sum_i b_i \exp(i\underline{Q}\,\underline{r}_i); \quad A_2(Q) = \sum_i b_i \exp[i\underline{Q}(\Omega\underline{r}_i + \underline{R})] \qquad (4.28)$$

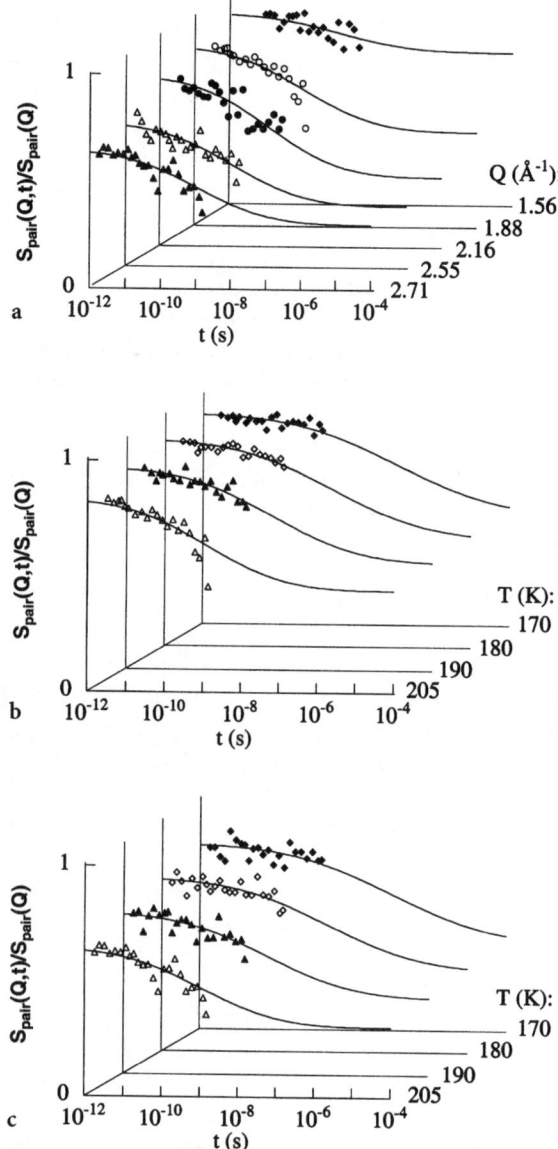

Fig. 4.26 PB NSE spectra in the β-relaxation regime: **a** at 205 K for the Q-values indicated; **b** at $Q=1.88$ Å^{-1} and **c** at 2.71 Å^{-1} for the temperatures indicated. *Solid lines* are the fitting curves obtained in the incoherent approximation for the inelastic part (jump distance $d_\beta=1.5$ Å). (Reprinted with permission from [133]. Copyright 1996 The American Physical Society)

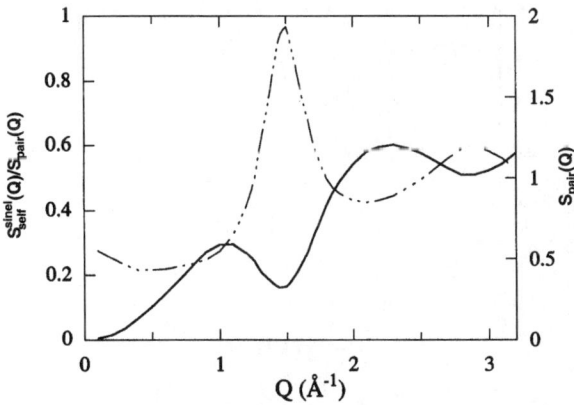

Fig. 4.27 Q-dependence of the amplitude of the relative quasi-elastic contribution of the β-process to the coherent scattering function $S^{inel}(Q)/S(Q)$ obtained for PB from the hopping model (*solid line*) with $d_\beta=1.5$ Å. The static structure factor $S(Q)$ at 160 K [123] is shown for comparison (*dashed-dotted line*)

If the object is embedded into a matrix with the same average scattering properties as the considered jumping unit, then the scattering contrast from the average size of the object is matched by the matrix and the corresponding forward scattering is suppressed. It can be shown [133] that the dynamic structure factor for an object embedded in a matrix, which performs jumps in a two level system, can be obtained as:

$$S^\beta_{pair}(Q,t) = \frac{1}{4}\langle[A_1(Q)+A_2(Q)]^2\rangle - \langle A(Q)\rangle^2 + \langle A(Q)\rangle^2 S_c(Q)$$
$$+ \frac{1}{4}\langle[A_1(Q)-A_2(Q)]^2\rangle \exp\left(-\frac{2t}{\tau}\right) \quad (4.29)$$
$$= S^{el}_{pair}(Q) + S^{inel}_{pair}(Q)\exp\left[-\frac{2t}{\tau}\right]$$

For reorientational motions the hole in the embedding medium does not change and Eq. 4.29 is valid for arbitrary reorientations. In the case of translational jumps the quantity $\frac{1}{4}\langle[A_1(Q)+A_2(Q)]^2\rangle-\langle A(Q)\rangle^2$ representing the structure factor for $t\to\infty$ may formally lead to negative intensity – if the separation between initial and final state is sufficiently large $\langle[A_1(Q)+A_2(Q)]^2\rangle$ may decay faster with Q than $\langle A(Q)\rangle^2$. In the taken approximation the requirement of overall positive intensity has to be fulfilled by the centre of mass correlation term $\langle A(Q)\rangle^2 S_c(Q)$, restricting possible translational jumps to those which are compatible with the translational correlations in the system.

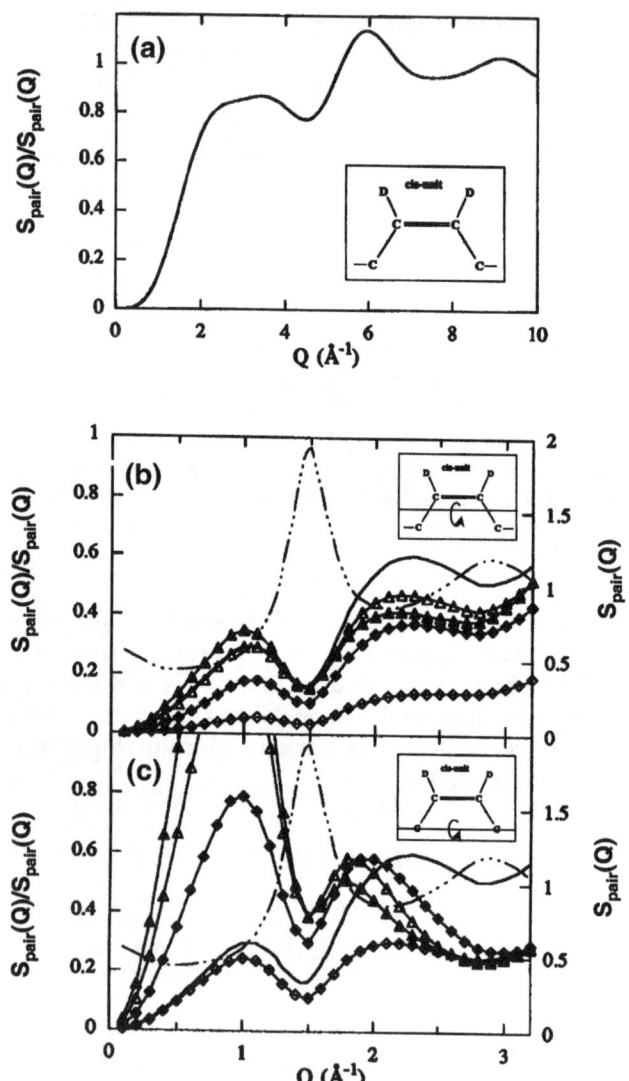

Fig. 4.28 a Form factor associated to the *cis*-unit of PB, which is schematically represented in the *inset*. b and c show the Q-dependence of the amplitude of the relative quasi-elastic contribution of the β-process to the coherent scattering function obtained for rotations of the *cis*-unit around an axis through the centre of mass of the unit and through the main chain, respectively, for different angles: 30° (*empty diamond*), 60° (*filled diamond*), 90° (*empty triangle*) and 120° (*filled triangle*). The static structure factor $S(Q)$ at 160 K [123] is shown for comparison (*dashed-dotted line*) (Reprinted with permission from [133]. Copyright 1996 The American Physical Society)

Using the above approach, an attempt to gain information on the molecular jump geometry was performed for PB. PB is built from basically two different rigid units, the *cis*- and the *trans*-units. The *cis*-group (see Fig. 4.28) carries the dielectric dipole – thus its motion is observed by dielectric spectroscopy. The form factor ($\langle A^2\rangle-\langle A\rangle^2$) of the *cis*-unit (Fig. 4.28a) shows a first intensity maximum around $Q=3$ Å^{-1}, where a high contribution of the β-process to the structure factor is experimentally observed. A possible elemental process for the β-relaxation could be the rotation of a *cis*-unit around a given axis parallel to the double bond and in the plane defined by the C-CD=CD-C rigid unit (see Fig. 4.28). The relative inelastic contribution $S_{\text{pair}}^{\text{inel}}(Q)/S_{\text{pair}}(Q)$ calculated for different rotation angles around an axis passing through the centre of mass is depicted in Fig. 4.28b and through the C-atoms along the main chain in Fig. 4.28c. For rotations of 60° and 120° around the centre of mass axis, the obtained coherent inelastic form factors follow qualitatively the result of the incoherent approximation for 1.5 Å jumps. Thus, such rotational jump processes are well compatible with the observed NSE spectra. We note that the form factor for a *trans*-unit is very similar and that analogous considerations hold. The picture is quite different for rotations around an axis through the main chain C-atoms. In this case strong inelastic scattering is predicted to occur around 1 Å^{-1}. For rotational angles >70° the consistency condition of Eq. 4.29 cannot be fulfilled, leading to negative elastic intensities. The corresponding large displacements are inconsistent with the approximation used and also are not reasonable. The NSE data thus indicate that rotational processes around the centre of mass axis of the *cis*-units are likely to be the motional mechanism behind the β-relaxation.

Finally, recently depolarized light scattering spectra [191] display an additional process that shows a much faster characteristic time and a much weaker temperature dependence than the dielectric β-relaxation (more than three orders of magnitude faster time at ~200 K and an activation energy of 0.16 eV, about half of the dielectric value). Also atomistic simulations on PB have indicated hopping processes of the *trans*-double bond [192, 193] with an associated activation energy of ~0.15 eV. Whether these observations may be related with the discrepancy in the apparent time scale of the NSE and dielectric experiments remains to be seen.

4.3
$\alpha\beta$-Merging

At the merging temperature T_m the α-relaxation time matches that of the β-relaxation. Around this temperature, the dynamic structure factor has to be generalized, in order to include also the segmental diffusion process underlying the α-process. The β-process can be considered as a local intrachain relaxation process, which takes place within the fixed environment set by the other chains. When the segmental diffusion reaches the timescale of the local relaxation, given atoms and molecular groups will noticeably participate simultaneously in both motional mechanisms: the intrachain β-relaxation and the interchain α-relaxation. The simplest way to approach the combination of the two processes is to assume that they are statistically independent. Then the self-correlation function describing the motion of an atom, which takes part in the statistically independent α- and β-processes, can be written as a convolution product of the corresponding self-correlation functions:

$$G_{\text{self}}^{\alpha\beta}(\underline{r},t) = \int G_{\text{self}}^{\beta}(\underline{r}',t)\, G_{\text{self}}^{\alpha}(\underline{r}-\underline{r}',t)\, d\underline{r}' \tag{4.30}$$

The self (incoherent) structure factor, which is given by the Fourier transformation of $G_{\text{self}}^{\alpha\beta}(\underline{r},t)$, then becomes the direct product of the structure factors corresponding to the two processes:

$$S_{\text{self}}^{\alpha\beta}(\vec{Q},t) = S_{\text{self}}^{\beta}(\vec{Q},t)\, S_{\text{self}}^{\alpha}(\vec{Q},t) \tag{4.32}$$

However, in the coherent case the derivation of a similar expression is not straightforward, because the correlations between all the pairs of scatters (j,i) have to be taken into account. One possibility is to follow the procedure described in [133], which is based on a generalization of the Vineyard approximation [194]. The dynamic structure factor of the combined process can be written as:

$$S_{\text{pair}}^{\alpha\beta}(Q,t) = S_{\text{pair}}^{\beta}(Q,t)\, S_{\text{self}}^{\alpha}(Q,t) \tag{4.33}$$

where $S_{\text{pair}}^{\beta}(Q,t)$ is the coherent structure factor of the β-process. In this approach we obtain an expression for $S_{\text{pair}}^{\alpha\beta}(Q,t)$ connecting the coherent structure factor of the β-relaxation with the relaxation function of the α-process, where in the language of the Vineyard approximation $S_{\text{pair}}^{\beta}(Q,t)$ takes the role of $S_{\text{pair}}(Q)$.

Such an approach for the first time was applied to NSE data from PB in [133]. It was assumed that $S_{\text{pair}}^{\beta}(Q,t)$ was known from the study performed below T_m (i.e. by using the dielectric distribution of activation energies and a jump distance of 1.5 Å) and that the values of the parameters involved could be extrapolated to higher temperatures. For $S_{\text{self}}^{\alpha}(Q,t)$, the spectral shape and the temperature dependence of the characteristic times were assumed to be those deduced from the inspection of the pair correlation function at the first maximum (KWW functional form (Eq. 4.9) with β=0.41 and $\tau_{\text{self}}(Q,T) \approx \tau_s(T) \approx \tau_\eta(T)$) (see Sect. 4.1, [133]). The Q-dependence of $\tau_{\text{self}}(Q)$ was obtained from the fitting of the NSE curves to the theoretical function. It is noteworthy that the only free parameters in the fitting procedure were $\tau_{\text{self}}(Q)$ and amplitude factors which account for faster processes like phonons, that are not visible in the NSE window but contribute to the total amplitude. Figure 4.29 shows resulting fit curves for several temperatures for Q values around the first maximum, minimum and second maximum of $S_{\text{pair}}(Q)$. The excellent agreement between the model scattering function and the experiment strongly supports the hypothesis that the α- and β-relaxations behave independently of each other. $\tau_{\text{self}}(Q)$ approximately follows a power law in Q as one would expect for the self-correlation function [133].

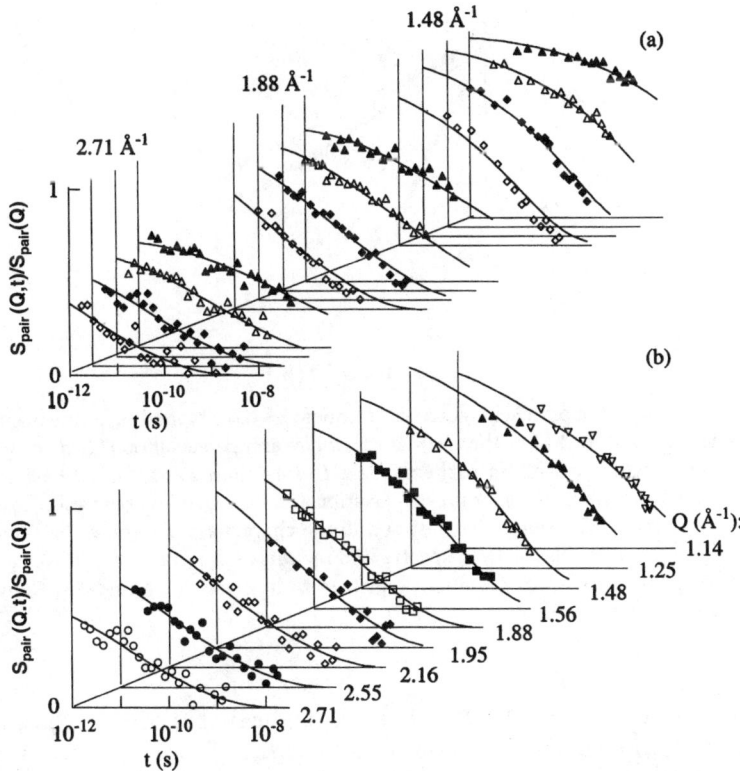

Fig. 4.29 a PB NSE spectra for the Q-values indicated at 220 K (*filled circle*), 240 K (*empty triangle*), 260 K (*filled diamond*) and 280 K (*empty diamond*). b at 260 K for the Q-values indicated. *Solid lines* are fitting curves (see text). (Reprinted with permission from [133]. Copyright 1996 The American Physical Society)

The same approach to describe the $\alpha\beta$-merging was further applied to PIB [125]. In contrast to PB, the neutron and dielectric prefactors for the dielectric β-relaxation timescale agree closely. The relaxation map for this polymer, displayed in Fig. 4.30, shows two processes that can be classified as secondary relaxations: the dielectrically determined β-relaxation and the so-called δ-process, characterized from NMR measurements and interpreted as a methyl group rotation [196]. We note that both relaxations show very similar activation energies, though the dielectric process cannot be understood in terms of a simple methyl rotation – this is dielectrically inactive. Thorough NSE measurements on the PIB dynamic structure factor in the supercooled liquid state were performed in a wide Q-range including the first and second maxima [125]. In the temperature region investigated (270 K$\leq T \leq$390 K) contributions from both, the α and the β-relaxations, were expected (see Fig. 4.30).

For this polymer the influence of the secondary relaxation on the dynamic structure factor seems to be much weaker than in the case of PB. Figure 4.31b

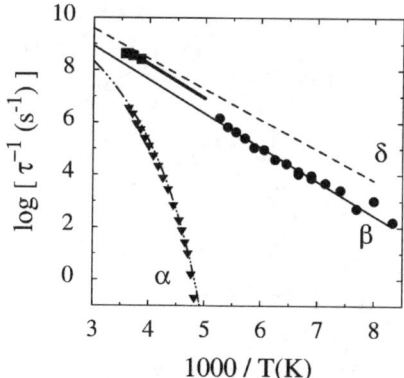

Fig. 4.30 Arrhenius plot of the characteristic frequencies (corresponding to the maximum of the dielectric loss) for PIB of the α- (*filled triangle*) and β-relaxation (*filled circle*). The *solid line* represents a fit with an Arrhenius law. *Dashed-dotted* and *dashed lines* are the temperature laws shown in [135] for the α-relaxation and the secondary relaxation observed by NMR respectively. The *squares* correspond to the characteristic rates of the β-process obtained from the quasi-elastic IN16 spectra and the *thick solid line* shows those deduced from the analysis of the elastic intensities (Reprinted with permission from [195]. Copyright 1998 American Chemical Society)

shows that for three characteristic Q-values (at the first maximum Q_{max}, at the minimum Q_{min} and at the second maximum of $S_{pair}(Q)$, see Fig. 4.31a) a good superposition of the spectra at different temperatures can be achieved by using the same shift factors. These were obtained from rheological measurements [34]. The only signature of contributions from processes other than the structural relaxation is the larger stretching observed at Q-values beyond the first maximum: the β-parameter describing the master curves in Fig. 4.31b decreases from 0.55 at Q_{max} to 0.46 at Q_{min} and 0.40 at the second maximum. Following a similar procedure as for PB, the NSE data were interpreted in terms of statistically independent α- and β-relaxations, assuming for the later the same distribution of activation energies as that deduced from dielectric spectroscopy [125]. Also, the prefactor of the timescale was fixed to the dielectric result. The good agreement achieved can be appreciated for several spectra at 365 K (Fig. 4.32a) and at Q_{min} (Fig. 4.32b). As discussed above, in this Q-region the $S_{pair}(Q)$ renormalization enhances to a maximum degree the relative contribution of the localized processes to the dynamic structure factor. For PIB the proposed scenario also accounts well for the experimental behaviour. The relative contribution of the secondary process to the normalized dynamic structure factor is found to be only some 10% at most. The involved jump distances are found to be in the range d_β=0.5–0.9 Å (see Fig. 4.33) [125]. This result relates well to molecular displacements expected for substate transitions in the double minimum potential for the split conformational states of PIB [197, 198].

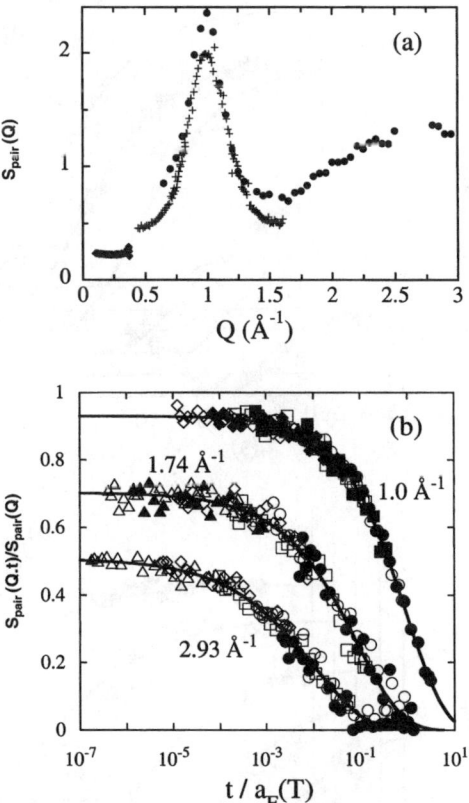

Fig. 4.31 a Static structure factor of PIB. Results from small angle scattering (*filled diamond*), NSE (*plus*) and diffraction (*filled circle*) [147]. b Scaling representation of the PIB NSE spectra at $Q=1.0$ Å^{-1} (Q_{max}), 1.74 Å^{-1} and 2.93 Å^{-1} using the shift factors given by Ferry [34]. The reference temperature is 390 K. Symbols correspond to different temperatures: 270 K (*empty triangle*), 280 K (*filled triangle*), 300 K (*empty diamond*), 32 K (*filled diamond*), 335 K (*empty square*), 350 K (*filled square*), 365 K (*empty circle*) and 390 K (*filled circle*). *Solid lines* are fits to KWW functions (b Reprinted with permission from [125]. Copyright 1998 American Chemical Society)

At lower temperatures – where the α-relaxation contribution in the dynamic window could be neglected – a deeper insight into the β-process was achieved by later backscattering (BS) experiments on the fully protonated sample, i.e. accessing the self-motions of protons [195]. For the interpretation the activation energy distribution was fixed to that obtained from the dielectric spectroscopy results. The resulting β-time scale from the BS data is shown by the squares in Fig. 4.30. It agrees with the dielectric data within a factor of about 2. Thus both processes can be considered as identical. We note moreover that the BS results are very close to the δ-process, indicating that this relaxation should also reflect the same motion.

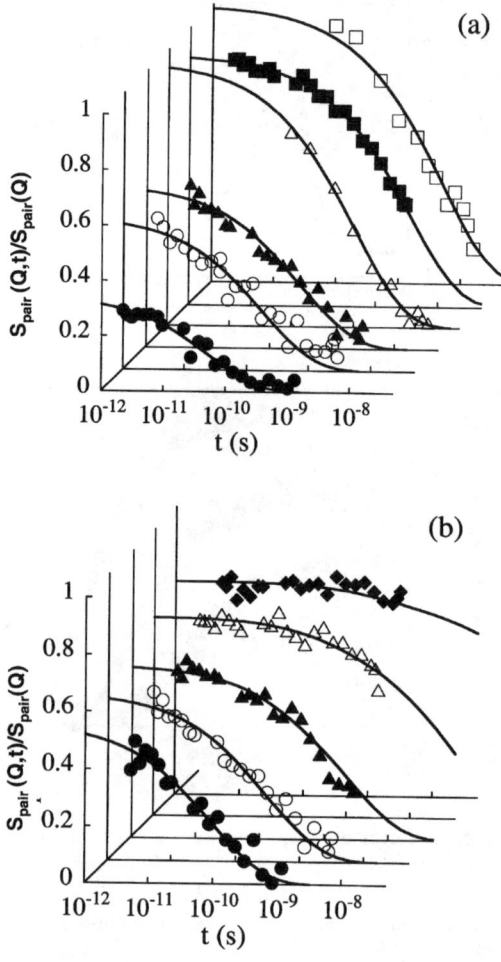

Fig. 4.32 PIB NSE spectra **a** at 365 K and different Q-values: 0.8 Å^{-1} (*empty square*), 1.0 Å^{-1} (*filled square*), 1.2 Å^{-1} (*empty triangle*), 1.74 Å^{-1} (*filled triangle*) 2.4 Å^{-1} (*empty circle*) and 2.93 Å^{-1} (*filled circle*). **b** at 1.74 Å^{-1} and different temperatures: 270 K (*filled diamond*), 300 K (*empty triangle*), 335 K (*filled triangle*), 365 K (*empty circle*), and 390 K (*filled circle*). *Solid lines* are the fit results with the assumption of statistically independent α- and β-relaxations [125] (see text) (Reprinted with permission from [125]. Copyright 1998 American Chemical Society)

Valuable information on the geometry of the proton motion is offered by the Q-dependence of the elastic incoherent intensity (EISF) (see Fig. 4.34). For two site jumps this intensity is described by:

$$S_{\text{self}}(Q, \omega \approx 0) = \frac{1}{2}(1 + A) + \frac{1}{2}(1 - A)\frac{\sin Qd}{Qd} \quad (4.34)$$

4 Local Dynamics and the Glass Transition

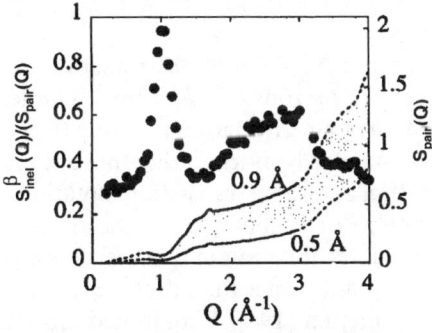

Fig. 4.33 Q-dependence of the relative quasi-elastic contribution from the β-process to the coherent scattering function of PIB for jump distances of 0.5 Å and 0.9 Å (*lines*). The static structure factor $S(Q)$ is shown for comparison (*filled circle*). (Reprinted with permission from [125]. Copyright 1998 American Chemical Society)

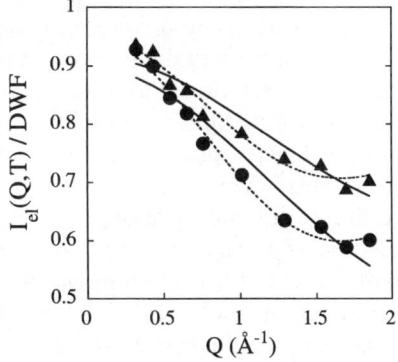

Fig. 4.34 Comparison between the descriptions of the elastic intensity at 260 K (*filled triangle*) and 280 K (*filled circle*) in terms of the EISF corresponding to a methyl-group rotation (*solid lines*) and to a 2-site jump (*dashed lines*). (Reprinted with permission from [195]. Copyright 1998 American Chemical Society)

where A is the non-resolved part of the quasielastic spectrum. The BS PIB data reveal a jump distance d_β=2.7 Å. This result rules out a simple methyl group rotation to be at the origin of the motion observed for the protons, as was invoked by NMR spectroscopy. This becomes clear in Fig. 4.34, where the expectation for such a rotation is compared with the experimental results. From these findings it becomes clear that the δ-process cannot be understood only in terms of a methyl-group rotation as proposed by NMR but must be related to a conformational rearrangement. This would explain its visibility in dielectric spectroscopy and the large motional amplitude derived from the BS experiment. Finally, a conformational process also makes plausible the observation of this process in Brillouin scattering, as reported by Patterson [199].

It would be interesting to understand how methyl-group reorientation alone couples to sound waves.

The value of the jump distance in the β-relaxation of PIB found from the study of the self-motion of protons (2.7 Å) is much larger than that obtained from the NSE study on the pair correlation function (0.5–0.9 Å). This apparent paradox can also be reconciled by interpreting the motion in the β-regime as a combined methyl rotation and some translation. Rotational motions around an axis of internal symmetry, do not contribute to the decay of the pair correlation function. Therefore, the interpretation of quasi-elastic coherent scattering appears to lead to shorter length scales than those revealed from a measurement of the self-correlation function [195]. A combined motion as proposed above would be consistent with all the experimental observations so far and also with the MD simulation results [198].

It is noteworthy that the neutron work in the merging region, which demonstrated the statistical independence of α- and β-relaxations, also opened a new approach for a better understanding of results from dielectric spectroscopy on polymers. For the dielectric response such an approach was in fact proposed by G. Williams a long time ago [200] and only recently has been quantitatively tested [133, 201–203]. As for the density fluctuations that are seen by the neutrons, it is assumed that the polarization is partially relaxed via local motions, which conform to the β-relaxation. While the dipoles are participating in these motions, they are surrounded by temporary local environments. The decaying from these local environments is what we call the α-process. This causes the subsequent total relaxation of the polarization. Note that as the atoms in the density fluctuations, all dipoles participate at the same time in both relaxation processes. An important success of this attempt was its application to PB dielectric results [133] allowing the isolation of the α-relaxation contribution from that of the β-processes in the dielectric response. Only in this way could the universality of the α-process be proven for dielectric results – the deduced temperature dependence of the timescale for the α-relaxation follows that observed for the structural relaxation (dynamic structure factor at Q_{max}) and also for the timescale associated with the viscosity (see Fig. 4.8). This feature remains masked if one identifies the main peak of the dielectric susceptibility with the α-relaxation.

4.4
Mode Coupling Theory

The only currently existing theory for the glass transition is the mode coupling theory (MCT) (see, e.g. [95, 96, 106]). MCT is an approach based on a rather microscopic description of the dynamics of density fluctuations and correlations among them. Although the theory was only formulated originally for simple (monatomic) fluids, it is believed to be of much wider applicability. In this review we will only briefly summarize the main basis and predictions of this theory, focusing on those that can be directly checked by NSE measurements.

4 Local Dynamics and the Glass Transition

MCT considers density fluctuations as the most important low-frequency process and describes the glass transition as an essentially dynamic phenomenon. Ergodicity breaking is understood as a result of strong non-linear coupling between the density fluctuations. At a critical temperature T_c structural arrest takes place. The physical interpretation behind is the "cage effect": particles are constrained in their motion due to their neighbours that form a "cage" in which each particle is more or less confined. As the system gets denser (or the temperature gets lower) structural arrest occurs because particles can no longer leave their cage at finite time. From this microscopic point of view, the density fluctuations are of prime interest. They are directly revealed by the normalized dynamic structure factor $S_{\text{pair}}(Q,t)/S_{\text{pair}}(Q)$ that is experimentally accessible by NSE. The theory results in a two-step relaxation behaviour leading to:

$$\frac{S_{\text{pair}}(Q,t)}{S_{\text{pair}}(Q)} = f_Q \Phi(t) + h_Q F\left(\frac{t}{\tau_c}\right) \tag{4.35}$$

Above T_c the first component $f_Q\Phi(t)$ relates to the structural relaxation while below T_c it measures the amount of structural arrest. The second part describes fast motional processes (that would take place in the picosecond range, not accessible by NSE) not related to transport phenomena. τ_c is the characteristic time of such fast microscopic dynamics. Concerning the structural relaxation, the following predictions are made:

i. All structural relaxations follow a common scaling law of the form $\Phi(t)= \Phi(t/\tau_s)$. Thereby the scale is universal.
ii. Stretching of $\Phi(t)$.
iii. Close to T_c the α-relaxation timescale diverges with a power law:

$$\tau_s \sim (T_c - T)^{-\gamma} \tag{4.36}$$

with an exponent $\gamma>1.76$.
iv. The non-ergodicity parameter f_Q measuring the degree of correlation remaining at infinite time in the non-ergodic state, $f_Q=\lim_{t\to\infty}S_{\text{pair}}(Q,t)/S_{\text{pair}}(Q)$ increases below T_c in the way

$$f_Q(T) = \begin{cases} f_Q^c & T > T_c \\ f_Q^c + h_Q(T_c - T)^{1/2} C_1 & T < T_c \end{cases} \tag{4.37}$$

where C_1 is a constant [95].
v. Qualitatively, the Q-dependence of f_Q resembles that of $S_{\text{pair}}(Q)$. However, no analytical expression is available.
vi. The amplitude and the timescale of the structural relaxation are related through:

$$\tau(Q) \sim \left(\frac{f_Q}{h_Q}\right)^{1/b} \tag{4.38}$$

where the so called von Schweidler exponent b is expected to be slightly larger than the Kohlrausch exponent β.

Since MCT is a quite simplified theory (originally elaborated for hard spheres) it has not been checked extensively with polymers. Moreover, from an experimental point of view the full dynamic range of interest for the theory is not easy to cover. To follow the two steps in the dynamic structure factor it is necessary to combine NSE with time-of-flight (TOF) measurements. This implies the use of Fourier transformation of the TOF data to the time domain. Therefore many of the existing investigations of MCT on polymers rest on NSE measurements, in particular on the observation of the decay of the dynamic structure factor at the intermolecular peak through the structural relaxation. It is remarkable that the main phenomenological features observed for the structural relaxation in polymers (time–temperature superposition, stretching and scaling with the timescale associated to the viscosity) follow directly from the mathematical properties of the MCT. These findings support the theory; however, as discussed above, the universality of the timescale might not be fulfilled for some correlators. Concerning the other specific MCT predictions related to the Q- and T- dependencies of the non-ergodicity parameter, there exist very few NSE experiments on polymers devoted to check them. These have been performed on PB [124, 204] and PI [8]. Apart from the features commented on already (stretching, scaling etc.) both polymers displayed a non-ergodicity parameter $f_Q(T)$ that was very much compatible with the MCT predictions. This can be realized from Figs. 4.3, 4.35, 4.36 and 4.37. The temperature

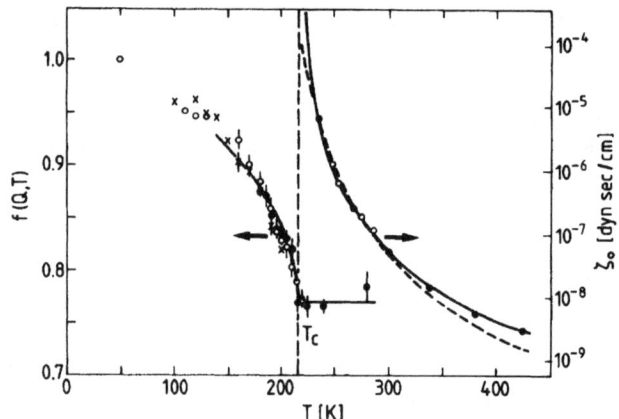

Fig. 4.35 Right-hand side: Monomeric friction coefficients derived from the viscosity measurements on PB [205]. The open and solid symbols denote results obtained from different molecular weights. Solid line is the result of a power-law fit. Dashed line is the Vogel-Fulcher parametrization following [205]. Left hand side: Temperature dependence of the non-ergodicity parameter. The three symbols display results from three different independent experimental runs. Solid line is the result of a fit with (Eq. 4.37) (Reprinted with permission from [204]. Copyright 1990 The American Physical Society)

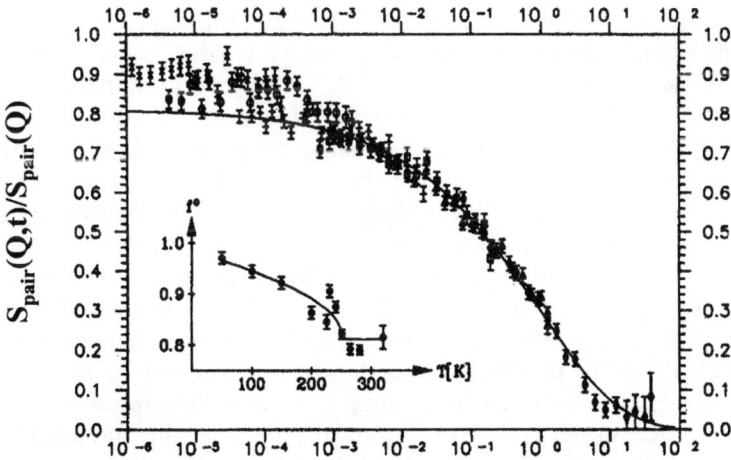

Fig. 4.36 Scaling representation of NSE data (density correlation function) corresponding to PI at $Q=1.92$ Å^{-1} [second maximum of $S(Q)$]. Times have been divided by the KWW time τ_{pair} to obtain a master curve. $T=230$ (*cross*), 240 (*empty circle*), 250 (*plus*), 264 (*empty square*), 280 (*empty triangle*), 320 K (*empty diamond*). The *solid line* indicates the fit with the KWW law for 250 K$\leq T \leq$320 K resulting in the parameters $f_Q^0 = 0.856\pm0.006$, $\beta=0.45\pm0.013$. *Insert*: Temperature dependence of $f_Q(T)$, the *solid line* denotes the prediction of MCT (Eq. 4.37) (Reprinted with permission from [8]. Copyright 1992 Elsevier)

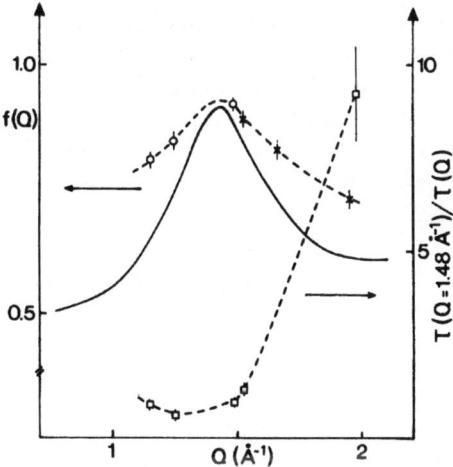

Fig. 4.37 Q-dependence of the non-ergodicity parameter $f_Q(T)$ (*empty circle*, $\lambda=6$ Å, *cross*, $\lambda=4.55$ Å) and the ratio of the relaxation rates $\tau_{\text{pair}}(Q=1.48$ Å$^{-1})/\tau_{\text{pair}}(Q)$ obtained from measurements on PB. The solid line represents the static structure factor $S(Q)$ at 230 K. (Reprinted with permission from [124]. Copyright 1988 The American Physical Society)

dependence of $f_Q(T)$ measured at the first minimum of $S_{pair}(Q)$ follows well the square root singularity of Eq. 4.37, allowing the determination of T_c: T_c= 214±3 K for PB (Figs. 4.3 and 4.35) and T_c~240 K for PI [8]. In the case of PI, the thermal evolution of the non-ergodicity factor has also been investigated at Q=1.92 Å^{-1}, in the region of second maximum of the static structure factor. The results were also compatible with the proposed square root singularity (see Fig. 4.36) [8]. Finally, for PB the Q-dependence of both the non-ergodicity parameter and the timescale characteristic for the slow decay of the dynamic structure factor was investigated in [124]. The experimental results provide strong support to the theory since the proposed modulation with the static structure factor becomes clearly evident for both magnitudes (see Fig. 4.37). A recent NSE study on PIB has also allowed this property to be scrutinized [147]. The results will be discussed in detail in the next chapter.

There exists an extensive work devoted to investigate the validity of MCT in a large variety of systems accumulated during past years (see, e.g. [206] and references therein). The agreement found between experiments or computer simulations and MCT seems to be rather good, though the occurrence of thermally activated ("hopping") processes had to be included in a more extended formulation of the theory in order to account for the experimental behaviour observed below T_c. In the case of polymers, further scrutiny of MCT has been realized by computer simulations of bead and spring models (see [207] and references therein, and [208]). In these simulations the predictions seem to be rather well fulfilled. Also, for a united atom model for a polymer the incoherent scattering function follows the MCT well [209]. Fully atomistic molecular dynamics simulations in polymers have not been yet used for this purpose, though it would be highly desirable. As will be shown in the next chapter, the dynamics of the atoms within the Gaussian blobs forming the Gaussian chain cannot be captured by bead and spring models. The dynamics involved in the decaging process, which is at the origin of the structural relaxation in real polymers, is just taking place within such Gaussian blobs.

5
Intermediate Length Scales Dynamics

Up to this point the large scale dynamics and the local scale dynamics have been dealt within an independent way. This is possible because at large length scales, where the detailed microstructure of the macromolecule is not important, the properties are determined by the chain character of the structural units. At local scales, some general features of glass-forming systems dominate, with the structural relaxation as main dynamic process. However, one of the big challenges for polymer physics today is to bridge the scales and to understand how the chain-specific properties like the α-relaxation cross over to the universal large scale dynamics. The investigation of polymer melt dynamics at length scales of several times the intermolecular distances – the so-called intermediate length scale regime – is then imperative. As we will see, this almost unexplored area contains unexpected phenomena.

Neutron scattering, and in particular NSE, is again the right tool for such studies. From the point of view of the glass-forming system, the question to answer is: how do the self- and collective motions evolve when longer and longer distances are explored and the hydrodynamic limit is approached? From the point of view of the macromolecular nature, NSE measurements of the single chain dynamic structure factor can be used to reveal how the single chain dynamics develop when the system is observed at shorter and shorter length scales, where finally the detailed nature of the monomer begins to play an important role.

We want to emphasize the difficulties posed by the study of the dynamics at intermediate length scales. From an experimental point of view studying the pair correlations, we face very weak intensities and possibly important multiple scattering contributions [210]. For these reasons, the intermediate length scale dynamics remains almost unexplored up to date. But even having experimental data at hand, the interpretation of the results is by no means trivial. As we will see, some theoretical approaches are at hand for explaining the development of the chain dynamics towards shorter length scales. However, none of them includes the fact that the dynamics at local scales has to fulfil the general behaviour of glass-forming systems, as for instance the stretching of the relaxation function. Moreover, as far as we know, there is no theory currently available which allows us to connect the three experimental approaches to the dynamics at intermediate length scales (self-, pair and single chain dynamics). In this sense, the accumulation of experimental data, even in a phenomenological framework, is of importance for gaining insight into this unexplored field.

5.1
Chain-Rigidity and Rotational Transitions as Rouse Limiting Processes

From a conceptual point of view the Rouse model is limited towards large scales by the onset of topological interactions (the associated confinement and reptation mechanism have been discussed in Chap. 4) and towards smaller scales as well. There, the simplifying assumptions of the Rouse model cease to be valid and the local chain structure comes into play. Locally the chain is stiff; this rigidity leads to deviations from the Rouse model. Furthermore, as a consequence of the rotational potentials local relaxation mechanisms across the rotational barriers come into play leading to an internal viscosity. Even more locally, we approach the regime of the elemental relaxation processes, the α-process and possibly higher order relaxations.

5.1.1
Mode Description of Chain Statistics

In a real chain segment-segment correlations extend beyond nearest neighbour distances. The standard model to treat the local statistics of a chain, which includes the local stiffness, would be the rotational isomeric state (RIS) [211] formalism. For a mode description as required for an evaluation of the chain motion it is more appropriate to consider the so-called all-rotational state (ARS) model [212], which describes the chain statistics in terms of orthogonal Rouse modes. It can be shown that both approaches are formally equivalent and only differ in the choice of the orthonormal basis for the representation of statistical weights. In the ARS approach the characteristic ratio C_∞ of the RIS-model becomes mode dependent.

As Allegra et al. [213, 214] have shown, for polyolefines it can be well approximated by:

$$C(q) = \frac{2(1-c)^{1/2}(C_\infty - c)}{(C_\infty - c)^2 + (1-c)^{1/2} - [(C_\infty - c)^2 - (1-c)^{1/2}]\cos q + c} \quad (5.1)$$

with $c \cong 0.4$, C_∞ the characteristic ratio of the particular polyolefine, and $q=p\pi/N$ (p: 1...N). The mode-dependent characteristic ratios are bell-shaped curves centred around $q=0$ with $C(0) = C_\infty$. If $\tilde{\underline{\ell}}(q)$ is the Fourier transformed of the segment vector $\underline{r}(n)=\underline{R}(n)-\underline{R}(n-1)$ (see Eq. 3.3), then $C(q)$ connects to the segment-segment correlation function:

$$\frac{1}{N}\langle \underline{\ell}(q) \underline{\ell}^*(q) \rangle = C(q) \ell_0^2 \quad (5.2)$$

The higher the mode number, the smaller becomes the square of the Fourier components. Using periodic boundary conditions and considering that a segment vector is given by the difference of adjacent position vectors the statistical average $\langle \tilde{R}(q)^2 \rangle$ in Fourier space becomes:

$$\langle \tilde{R}^2(q) \rangle = \frac{\langle \tilde{\ell}^2(q) \rangle}{4\sin^2\left(\frac{q}{2}\right)} = \frac{N\ell_0^2 C(q)}{4\sin^2\left(\frac{q}{2}\right)} \quad (5.3)$$

5 Intermediate Length Scales Dynamics

With this equation in place we now may evaluate the statistical average for any distance $k=|h-j|$ along the chain, where h and j denote the position of chain segments. Fourier transforming ($\underline{r}\{n\}$) (Eq. 3.3) we obtain a probability distribution in Fourier space:

$$\tilde{P}rob(\{\tilde{\underline{\ell}}(q)\}) = \prod_q \left(\frac{3}{2\pi NC(q)\ell_0^2}\right)^{3/2} \exp\left[-\sum_q 3\frac{\tilde{\underline{\ell}}(q)\,\tilde{\underline{\ell}}^*(q)}{2NC(q)\ell_0^2}\right] \quad (5.4)$$

Like in the Rouse model from the probability distribution $\tilde{P}rob$ the free energy is obtained by taking the logarithm and finally the force exerted on a segment h (x-component) follows by taking the derivative of the free energy:

$$f_x(h) = -\frac{\partial F}{\partial x(h)} = -\frac{3k_BT}{N\ell_0^2}\sum_{\{q\}}\frac{4\sin^2\left(\frac{q}{2}\right)}{C(q)}\tilde{x}(q)\,e^{-iqh} \quad (5.5)$$

The Fourier transformed position coordinate $\tilde{x}(q)$ thereby is taken from Eq. 3.8. Having evaluated the force we now can reformulate the Rouse equation (Eq. 3.6) introducing the new force term of Eq. 5.5. The Fourier-transformed Rouse equation then reads:

$$\zeta_0\dot{\tilde{x}}(q) - \frac{3k_BT}{\ell_0^2}\frac{4\sin^2\left(\frac{q}{2}\right)}{C(q)}\tilde{x}(q) = \tilde{F}_x(q,t) \quad (5.6)$$

where $\tilde{F}_x$ denotes the x-component of the appropriate Fourier component of the random force. Comparing Eq. 5.6 with Eq. 3.6 we realize that the mode-dependent characteristic ratio $C(q)$ leads to a stiffening of the chain for higher q, where $C(q)$ drops (Eq. 5.1) and consequently the spring constant increases. As a result the characteristic relaxation times are shortened compared to Eq. 3.12. We have (second part is valid for $p\ll N$):

$$\frac{1}{\tau_q} = \frac{12k_BT\sin^2\left(\frac{q}{2}\right)}{\zeta_0\ell_0^2 C(q)} \approx \frac{3k_BT}{\zeta_0\ell_0^2 C(q)}q^2 \quad (5.7)$$

The mean square displacements are calculated in the same spirit as for the simple Rouse model. If for simplicity we consider the periodic chain transform we get:

$$B(n,m,t) = B(k,t) = \frac{6k_BT}{N_0\zeta_0}\sum_{\{q\}}\tau_q\left[1-\cos(qk)\exp\left(-\frac{t}{\tau_q}\right)\right] \quad (5.8)$$

where $k=|n-m|$. We note that because of the shorter relaxation times the weight at which the contribution of an eigenmode q appears in the mean square displacement is reduced compared to the simple Rouse model. According to the equipartition theorem a stiffer spring leads to a smaller fluctuation amplitude.

The dynamic structure factors now may be evaluated following Eq. 3.19 as above.

5.1.2
Effect of Bending Forces

While we started from an ARS description of the chain statistics, we now coarse grain and introduce the chain stiffness more generally in terms of a local rigidity which is exhibited by locally stiff chain molecules. Harnau et al. [215, 216] have shown that the partition function for such a chain can be formulated by a maximum entropy principle. Its evaluation leads to the correct average static properties of the Kratky-Porod wormlike chain. Here we consider such a chain and describe it by a continuous chain contour coordinate $-L/2 \leq s \leq L/2$, where L is the overall contour length and s the contour coordinate. L relates to the number of chain bonds N and the bond length ℓ_0 by $L=N\ell_0 \sin(\theta/2)$, where θ is the bond angle. The stiffness is introduced into the continuous form of the Rouse Eq. 3.7 by a fourth order derivative with respect to the contour coordinate:

$$\gamma \frac{\partial}{\partial t} \underline{R}(s,t) + \varepsilon \frac{\partial^4}{\partial s^4} \underline{R}(s,t) - 2\nu \frac{\partial^2}{\partial s^2} \underline{R}(s,t) = \underline{f}(s,t) \tag{5.9}$$

with $\gamma = \dfrac{\zeta}{\ell_0 \sin\theta/2}$ the friction per unit length, $\varepsilon = \dfrac{3k_B T}{4\tilde{p}}$ and $\nu = \dfrac{3\tilde{p}k_B T}{2}$; $\tilde{p}$ relates to the persistence length of the chain and may be connected to C_∞. e.g. by the expression for the end-to-end distance of a wormlike chain $\langle R_E^2 \rangle = \dfrac{L}{\tilde{p}} - \dfrac{(1-e^{-2\tilde{p}L})}{2\tilde{p}^2}$ and the corresponding expression for a RIS-chain. For large N, $\tilde{p} = \dfrac{\sin(\theta/2)}{C_\infty \ell_0}$ holds. With the appropriate boundary conditions of force-free ends a normal mode analysis may be performed. From that analysis characteristic relaxation times are obtained as:

$$\frac{1}{\tau_p^b} = \frac{3k_B T}{\gamma} \left(\tilde{p}\alpha_p^2 + \frac{\alpha_p^4}{4\tilde{p}} \right) \tag{5.10}$$

The coefficients α_p are obtained from transcendent equations (Eq. 2.25 in [216]). For large $L\tilde{p}$ the coefficients α_p are very close to $\alpha_p \cong p\pi/L$. With this approximation Eq. 5.10 becomes:

$$\frac{1}{\tau_p^b} = \frac{W\pi^2}{N^2} \left[p^2 + \frac{\pi^2 C_\infty^2 p^4}{4N^2 \sin(\theta/2)} \right] = \frac{W\pi^2}{N^2} (p^2 + \alpha_B p^4) \tag{5.11}$$

5 Intermediate Length Scales Dynamics

Equation 5.11 in its first part agrees with the Rouse result of Eq. 3.12. The term proportional to p^4 is the correction due to rigidity effects. For higher mode numbers p the relaxation rates $(\tau_p^b)^{-1}$ are increasingly accelerated. Furthermore, because of the equipartition theorem (see e.g. Eq. 3.13) the amplitude of the corresponding correlator will diminish. We notice that under these conditions the eigenfunctions of Eq. 5.9 also become very similar to those of the Rouse problem. In a first order approximation therefore, the mean square displacements (Eq. 3.17) may be calculated by replacing the characteristic relaxation times $\tau_p = \tau_R/p^2$ by τ_p^b. With these new expressions for the msd we then may evaluate the dynamic structure factor following the prescription of Eq. 3.19. Using the same approach as for the Rouse model, we also may calculate the full solutions.

5.1.3
Internal Viscosity Effects

The description of the chain dynamics in terms of the Rouse model is not only limited by local stiffness effects but also by local dissipative relaxation processes like jumps over the barrier in the rotational potential. Thus, in order to extend the range of description, a combination of the modified Rouse model with a simple description of the rotational jump processes is asked for. Allegra et al. [213, 214] introduced an internal viscosity as a force which arises due to a transient departure from configurational equilibrium, that relaxes by reorientational jumps. Thereby, the rotational relaxation processes are described by one single relaxation rate τ_a. From an expression for the difference in free energy due to small excursions from equilibrium an explicit expression for the internal viscosity force in terms of a memory function is derived. The internal viscosity force φ_k acting on the k-th backbone atom becomes:

$$\varphi_k(t) = \mp \frac{iV}{N} \int_{-\infty}^{t} \exp\left[-\frac{(t-t')}{\tau_a}\right] \times \left[\sum_{\{q\}} \frac{3k_B T}{C(q)\ell_0^2} \tilde{x}(q,t') \sin(q)\, e^{-iqk}\right] dt' \quad (5.12)$$

Thereby, V is a numerical constant in the order of 1. The sign describes the direction of the net force, which is dictated by the direction of strain propagation along the chain. After Fourier transformation Eq. 5.12 may be introduced into the modified Rouse equation (Eq. 5.6) yielding:

$$\frac{12 k_B T}{C(q)\ell_0^2} \sin^2\left(\frac{q}{2}\right) \tilde{x}(q,t) \pm iV \frac{3k_B T}{C(q)\ell_0^2} \sin(q)$$

$$\int_{-\infty}^{t} dt' \times \exp\left[-\frac{(t-t')}{\tau_a}\right] \dot{\tilde{x}}(q,t') + \zeta_0 \dot{\tilde{x}}(q,t) = \tilde{F}_x(q,t) \quad (5.13)$$

The homogeneous Eq. 5.13 is solved by a decaying travelling wave:

$$\tilde{x}(q,t) = \tilde{x}_o(q) \exp\left\{\left[i\omega(q) - \frac{1}{\tau(q)}\right]t\right\} \tag{5.14}$$

with

$$\omega(q) + \frac{i}{\tau(q)} = \frac{A(q) \mp \sqrt{A^2(q) + B(q)}}{2\zeta_0 \tau_a} \tag{5.15a}$$

where

$$A(q) = V\frac{3k_BT}{\ell_0^2}\tau_a \sin(q) - i\left[\zeta_0 + 2\frac{12k_BT}{\ell_0^2 C(q)}\tau_a \sin^2\left(\frac{q}{2}\right)\right] \tag{5.15b}$$

$$B(q) = 8\frac{12k_BT}{\ell_0^2 C(q)}\zeta_0\tau_a \sin^2\left(\frac{q}{2}\right) \tag{5.15c}$$

The eigenvalues Eq. 5.15a lead to two branches of the relaxation time τ and the propagation frequency ω. For experimental parameters appropriate for PIB, Fig. 5.1 displays the dependence of the relaxation times on the mode number

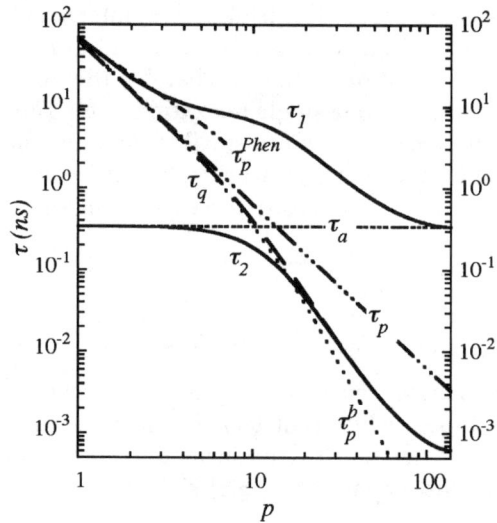

Fig. 5.1 Mode number dependence of the relaxation times τ_1 and τ_2 (*solid lines*). The *dashed-dotted line* shows the relaxation time τ_p in the Rouse model (Eq. 3.12). The *horizontal dashed line* displays the value of τ_a. The *dashed* and the *dotted lines* represent the relaxation time when the influence of the chain stiffness is considered: mode description of the chain statistics τ_q (*dashed*, Eq. 5.11) and bending force model τ_p^b (*dotted*, Eq. 5.7). The behaviour of the relaxation time τ_p^{phen} used in the phenomenological description is also shown for the lowest modes (see text). (Reprinted with permission from [217]. Copyright 1999 American Institute of Physics)

$p=qN/\tau$. For comparison also the solution τ_q of the ARS-Rouse equation Eq. 5.7 is displayed. While at low p the solution for the longer relaxation time τ_1 coincides with that of the modified Rouse equation, at higher p such a coincidence is observed for the shorter relaxation time τ_2. In between both modes repel each other and for small and large p reach the value of τ_A. The multidot-dashed line represents the Rouse model. We realize that for small p the Rouse relaxation times agree well with τ_q, while for higher mode number the relaxation times are significantly decreased due to the stiffness. Finally we also include the relaxation times of the bending model τ_p^b (dotted line). These times decrease even more strongly than those of the ARS-model.

The mean square segment displacements, which are the key ingredient for a calculation of the dynamic structure factor, are obtained from a calculation of the eigenfunctions of the differential Eq. 5.13. After retransformation from Fourier space to real space $B(k,t)$ is given by Eq. 41 of [213]. For short chains the integral over the mode variable q has to be replaced by the appropriate sum. Finally, for observation times $t \gg \tau_a$ the mean square displacements can be expressed in terms of eigenvalue branch τ_1 only. Thereby a significant simplification of the expression for the mean square displacements is achieved (see [213]).

5.1.4
Hydrodynamic Interactions and Internal Viscosity Effects

In order to access more directly the internal relaxation time τ_a, which is suspected to relate to the intrachain rotational potential, it is advantageous to investigate polymer solutions where the interchain interactions can be removed by dilution. The interpretation of solution results requires the consideration of hydrodynamic interactions between the chain segments, which are mediated by the solvent. In this section, we will introduce these interactions briefly in terms of the Zimm model [218], which considers the forces exerted on other chain segments by a moving segment of the same chain due to the induced motion of the surrounding fluid. If a force $f(m,t)$ acts on the solvent at position "m" along the chain, the resulting force at position "n" is given by:

$$\underline{f}^{\text{hyd}}(n,m) = \underline{\underline{H}}(\underline{r}_n - \underline{r}_n)\underline{f}(m) \tag{5.16a}$$

where

$$\underline{\underline{H}}(r) = \frac{1}{8\pi\eta_s r}\left(1 - \frac{\underline{r} \otimes \underline{r}}{r^2}\right) \tag{5.16b}$$

is the Oseen tensor. It is derived from the linearized hydrodynamic equations (Stokes approximation) and η_s is the solvent viscosity. With $\underline{\underline{H}}_{nm} \equiv (\underline{r}_n - \underline{r}_m)$ inserting Eq. 5.16a into the Rouse Eq. 3.7 we arrive at:

$$\zeta_0 \frac{\partial \underline{R}_n}{\partial t} = \sum_m \underline{\underline{H}}_{nm} \frac{k_B T}{\ell^2}\frac{\partial^2 R_m}{\partial m^2} + \underline{f}(n,t) \tag{5.17}$$

Since Eq. 5.16b assumes point-like interactions, the self term diverges and has to be replaced by the friction of the bead with the embedding medium $H_{nn}=1/\xi$ – the only friction present in the Rouse model – with $\xi=6\pi\eta a$, where a stands for the effective bead radius. More general the ratio of the diagonal (Rouse-like) friction and the solvent mediated interaction strength is expressed by the draining parameter $B=(\xi/\eta)/(6\pi^3\ell^2)^{1/2}$ [214]. The Rouse model has $B=0$ whereas the assumption (segment=sphere) of $\ell=2a$ and $\xi=6\pi\eta a$ leads to $B=0.69$. The form of Eq. 5.17 yields a non-linear Langevin equation for which no analytic solution can be given. Only replacing H_{nm} by its average over the equilibrium chain configurations restores linearity.

For a Gaussian chain we have:

$$\langle H_{nm}\rangle = \sqrt{\frac{6}{\pi}}\frac{1}{6\pi\eta}\langle r^2(n,m)\rangle^{-1/2} \tag{5.18}$$

In the ARS-picture again we have to consider the q-dependence of $C(q)$ (Eqs. 5.2 and 5.3). After Fourier transformation and considering again periodic bounding conditions, Eq. 5.17 yields characteristic relaxation rates:

$$\frac{1}{\tau_q^Z} = \frac{3k_BT}{\ell_0^2 \zeta(q)C(q)}\sin^2\left(\frac{q}{2}\right) \tag{5.19}$$

They differ from (Eq. 5.7) by a mode-dependent friction coefficient.

$$\frac{\zeta_0}{\zeta(q)} = 1 + \frac{\zeta}{3\pi\eta_s}\sqrt{\frac{6}{\pi}}\sum_{k=1}^{N-1}\left(1-\frac{k}{N}\right)\frac{1}{\ell_0\sqrt{C(k)}}\frac{\cos qk}{\sqrt{k}} \tag{5.20}$$

In the limit of $C(q)=C_\infty$ Eq. 5.20 may also be formulated in terms of the draining parameter. In this case for numerical purposes the sum in Eq. 5.20 is well approximated by $\sqrt{\frac{2\pi}{q}}\,[1-2\,FresnelC(\sqrt{1.06q/\pi})]$ with $FresnelC(x)=\int_0^x\cos\left(\frac{\pi t^2}{2}\right)dt$. In order to include the hydrodynamic interaction into the equations for the internal viscosity effect ζ_0 in Eq. 5.14 and Eq. 5.15 has to be replaced by $\zeta(q)$. The modified eigenvalues of Eq. 5.18 introducing $\zeta(q)$ are displayed in Fig. 5.2 anticipating the experimental parameters from a solution of PIB in toluene (see later) [186]. For comparison also the dispersion of the Rouse-Zimm relaxation τ_q^Z (Eq. 5.19) without the influence of the local relaxation process is shown. For small q the Rouse-Zimm dispersion is close to the branch τ_1 while at large q, τ_2 coincides with τ_q^Z. In between both modes repel each other and for small or large q reach the value of τ_a. For intermediate scale motion q/π-values in the range above 0.01 are important. The main effect of the

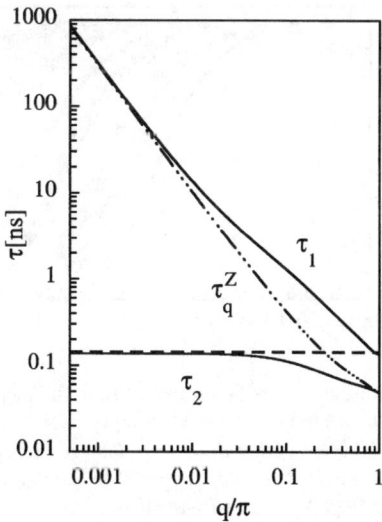

Fig. 5.2 Mode number dependence of the relaxation times τ_1 and τ_2 (*solid lines*) found for PIB in dilute solution at 327 K. The *dashed-dotted line* shows the relaxation time τ_q^Z of the Rouse-Zimm model. The *horizontal dashed line* displays the value of τ_a. (Reprinted with permission from [186]. Copyright 2001 American Chemical Society)

presence of the local relaxation process is a general slowing down of the chain relaxation – τ_1 bends upwards!

5.1.5
NSE Results on Chain Specific Effects Limiting the Rouse Dynamics

Recently a very detailed study on the single chain dynamic structure factor of short chain PIB (M_n=3870) melts was undertaken with the aim to identify the leading effects limiting the applicability of the Rouse model toward short length scales [217]. This study was later followed by experiments on PDMS (M_n=6460), a polymer that has very low rotational barriers [219]. Finally, in order to access directly the intrachain relaxation mechanism experiments comparing PDMS and PIB in solution were also carried out [186]. The structural parameters for both chains were virtually identical, R_g=19.2 (21.3 Å). Also their characteristic ratios C_∞=6.73 (6.19) are very similar, i.e. the polymers have nearly equal contour length L and identical persistence lengths, thus their conformation are the same. The rotational barriers on the other hand are 3–3.5 kcal/mol for PIB and about 0.1 kcal/mol for PDMS. We first describe in some detail the study on the PIB melt compared with the PDMS melt and then discuss the results.

Figure 5.3 presents NSE results obtained on PIB at 470 K together with a fit with the Rouse dynamic structure factor Eq. 3.19. The Rouse model provides a good description of the spectra for $Q \leq 0.15$ Å^{-1}. In this range, the elementary

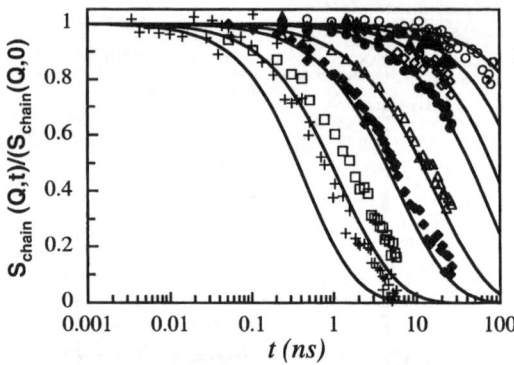

Fig. 5.3 Single chain dynamic structure factor from PIB in the melt at 470 K and $Q=0.04$ Å^{-1} (*empty circle*), 0.06 Å^{-1} (*filled triangle*), 0.08 Å^{-1} (*empty diamond*), 0.10 Å^{-1} (*filled circle*), 0.15 Å^{-1} (*empty triangle*), 0.20 Å^{-1} (*filled diamond*), 0.30 Å^{-1} (*empty square*), and 0.40 Å^{-1} (*plus*). The *solid lines* show the fit of the Rouse model to the data. (Reprinted with permission from [217]. Copyright 1999 American Institute of Physics)

Rouse rate $W\ell^4$ at 470 K is 8100 Å^4/ns. The corresponding mode-dependent characteristic times are represented by the dashed-dotted line in Fig. 5.1. Taking into account the relationship between D_R and $W\ell^4$, the corresponding diffusion coefficient D_R becomes $D_R = 1.22 \pm 0.17 \times 10^{-11}$ m^2/s.

While the NSE results show that, within the experimental accuracy, in the range $Q \leq 0.15$ Å^{-1} the Rouse model gives a good account for the internal modes as well as for the diffusion of the chain centre of mass, it is also clear that for higher Q-values the experimental structure factors decay significantly more slowly than the Rouse model would require. These deviations are quantified in fitting the Rouse model to the different spectra separately. This procedure results in a strong dispersion of the elementary Rouse rate. The values determined for $W\ell^4$ at $Q \geq 0.15$ Å^{-1} follow a Q-dependence, which can be described by the power law:

$$W\ell^4 \sim Q^{-0.84} \tag{5.21}$$

Note that another origin of the slowing down of the chain relaxation compared to the Rouse prediction could be a reduction of the weights of the higher modes, which in the Rouse model are proportional to p^{-2} (see Eq. 3.19).

5.1.5.1
Chain Stiffness

As explained above, towards shorter scales a more realistic description of the chain dynamics must include the stiffness of the chain. The influence of the stiffness that can be expected from the characteristic ratio of PIB was calculated according to both the ARS model and bending force models. For the mode description of the chain statistics $C(p)$ (Eq. 5.1) with $C_\infty=6.73$ and $c=0.4$ has to be considered. The chain-bending model was investigated both in its approximated version (Eq. 5.11) with $\alpha_B=0.013$ and in terms of its full solution. Within the accuracy of the calculation both expressions are indistinguishable and lead to the same structure factors. The obtained relaxation times τ_p^h at 470 K show the mode number dependence depicted in Fig. 5.1. For p values lower than ≈20 they coincide and for higher number modes the values resulting from the bending force model (Eq. 5.11) are faster than those obtained in the other framework (Eq. 5.7). The predicted dynamic structure factors are shown in Fig. 5.4. The dashed-dotted lines correspond to the bending force approach, and the solid lines to the mode description of the chain statistics. These structure factors hardly agree better with the experimental data than those obtained from the Rouse model (Fig. 5.3). On the other hand, the prediction of the bending force model calculated in terms of its full solution is almost indistinguishable from that of the approximate version. It follows that the stiffness corrections to the chain motion are clearly not enough to account for the experimentally observed slowing down of the relaxation. Recent atomistic simulations on C_nH_{2n+2} [53] confirm this result.

We would like to mention that a fit with the bending force model to the data allowing the variation of the parameter α_B leads to a very good agreement with the experimental spectra. However, the resulting value for α_B was 0.20, about 15 times larger than that deduced from chain structure, and therefore clearly unphysical [219]. As pointed out before, in the decay of $S_{\text{chain}}(Q,t)$ each mode contributes with a weight dictated by equipartition. Since this weight is inversely proportional to $(p^2+\alpha_B p^4)$, a large value of α_B drastically reduces the contribution of the higher modes to the relaxation. Therefore, a good descrip-

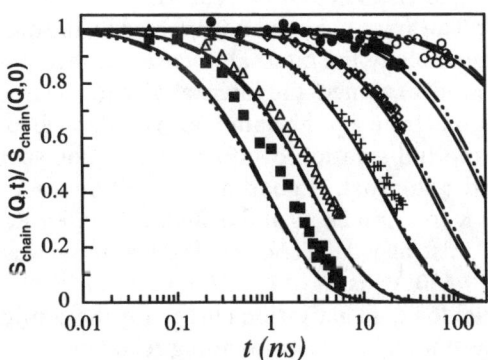

Fig. 5.4 Comparison between the predictions of the bending force model (Eq. 5.11) (*dashed-dotted lines*) and the mode description (*solid lines*) with the experimental NSE data obtained at 470 K on a PIB melt. The Q-values shown are: 0.03 Å^{-1} (*empty circle*), 0.06 Å^{-1} (*filled circle*), 0.10 Å^{-1} (*empty diamond*), 0.15 Å^{-1} (*plus*), 0.25 Å^{-1} (*empty triangle*), and 0.35 Å^{-1} (*filled square*). (Reprinted with permission from [217]. Copyright 1999 American Institute of Physics)

tion of the experimental data by the bending force model in fitting α_B is achieved by basically suppressing the contribution of the higher modes in an artificial way. A value of 0.2 for α_B demands a characteristic ratio C_∞=26.4, completely incompatible with the actually observed chain dimensions. The amplitudes are not affected by the damping.

5.1.6
Internal Viscosity

The Rouse analysis leads to the observation of a strong Q-dispersion of the elemental rate $W\ell^4$, which can be parametrized by the power law given by Eq. 5.21. Moreover, this result cannot be explained only by the stiffness of the polymer. Thus, we are forced to assume that some damping mechanism of the Rouse modes is at the origin of the apparent Q-dependence of W. Taking Eq. 5.21 as a starting point, and considering an intuitive parallelism between Q-vector and mode number p, we tentatively assume a damping mechanism that depends on the mode number in form of a power law:

$$\frac{1}{\tau_p^{\text{Phen}}} = \frac{W\pi^2}{N^2} (p^2 + \alpha_B p^4)\, p^{-\gamma} \qquad (5.22)$$

where γ is a parameter describing the strength of the damping. These defined frequencies contain the effect of both the stiffness and an internal viscosity. In terms of the friction coefficient Eq. 5.22 proposes an effective mode-dependent friction $\zeta_p \sim p^\gamma$. With such an approach for γ=0.67, an excellent description of the experimental results over the whole Q-range can be achieved. The phenomenological Ansatz appears to demonstrate that a mode-damping mechanism is a possible answer to the discrepancies. Thereby the retardation of the modes increases with the mode number. The amplitudes are not affected by the damping.

Beyond phenomenology, the internal viscosity model of Allegra et al. [214] involves a physical idea, namely the internal viscous resistance of a chain to equilibrate deviations from equilibrium. This was described above. Figure 5.5 presents a fit of the full solution of this model to the spectra allowing the Rouse rate W and τ_a to float. A good agreement between theory and data was found at 470 K for τ_a=0.28 ns and a slightly revised Rouse frequency of $W\ell^4$=6,900Å^4 ns^{-1}. Also at 390 K a reasonable description was possible yielding τ_a=2.5 ns and $W\ell^4$=650 Å^4/ns.

Using the results for τ_a an activation energy for the configurational change of about 0.43 eV is found. The corresponding relaxation times τ_1 and τ_2 of the model at 470 K are displayed as solid lines in Fig. 5.1. We note that for low mode numbers the phenomenological time τ_p^{Phen}, which is also presented, agrees quite well with τ_1. Thus, the Allegra et al. model is able to underpin the phenomenological result in interpreting the "damping" in terms of the hybridization of a local and a dispersive mode. That it is indeed an intrachain damping mechanism that affects the Rouse models is demonstrated very decisively by compar-

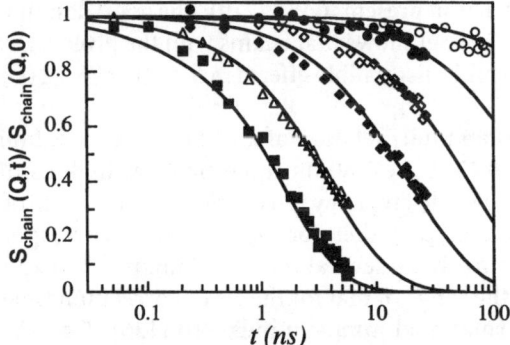

Fig. 5.5 Fitting curves in the frame of the model proposed by Allegra et al. [213, 214] for different Q-values at 470 K. The symbols correspond to: $Q=0.03$ Å^{-1} (*empty circle*), 0.06 Å^{-1} (*filled circle*), 0.10 Å^{-1} (*empty diamond*), 0.15 Å^{-1} (*filled diamond*), 0.25 Å^{-1} (*empty triangle*), 0.35 Å^{-1} (*filled square*). (Reprinted with permission from [217]. Copyright 1999 American Institute of Physics)

ison with the single chain dynamic structure factor of a PDMS melt of similar chain contour length, a system with basically no torsional barriers.

Figure 5.6 shows that the PDMS data perfectly match the prediction of the simple Rouse model up to the highest Q-values, whereas the PIB data show severe deviations from the Rouse model (Fig. 5.3) and the stiff chain model (Fig. 5.4). From the fact that two polymers with very similar structural parameters but strongly different torsional barriers display completely different relaxation behaviour the conclusion is compelling that there must be an addi-

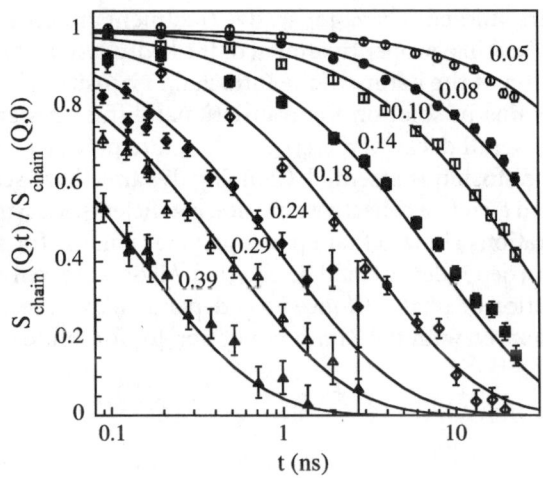

Fig. 5.6 Single chain dynamic structure factor measured for PDMS chains at 373 K in the melts compared to the standard Rouse model (*lines*) at the Q-values (Å^{-1}) indicated. (Reprinted with permission from [186]. Copyright 2001 American Chemical Society)

tional (internal) friction present in PIB [219]. The good description of the PDMS data by the Rouse model shows that chains with the given C_∞ values are not yet stiff enough to exhibit discernible effects in $S_{chain}(Q,t)/S_{chain}(Q)$ due to bending forces.

There remains an interpretation of τ_a to be found. τ_a exhibits an activation energy of about 0.43±0.1 eV, about three times as high as the C-C torsional barrier of 0.13 eV. The discrepancy must reflect the influence of the interactions with the environment and therefore τ_a appears to correspond to relaxation times most likely involving several correlated jumps. The experimental activation energy is in the range of that for the NMR correlation time associated with correlated conformational jumps in bulk PIB [136] (0.46 eV) and one could tentatively relate τ_a to the mechanism underlying this process (see later).

5.1.7
Polymer Solutions

To more deeply understand the leading mechanism coming into play upon leaving the universal Rouse regime, the same PIB molecule was studied in dilute toluene solution [186]. In dilute solution interchain friction effects are weak, and the nature of the intrachain relaxation parameter τ_a should reveal itself. The interpretation is complicated, however, because the dynamics of polymers in solution is affected by the hydrodynamic interaction between different moving entities [218, 220–222]. Since the description of this interaction within the Rouse-Zimm model is only well proven in the long wavelength limit and has its shortcomings at higher wave numbers, a comparative approach was taken in directly comparing the dynamics of PIB with that of PDMS. Chains of practical identical chain dimensions and identical translational diffusion coefficients were studied. The comparative treatment greatly reduces the uncertainties involved in a proper treatment of the hydrodynamic interaction and allows elucidation of the nature of the intrachain viscosity effect.

For small chains in solution the translational diffusion significantly contributes to the overall decay of $S_{chain}(Q,t)$. Therefore precise knowledge of the centre of mass diffusion is essential. Combing dynamic light scattering (DLS) and NSE revealed effective collective diffusion coefficients. Measurements at different concentrations showed that up to a polymer volume fraction of $\varphi=10\%$ no concentration dependence could be detected. All data are well below the overlap volume fraction of $\varphi^*=0.23$. Since no φ-dependence was seen, the data may be directly compared with the Zimm prediction [6] for dilute solutions:

$$D_Z = 0.196(0.203)\frac{k_B T}{\eta R_e} \tag{5.23}$$

where the two coefficients in front are valid for θ- and (good-) solvents. Figure 5.7 displays the dependence of $D(\varphi)$ on T/η. In accordance with Eq. 5.23 a very good linear relationship is observed. A comparison of the absolute values

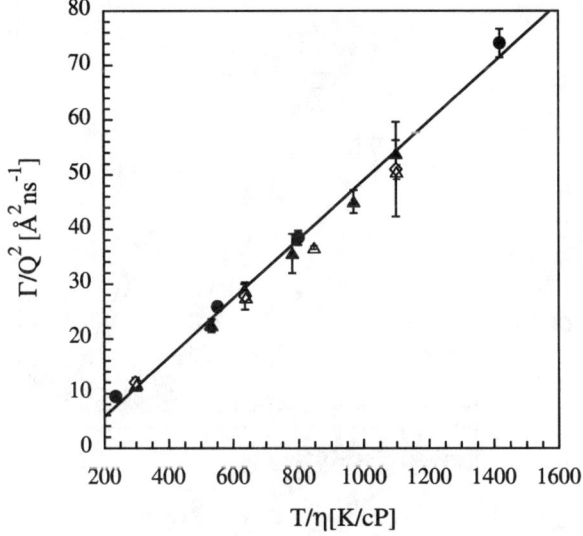

Fig. 5.7 Relaxation rate Γ divided by Q^2 for PDMS in toluene solution vs. the reduced variable T/η. Results correspond to NSE measurements: $\varphi=0.1$ (*filled circle*) and DLS for $\varphi=0.1$ (*filled triangle*), $\varphi=0.07$ (*empty triangle*) and $\varphi=0.02$ (*empty diamond*). *Solid line* shows a linear regression fit of the $\varphi=0.1$ NSE and DLS data. (Reprinted with permission from [186]. Copyright 2001 American Chemical Society)

with Eq. 5.23 is done best by looking on the slope of $D(\varphi)$ vs. T/η. Using R_e=52.2 Å (mean value in the explored temperature range, see later) Eq. 5.23 predicts a slope of 5.37×10^{-13} N/K which agrees very well with the measured value of 5.4×10^{-13} N/K.

Figure 5.8 presents typical spectra taken on both polymer solutions at 300 K (a) and 378 K (b). The PDMS data are represented by open symbols, while the PIB data are shown by full symbols. Let us first look at the data at 378 K. At $Q=0.04$ Å^{-1} ($QR_g\equiv0.8$) we are in the regime of translational diffusion, where the contributions of the intrachain modes amount to only $\approx 1\%$. There the spectra from both polymers are identical. Since both polymers are characterized by equal chain dimensions, the equality of the translational diffusion coefficients implies that the draining properties are also equal. In going to larger Q-values, gradually the spectra from the PIB-solutions commences to decay at later times. This effect increases with increasing Q and is maximal at $Q=0.4$ Å^{-1} (see Fig. 5.8a).

Figure 5.9 compares the effective diffusion coefficients $\Gamma(Q)/Q^2=D_{eff}$ from both polymers for two different temperatures. Again we realize the close agreement of the corresponding D_{eff} from both polymers in the low Q-regime. At higher Q on the other hand, where we are in the regime of the internal chain relaxations, significant differences are visible. Thus, without any sophisticated data evaluation, just from a qualitative inspection of the results the retardation

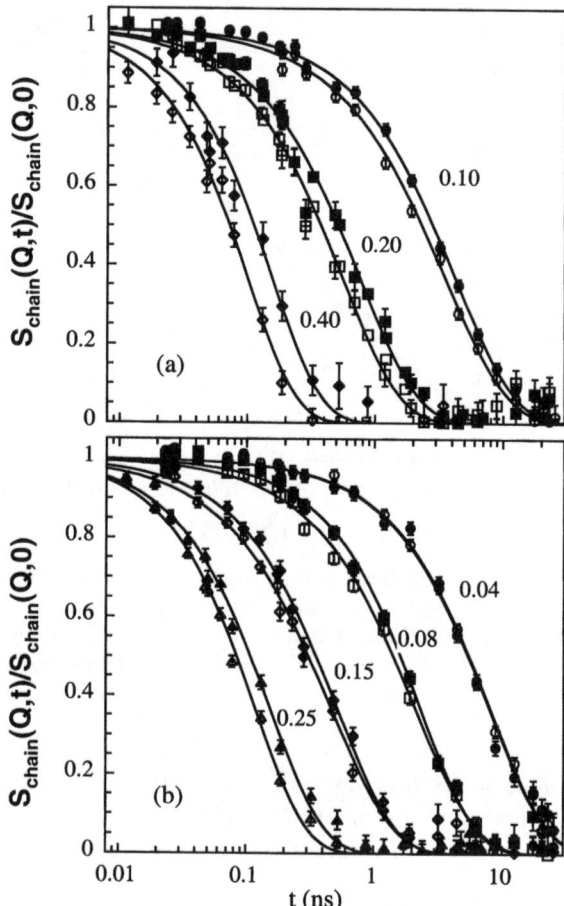

Fig. 5.8 Chain dynamic structure factor of PDMS (*empty symbols*) and PIB (*full symbols*) in toluene solution at 300 K (**a**) and 378 K (**b**). The corresponding Q-values [Å^{-1}] are indicated. *Lines* through the points represent the single exponential fits, which describe the data well. (Reprinted with permission from [186]. Copyright 2001 American Chemical Society)

of the intrachain relaxation in PIB compared to PDMS reveals itself. At larger Q, D_{eff} in PIB is reduced by about 50% at 300 K and by around 30% at 378 K. The effect is weaker than in the melt, where reductions by factors of three are found, but still significant.

The PDMS and the PIB chains consist of around 78 monomers rendering the usually applied long wavelength approximation: ($\tau_p^{-1} \sim p^{3/2}$) of the Zimm model invalid. For such short chains the Rouse-Zimm model needs to be evaluated according to Eq. 5.19, where the approximation by the Fresnel$C$-function was applied. The data were fitted keeping the translational diffusion coefficient fixed. The draining parameter B served as the only fitting parameter [$C(q) \equiv C_\infty$].

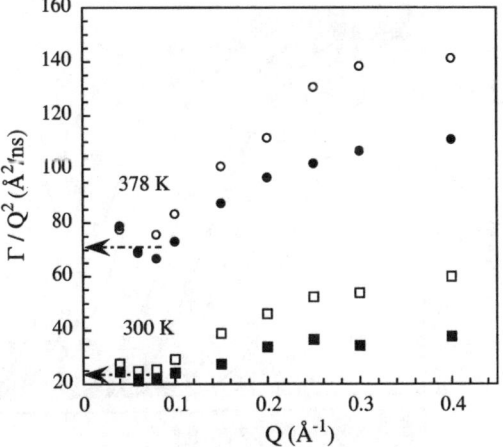

Fig. 5.9 Q-dependence of the decay rate Γ divided by Q^2 for PDMS (*empty symbols*) and PIB (*full symbols*) in solution. *Squares* correspond to 300 K and *circles* to 378 K. The *dashed arrows* show the values obtained by DLS for PDMS. (Reprinted with permission from [186]. Copyright 2001 American Chemical Society)

With this approach an excellent description of all PDMS spectra was reached (see Fig. 5.10).

In these fits the hydrodynamic interaction is captured by the draining parameter. Depending on temperature, values for B between 0.38 (251 K) and 0.30 (378 K) were found. Compared to NSE-results on long PDMS chains these B-values are low. For example, for PDMS (M_w=60,000) in the θ-solvent bromobenzene at 357 K B=0.60 was found [223]. Experiments in toluene gave $B\approx 0.50$ at 373 K [221]. In particular, a comparison with the earlier study of long chains in toluene at 373 K with our results at 378 K shows that the observed small draining parameters seem to be a chain length effect: since the hydrodynamic flow field decays only weakly with distance ($1/r$) the development of full draining behaviour appears to need a large number of monomers within a chain.

5.1.8
Intrachain Viscosity Analysis of the PIB-Data

The Allegra approach is based on the physical idea of an intrachain viscous resistance to configurational equilibration after an excursion from equilibrium. In solution the corresponding relaxation time should relate to jumps across the rotational barriers within the chain. For the PIB analysis the results of the Rouse-Zimm evaluation of PDMS were taken as an input. The analysis with the full solution of the Allegra model was performed fixing the draining parameters to the PDMS values. Figure 5.10b displays the results of such a fit at 327 K with τ_a being the only fit parameter – the diffusion coefficient was extracted

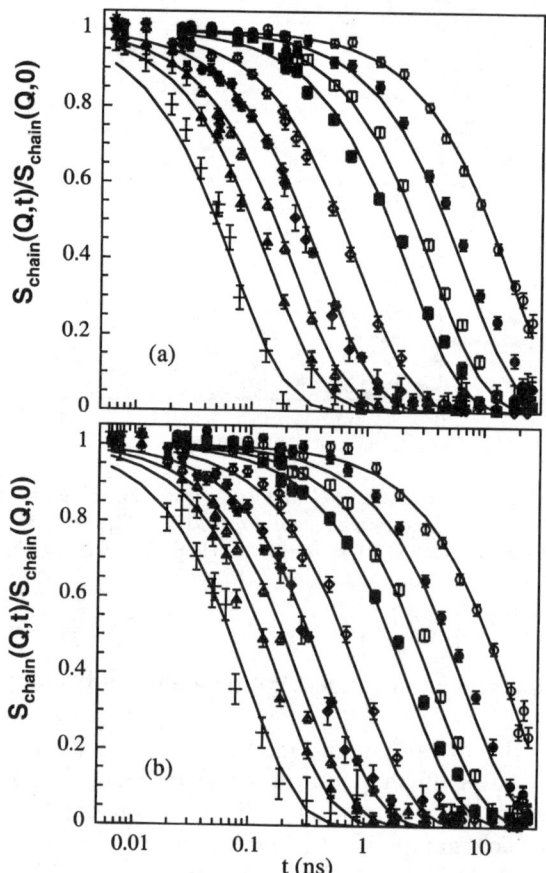

Fig. 5.10 Chain dynamic structure factor of PDMS (**a**) and PIB (**b**) in toluene solution at 327 K at the Q-values: 0.04 Å^{-1} (*empty circle*), 0.06 Å^{-1} (*filled circle*), 0.08 Å^{-1} (*empty square*), 0.10 Å^{-1} (*filled square*), 0.15 Å^{-1} (*empty diamond*), 0.20 Å^{-1} (*filled diamond*), 0.25 Å^{-1} (*empty triangle*), 0.30 Å^{-1} (*filled triangle*), 0.40 Å^{-1} (*plus*). *Solid lines* correspond to fitting curves: Rouse-Zimm model for PDMS and Rouse-Zimm with intrachain viscosity for PIB (see the text). (Reprinted with permission from [186]. Copyright 2001 American Chemical Society)

from the low Q-results and kept fixed. The fits are excellent for all temperatures. The resulting values for τ_a are displayed in Fig. 5.11.

$\tau_a(T)$ follows an activated behaviour with an activation energy $E=3.1\pm 0.3$ kcal/mol:

$$\tau_a = 1.27 \times 10^{-12} \exp\left[\frac{E}{RT}\right] \text{[sec]} \tag{5.24}$$

The preexponential factor of about 1 ps lies well in the microscopic range. The true activation energy might be somewhat higher because all the evaluation

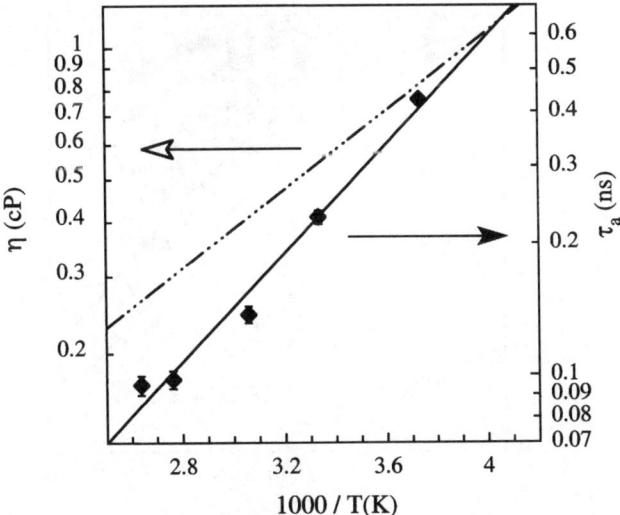

Fig. 5.11 T-dependence of the solvent viscosity (*dashed line*) and the characteristic time τ_a deduced for the conformational transitions in PIB (*filled diamond*). The *solid line* through the points corresponds to the fit to an Arrhenius law. (Reprinted with permission from [186]. Copyright 2001 American Chemical Society)

was performed relative to PDMS and any rotational barrier of PDMS would have to be added. For comparison we also display the temperature dependence of the solvent viscosity η_{tol}, which is characterized by an activation energy of 2 kcal/mol.

The activation energy of 3.1 kcal/mol lies well in the range of activation energies for torsional jumps for PIB found by other techniques [224] and simulation [197] and allows identification of the mechanism behind the intrachain friction as rotational jumps that change the chain conformation. The finding corroborates the physical picture behind the Allegra mechanism and shows that the intrachain viscosity appears to be the leading mechanism causing deviations from universal relaxation in going to local scales.

Summarizing, with the aid of NSE spectroscopy detailed information about the leading mechanism limiting the universal Rouse model towards short length scales for a typical polymer has evolved. Comparing PIB and PDMS relaxations in the melt led to the conclusion that, given identical static rigidity, the different dynamics must relate to an intrachain relaxation process. The solution data clearly demonstrate that (1) the internal viscosity model of Allegra et al. is able to quantitatively describe all solution results and (2) the relaxation mechanism behind the internal viscosity effect is indeed the jump across rotational barriers.

Not only the single chain dynamic structure factor of PIB but also other investigated polymers show deviations from Rouse dynamics when approaching local scales. As already discussed above, in the combined NSE and MD-simu-

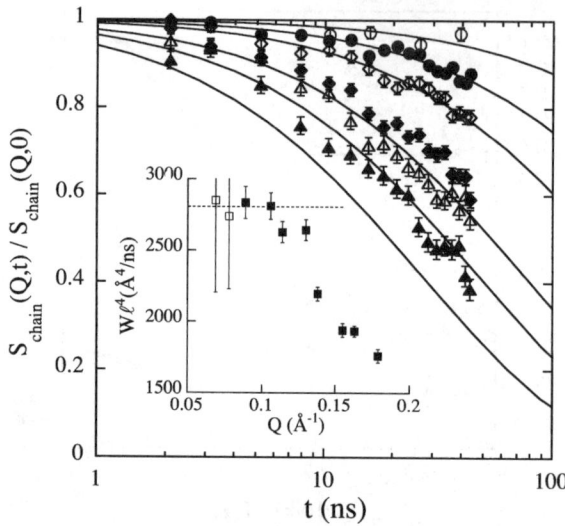

Fig. 5.12 $S_{chain}(Q,t)$ obtained for PVE at Q=0.07, 0.09, 0.11, 0.14, 0.16 and 0.18 Å^{-1} (*top to bottom*). The *solid lines* are the prediction of the Rouse model with $W\ell^4$=2800 Å^4/ns. *Inset*: Q-dependence of the parameter $W\ell^4$ as obtained from individual fits to the Rouse model. The *dotted line* shows the mean value for small Q. (Reprinted with permission from [39]. Copyright 2004 EDP Sciences)

lations study on PE [52] the comparison between Rouse predictions and the results of both techniques lead to a reasonable agreement only for Q-values up to 0.14 Å^{-1}, while for $Q\geq 0.18$ Å^{-1} the theoretical Rouse curves lie systematically below both the experimental and simulation data. Also, recent NSE measurements of the PVE single chain dynamic structure factor have revealed similar deviations. As can be seen in Fig. 5.12, the Rouse prediction fails at higher Q-values. In order to quantify the deviations, individual fits with the Rouse equation Eq. 3.24 to all data were performed. These fits reveal the Rouse variable $W\ell^4$. Its Q-dependence is displayed in the insert of Fig. 5.12. The validity of the Rouse model obviously requires $W\ell^4$=constant, a condition that is fulfilled only for $Q\leq Q_R$=0.11 Å^{-1} (dashed line). Towards higher Q the effective Rouse parameter decreases by nearly a factor of two. We note that PVE significantly shows a higher rigidity than PIB (C_∞^{PVE}=8.2) but the deviations from Rouse appear in a similar Q-range. This observation again underlines that the chain stiffness is not the main reason for the deviations from purely entropic dynamics at higher Q.

5.2
Collective Motions

Up to date the only published work on the dynamic structure factor of a glass forming system at intermediate length scales is a study on PIB [147]. The NSE measurements explored the Q-range $0.20\leq Q\leq 1.6$ Å^{-1} ($Q_{max}\approx 1.0$ Å^{-1}) for 335 K,

5 Intermediate Length Scales Dynamics

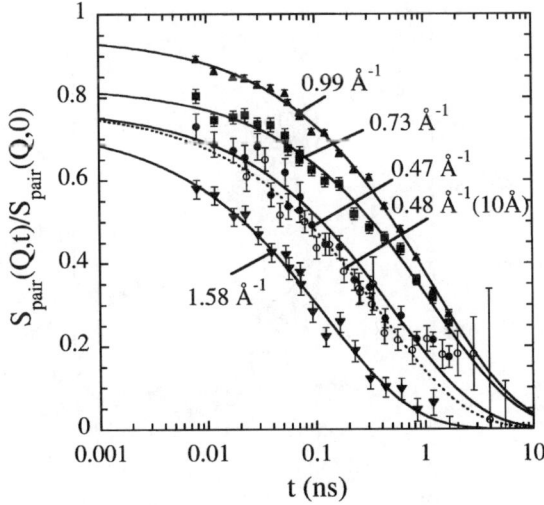

Fig. 5.13 Dynamic structure factor measured on deuterated PIB. The symbols (*full incoming wavelength $\lambda=6$ Å, empty $\lambda=10$ Å*) correspond to the different Q-values indicated. *Lines* are the resulting KWW fit curves (Eq. 4.21) (*solid $\lambda=6$ Å, dotted $\lambda=10$ Å*). (Reprinted with permission from [147]. Copyright 2002 The American Physical Society)

365 K and 390 K ($T_g=205$ K). As can be seen in Fig. 4.32a, the dynamics were accessed from the intermediate length scales region to the valley between the first and the second peaks of $S_{pair}(Q)$ across Q_{max}. Figure 5.13 shows some of the spectra recorded at 390 K in different regions of the structure factor:

i. From the low Q plateau of $S_{pair}(Q)$, data at Q around 0.47 Å^{-1} are shown, which were taken with incident wavelengths of 6 and 10 Å. In spite of the low scattered intensity reasonable statistics were still achieved. Furthermore, it is satisfying to observe the good agreement between the results from the two different experimental setups.
ii. The slowest relaxation is observed for the spectrum at Q_{max} (where the structural relaxation is revealed).
iii. The fastest relaxation is detected at $Q=1.58$ Å^{-1}, close to the first minimum of $S_{pair}(Q)$.
iv. The spectrum at $Q=0.73$ Å^{-1} in the lower Q flank of $S_{pair}(Q)$ relaxes at a rate intermediate between that of the low Q plateau and that of the structure factor peak. Finally, we note that in all cases a nearly full relaxation of the dynamic structure factor within the instrumental time frame is observed.

One of the most serious difficulties for experimentally accessing the dynamics in the low Q plateau is the presence of multiple scattering [210]. In order to assess this effect, NSE spectra were taken above and below the multiple scattering threshold. For neutron wavelengths $\lambda \geq 14$ Å the first structure factor peak cannot be seen with neutrons and therefore multiple scattering contributions

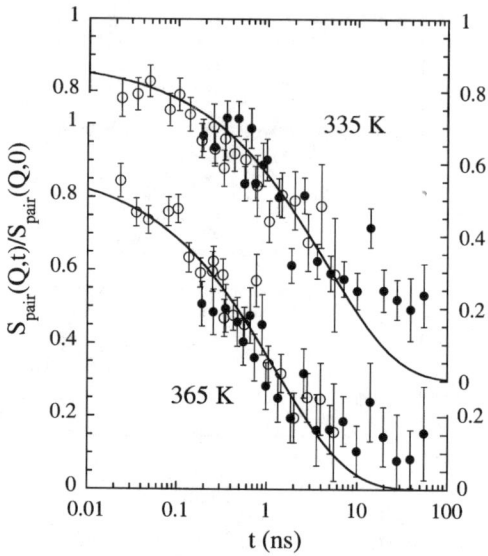

Fig. 5.14 Dynamic structure factor of PIB measured at $Q=0.2$ Å^{-1} with an incident wavelength of $\lambda=10$ Å (*open symbols*) and with $\lambda=15$ Å (*full symbols*) at the two temperatures indicated. (Reprinted with permission from [147]. Copyright 2002 The American Physical Society)

are expected to be minimal. Figure 5.14 displays NSE data taken in the low Q-regime at neutron wavelengths $\lambda=15$ Å and $\lambda=10$ Å respectively. Though the data have some statistical scatter – experiments at these long wavelengths at a cross section of ≈0.01 cm^{-1} are very difficult – in both cases (within the statistical error) the respective data sets taken above and below the multiple scattering threshold agree very well. Thus, for the intermediate length scale dynamics of PIB the multiple scattering effects on the dynamics may be ignored within the level of accuracy.

An analysis in terms of KWW functions (Eq. 4.21) revealed a Q- and T-independent spectral shape ($\beta \approx 0.55$) within the uncertainties. Taking into account the incoherent contributions to the signal [147], the amplitudes $f_Q(T)$ and characteristic times $\tau_{\text{pair}}(Q,T)$ were obtained. A well-defined maximum is observed for the amplitude at Q_{max} with a Q-dependence mirroring the structure factor (see Fig. 5.15). Reflecting the lower statistics of the data at low Q, the amplitude scatters around an average value that decreases with increasing temperature. Figure 5.16 presents the Q-dependent relaxation times for the different temperatures. For all three temperatures the same general behaviour is observed. The collective relaxation times display a low Q plateau like behaviour with some tendency of an increase towards the lowest Q. The low Q plateau is followed by an increase of the relaxation times to a maximum value which shifts from about $Q=0.8$ Å^{-1} at $T=390$ K to $Q=1$ Å^{-1} at $T=335$ K. Towards higher Q a strong decrease of the relaxation times is observed.

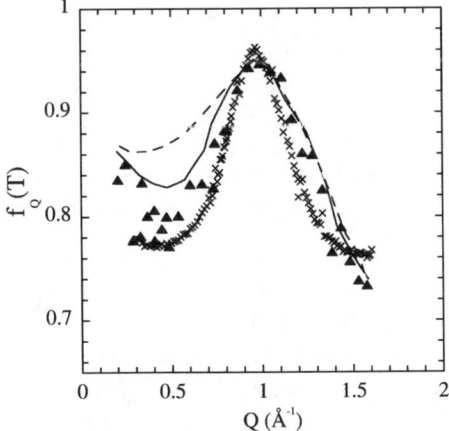

Fig. 5.15 Momentum transfer dependence of the amplitude of the KWW functions describing the dynamic structure factor. For 390 K all the values obtained are shown (*filled triangle*), while the *dashed and solid lines* represent the smoothed behaviour for 335 K and 365 K, respectively. The static structure factor is shown in arbitrary units for comparison (*cross*). (Reprinted with permission from [147]. Copyright 2002 The American Physical Society)

It is noteworthy that the peak height of $\tau_{pair}(Q)$ relative to the low Q-plateau increases with decreasing temperature. While at $T=390$ K the relative peak height amounts to about a factor of two, this changes to four at 335 K. From this qualitative observation we may conclude that the relaxation times in the peak region close to Q_{max} shift differently with temperature than in the low Q regime – a rather unexpected result! Figures 5.16b–c investigate the temperature dependence in more detail. In Fig. 5.16b the observed relaxation times are shifted with the rheological shift factor due to Ferry [34]. The single master curve obtained for $Q \geq 1$ Å^{-1} corroborates the earlier result [125] that at $Q \geq Q_{max}$ the temperature dependence of the collective response agrees well with the viscosity results. On the low Q-side, however, severe discrepancies evolve. Obviously the data in the intermediate length scales region do not follow the temperature dependence of the viscosity. They show a considerably weaker temperature dependence compatible with an apparent activation energy close to 0.43 eV (Fig. 5.16c).

The low-Q limit of the coherent scattering relates to the imaginary part of the susceptibility $\chi_{11}=1/C_{11}$, where C_{11} is the elastic modulus of the longitudinal sound waves [225, 147]. C_{11} relaxes with the stress relaxation time τ_M. From longitudinal sound wave damping in the MHz region [226], long-time stress-relaxation data [227], dynamic shear data from lower frequency [34] and light scattering Brillouin data [199] an apparent activation energy of 0.48 eV is deduced for the stress relaxation time in the T-range investigated by NSE. The interpolated values for τ_M, shown as arrows in Fig. 5.16a, agree nearly quantita-

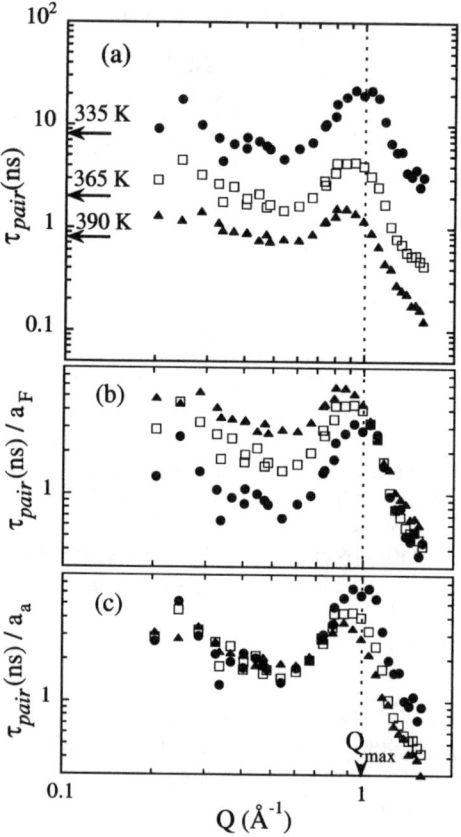

Fig. 5.16 Q-dependence of the characteristic times of the KWW functions describing the PIB dynamic structure factor at 335 K (*filled circle*), 365 K (*empty square*) and 390 K (*filled triangle*). **a** Shows the values obtained for each temperature. Taking 365 K as reference temperature, the application of the rheological shift factor to the times gives **b** and a shift factor corresponding to an activation energy of 0.43 eV delivers **c**. The *arrows* in **a** show the interpolated mechanical susceptibility relaxation times at the temperatures indicated. (Reprinted with permission from [147]. Copyright 2002 The American Physical Society)

tively with the pair relaxation times in the low Q-regime. This finding supports the consistency of the NSE results and corroborates the observation of an apparent activation energy ≈0.43 eV when approaching the hydrodynamic regime. While the low Q collective relaxation seems to follow a similar T-dependence as the stress relaxation, the structural relaxation (as observed at the structure factor maximum) clearly displays the stronger temperature dependence dictated by viscosity.

Further results on PI and PVE also point to different thermal behaviours for the structural relaxation and the dynamics at intermediate length scales [228, 229].

5.2.1
Application of Mode Coupling Theory

To rationalize experimental results in unexplored dynamic regimes constitutes a challenge for any theoretical approach. Does mode coupling theory account for the observed behaviour on the intermediate length scales dynamics of PIB? The experiments have revealed the Q- and temperature dependencies of two main parameters in the theory, the relaxation strength of the α-relaxation (non-ergodicity parameter) $f_Q(T)$ and the characteristic time of the dynamic structure factor τ_{pair}. $h_Q(T)=1-f_Q(T)$ is the contribution of the fast process to the dynamic structure factor. The strong modulation of $f_Q(T)$ following $S_{pair}(Q)$ (see Fig. 5.15) qualitatively agrees with MCT. For the Q-dependence of the characteristic time of the α-relaxation, MCT predicts the relation between amplitude and timescale given by Eq. 4.38. To investigate this prediction, the ratio between the timescales at Q_{max} at the different temperatures was fixed to that observed for the viscosity. Assuming $b=\beta$ the Q- and T-dependencies represented in Fig. 5.17 (lines) evolve. The agreement is good for the lowest temperature investigated. There, both, the modulation in the peak region as well as the plateau, are fairly reproduced by MCT predictions. However, the thermal evolution obtained from experiment and theory are clearly different. From Eq. 4.38 different apparent activation energies in the peak and in the low Q-regime are obtained – $f_Q(T)$ and $h_Q(T)$ vary with temperature only in the low Q-regime. There $f_Q(T)$ decreases with increasing temperature (see Fig. 5.15). This translates in an additional decrease of the predicted relaxation time with respect to its value at Q_{max}. Thus, though the relation between the timescale and the amplitude of the second relaxation step proposed by MCT (Eq. 4.38) reproduces the

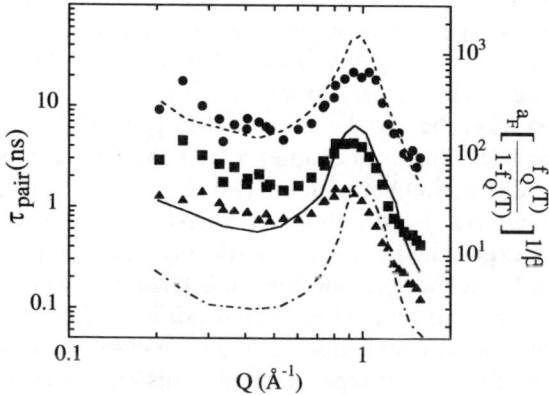

Fig. 5.17 Momentum transfer dependence of the characteristic times of the KWW functions describing PIB dynamic structure factor at 390 K (*filled triangle*), 365 K (*empty square*) and 335 K (*filled circle*). The *lines* display the prediction of MCT (Eq. 4.37) affected by the rheological shift factor for 335 K (*dashed*), 365 K (*solid*) and 390 K (*dashed-dotted*). (Reprinted with permission from [147]. Copyright 2002 The American Physical Society)

Q-dependence of τ_{pair} at the lowest temperature investigated, it fails to describe the temperature behaviour of the PIB collective response at intermediate length scales, at least in the temperature range investigated. MCT predicts universality of timescales. This prediction is incompatible with the observation of two different temperature dependencies for the structural and the stress relaxation.

5.3
Self-Motion

There is a fundamental question concerning the nature of the self-motion of protons in glass-forming polymers. In Sect. 4.1 we have shown that the existing neutron scattering results on the self-correlation function at times close to the structural relaxation time τ_s (Q-region $0.2 \leq Q \leq 1$ Å^{-1}) are well described in terms of sublinear diffusion, $\langle r^2(t)\rangle \propto t^\beta$. This qualification rests on the observation of $S_{self}(Q,t)$ with a KWW-like functional form and stretching exponents close to $\beta \cong 0.5$.

As predicted by mode coupling theory [95, 96, 106] in simple glass forming systems, the decaging process, which is considered to be at the origin of the α-relaxation, is followed by translational diffusion of the whole molecule [230]. For polymers, both molecular dynamics (MD) simulations and MCT calculations on coarse-grained polymer models (bead and spring models [207,208]) after the initial ballistic regime show the existence of a plateau in the early stages of $\langle r^2(t)\rangle$. This proves that the "cage effect" also exists for polymers. However, as $\langle r^2(t)\rangle$ increases above the plateau, instead of simple diffusion a subdiffusive regime ($\langle r^2(t)\rangle \approx t^x$, $x \approx 0.5$–0.6) is found. This difference to low molecular weight glass-forming systems was related to the connectivity of polymer chains. Though, in principle, these results from coarse grained bead and spring models appear to agree with NS results on real polymers, an important question arises. While in the case of simulations the subdiffusive regime was identified with the Rouse regime ($\langle r^2(t)\rangle \approx t^{0.5}$ for long chains), NS results suggest that this region cannot be explained only in terms of the Rouse regime. This follows from the observation that the Rouse model in general fails at intermediate Q-values, as has been shown in chapter 5.1. As the β-values in polymers are close to 0.5, the distinction between the two regimes $\langle r^2(t)\rangle \approx t^\beta$ and $\langle r^2(t)\rangle \approx t^{0.5}$ (Rouse) or even a proof that they both exist separately is very difficult. In fact, in the literature experimental neutron scattering results obtained in the high Q-regime can be found that were interpreted in terms of the Rouse model [231], even though the precondition of Gaussian beads is fulfilled at larger scales.

Only recently has the connection between the glassy α-relaxation and the Rouse regime been addressed experimentally. This was realized on poly(vinylethylene) (PVE), where for the first time it was possible to distinguish clearly two well-separated dynamic regimes of sublinear diffusion [39]. The experiment required the combination of NSE studies on $S_{chain}(Q,t)$ and $S_{self}(Q,t)$ pushing the experimental capabilities as much as possible to overlap the results in the intermediate length scale region. In Sect. 5.1 the single chain dynamic structure

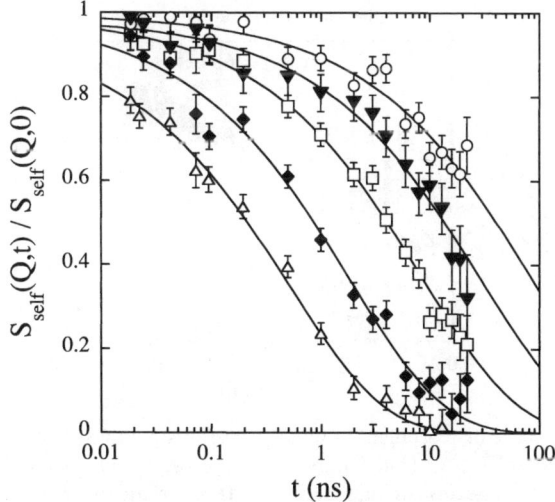

Fig. 5.18 $S_{self}(Q,t)$ of PVE measured at $Q=0.1, 0.15, 0.2, 0.3$ and 0.4 Å^{-1} (*top to bottom*). *Solid lines* are KWW fits with $\beta=0.5$. (Reprinted with permission from [39]. Copyright 2004 EDP Sciences)

factor results of this study have already been presented (see Fig. 5.12). Deviations from Rouse behaviour were observed for $Q>Q_R=0.11$ Å^{-1}. At Q_R the Rouse relaxation time is 100 ns and may be taken as the characteristic time for the self-motion at this Q-value, $t_R = \tau_{self}^{Rouse}(Q_R) \approx 100$ ns (see Eq. 3.18).

Turning to the self-motion of PVE protons, Fig. 5.18 shows the NSE data obtained for $S_{self}(Q,t)$. If indeed there existed a second Gaussian subdiffusive regime at length scales shorter than that of the Rouse dynamics, then following Eq. 3.26 also for $Q>Q_R$ it should be possible to construct a Q-independent $\langle r^2(t) \rangle$.

This is shown in Fig. 5.19. For $Q \geq 0.20$ Å^{-1} a nearly perfect collapse of the experimental data into a single curve is obtained. The time dependence of the determined $\langle r^2(t) \rangle$ can be well described by a power law with an exponent of 0.5 (solid line in the figure). This demonstrates that for times shorter than t_R there exists a second subdiffusive regime in PVE, in this case following the same power law in time as the Rouse motion. This subdiffusive regime is indeed the one relevant for the glass process in this polymer. Two timescales may be invoked as representative for defining the time range where the glassy dynamics evolves:

i. The time needed by a proton to move as far as the average interchain distance d_{chain}.
ii. The time τ_s of the structural relaxation.

From $S_{pair}(Q)$ measurements (see the insert in Fig. 5.20) $Q_{max}=0.9$ Å^{-1}, revealing $d_{chain}=2\pi/Q_{max} \approx 7$ Å. Such a distance is covered by the protons in about 1 ns, just in the middle of the time region where the sublinear diffusive regime has

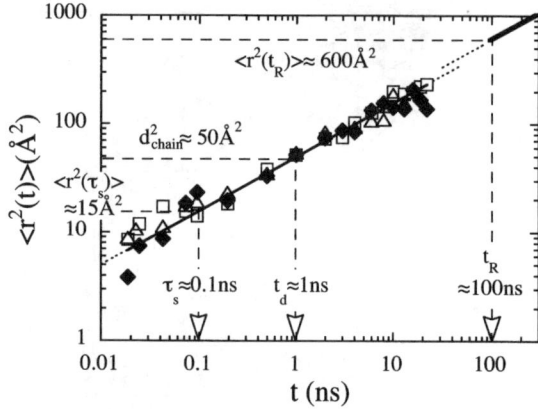

Fig. 5.19 Mean square displacement of the PVE protons obtained from $S_{self}(Q,t)$ at $Q>Q_R$: 0.2 Å^{-1} (*empty square*), 0.3 Å^{-1} (*filled diamond*) and 0.4 Å^{-1} (*empty triangle*). The *solid line* represents $\langle r^2(t)\rangle \approx t^{0.5}$. The *dashed lines* mark the characteristic times and lengths discussed in the text. The *thick solid line* in the *upper right corner* displays the mean squared displacement in the Rouse regime. The two *dotted lines* extrapolate into the crossover regime. (Reprinted with permission from [39]. Copyright 2004 EDP Sciences)

been experimentally obtained (see Fig. 5.19). On the other hand, $S_{pair}(Q_{max},t)$ decays at this temperature with $\tau_s=0.1$ ns, a value also belonging to the subdiffusive regime. For comparison, in Fig. 5.19 the mean squared displacement following from the Rouse motion (Eq. 3.17) is depicted in its range of validity (after t_R). Both regimes differ by a change in the prefactor governing the magnitude of $\langle r^2(t)\rangle$.

The crossover found between the Rouse and the subdiffusive glassy regime becomes most clear in the Q-dependence of the relaxation times. The correlation functions $S_{chain}(Q,t)$ and $S_{self}(Q,t)$ are different and therefore their direct comparison is not possible. However, we may easily relate the characteristic times:

i. A fit of the $S_{self}(Q,t)$ to Eq. 4.9 with $\beta=0.5$ (see solid lines in Fig. 5.18) reveals directly τ_{self} in both regimes.
ii. With Eq. 3.18 ($\tau_{self}^{Rouse}(Q)=9\pi/(W\ell^4 Q^4)$) the effective Rouse constants $W\ell^4(Q)$ determined from $S_{chain}(Q,t)$ (see inset of Fig. 5.12) can be directly transformed into $\tau_{self}(Q)$.

Figure 5.21 presents the Q-dependent relaxation times obtained in this way. Removing the Q^{-4} dependence expected for both the Rouse as well as for the glassy regime (Eq. 4.11) should lead to a plateau in the variable $\tau_{self}Q^4$. This is indeed observed in Fig. 5.21 for each regime: the plateau at $Q<0.11$ Å^{-1} corresponds to Rouse dynamics while the plateau at higher Q reflects the Gaussianity in the α-relaxation region. The two regimes are separated by a step occurring around $Q_{cross}\approx 0.13$ Å^{-1}. As a consequence of the steep dispersion in time, the gradual

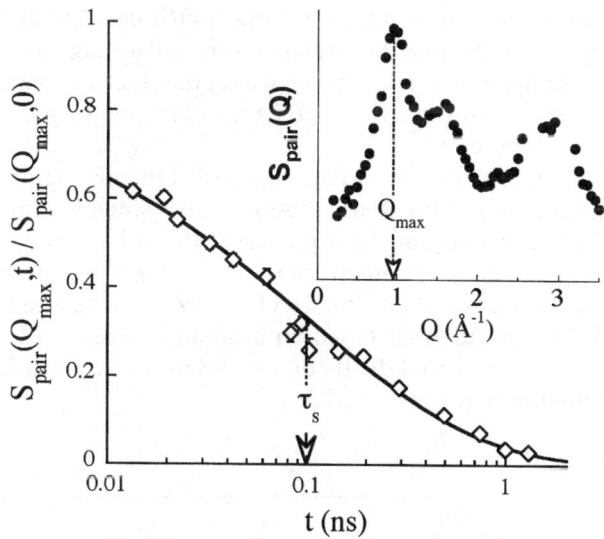

Fig. 5.20 Temporal evolution of the dynamic structure factor of PVE at the first maximum of the static structure factor at 418 K. The *solid line* represents a KWW fit (β=0.5). *Insert*: S(Q) of PVE measured at 320 K. (Reprinted with permission from [39]. Copyright 2004 EDP Sciences)

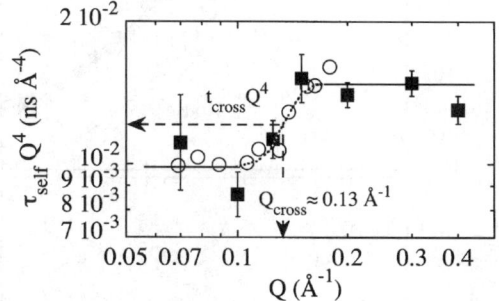

Fig. 5.21 Q-dependence of $\tau_{self}Q^4$ obtained from $S_{self}(Q,t)$ (*filled square*) and from the description of $S_{chain}(Q,t)$ in terms of the Rouse model (*empty circle*) for PVE. (Reprinted with permission from [39]. Copyright 2004 EDP Sciences)

crossover for $\langle r^2(t) \rangle$ (Fig. 5.19) here is compressed to a rather narrow step in $\tau_{self}(Q)Q^4$. Finally we note that the results from a direct measurement of $S_{self}(Q,t)$ in the low Q regime perfectly agree with those deduced from the Rouse description of $S_{chain}(Q,t)$, demonstrating the consistency of the data analysis performed.

The dynamic crossover occurs at $t_{cross} \approx 40$ ns. From $\langle r^2(t_{cross}) \rangle = 350$ Å^2 a crossover length $\ell_{cross} \approx 19$ Å is deduced. This length scale may be identified with the size of a Gaussian blob underlying the Rouse model, i.e. with the end-to-end distance of the chain sections within the blob. Considering C_∞^{PVE}=8.2, the Gaussian blob would be made up of ten monomers or 20 bonds. For distances along the chain larger than this size, the Gaussian bead and spring model works and we obtain Rouse dynamics. For smaller scales, the internal dynamics within the Gaussian blobs is observed. This dynamics manifests itself in the glassy dynamics of the α-process.

Thus for polymer melts there exists a generic dynamic regime in between the decaging process and the Rouse motion. This regime is characterized by sublinear diffusion underlying the α-process [148, 146] and is separated from the Rouse process. It evolves from the motion of polymer segments, which are subject to specific intra- and inter-molecular forces. Coarse grained MD-simulations and MCT calculations based on bead and spring models [207, 208] could not reveal the observed distinction between the subdiffusive α-regime and the genuine Rouse regime.

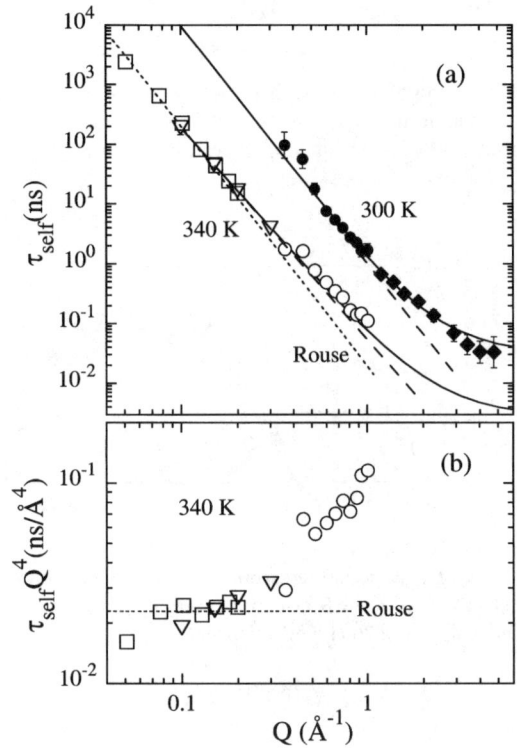

Fig. 5.22 a Q-dependence of τ_{self} obtained from KWW fits of $S_{self}(Q,t)$ of PId3 at 340 K (β=0.57) (*filled circle* FZ-Jülich, *empty circle* IN11C) and 300 K (β=0.50) (*empty diamond* IN11C, *filled diamond* IN13). τ_{self}^{Rouse} deduced for 340 K from the Rouse description of $S_{chain}(Q,t)$ [47] are also depicted (*empty square*). *Dotted line* shows the Rouse prediction, *dashed lines* the Gaussian extrapolations $\tau_{self} \propto Q^{-2/\beta}$ and *solid lines* the descriptions in terms of the anomalous jump diffusion model [9, 154]. **b** Q-dependence of $\tau_{self}Q^4$ for 340 K; symbols as above

In another well-investigated polymer, PI, such a crossover can also be visualized by combining NSE results on $S_{chain}(Q,t)$ and $S_{self}(Q,t)$ in a similar way as described above for PVE. In this case, the crossover takes place in a more subtle way. Its signature is a change of the value of the stretching parameter β and consequently a different exponent in the Q-dependence of the characteristic times in both regions, as can be seen in Fig. 5.22 [232]. Here also, deviations from Gaussian behaviour at $Q>1$ Å^{-1} are evident; they were attributed to a finite jump length of the α-process [154]. We note again the nearly perfect agreement between the data corresponding to measurements of different correlators, indicating the consistency of the measurements and the data evaluations performed. From this it can be deduced that for PI the crossover takes place at shorter length scales than for PIB and PVE.

5.4
Scenario for the Intermediate Length Scale Dynamics

5.4.1
Direct Comparison of the Correlators

It is highly interesting to compare the experimental results obtained at the same conditions for the different correlation functions accessible by NSE: $S_{self}(Q,t)$, $S_{pair}(Q,t)$ and $S_{chain}(Q,t)$. To date this can only be realized for PIB, and due to the inherent experimental difficulties, only in a narrow Q-range (0.3–0.4 Å^{-1}) at one temperature, 390 K. Figure 5.23 shows such a comparison for $Q=0.3$ Å^{-1}. We

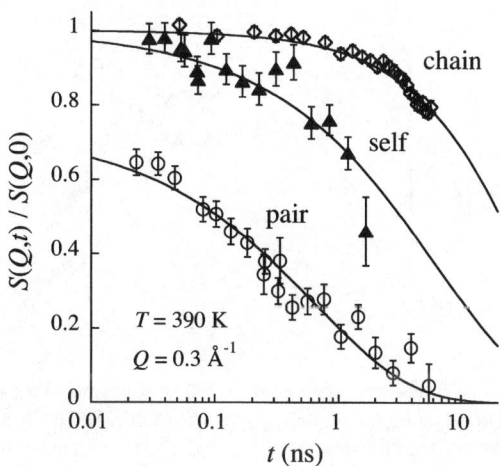

Fig. 5.23 Time evolution of the three functions investigated for PIB at 390 K and $Q=0.3$ Å^{-1}: pair correlation function (*empty circle*); single chain dynamic structure factor (*empty diamond*) and self-motion of the protons (*filled triangle*). Solid lines show KWW fitting curves. (Reprinted with permission from [187]. Copyright 2003 Elsevier)

observe that for the single chain structure factor the decay is the slowest and is closer to a single exponential function than for the other correlators. A phenomenological fit with a KWW law reveals a β-value of 0.82, while β=0.55 has been used for describing the self- and pair correlation functions (see solid lines in the figure). The timescales may be compared taking the average times $\langle \tau \rangle$ obtained through:

$$\langle \tau \rangle = \int_0^\infty e^{-(t/\tau_w)^\beta} dt = \frac{\Gamma(1/\beta)}{\beta} \tau_w \qquad (5.25)$$

where τ_w and β are the characteristic times and the shape parameter of the KWW functions. The resulting $\langle \tau \rangle$ are displayed in Fig. 5.24 as a function of Q. The characteristic times for the chain dynamics (black dots) are significantly slower than those corresponding to the pair correlation function (black triangles). This can be understood since, though both functions are determined by the relative positions of pairs of atoms, in the case of the chain dynamic structure factor these atoms are restricted to belonging to the same chain. Therefore, fluctuations can only be equilibrated by transport of segments over larger distances under the further restriction of the chain connectivity. In the collective case at low Q, density fluctuations are observed that may be relaxed by cooperative small amplitude motions of the chain ensemble. Thus, the relaxation of S_{chain} will be slow, while S_{pair} may relax significantly faster at low Q.

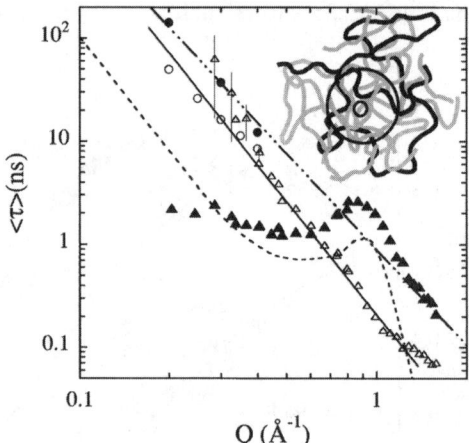

Fig. 5.24 Q-dependence of the average times at 390 K corresponding to: $S_{\text{self}}(Q,t)$ (*empty triangle*), $S_{\text{pair}}(Q,t)$ (*filled triangle*), and $S_{\text{chain}}(Q,t)$ (*filled circle*). The deduced values for the self-correlation obtained from the description of $S_{\text{chain}}(Q,t)$ in terms of the model proposed by Allegra et al. are also shown (*empty circle*). The *solid line* represents a $Q^{-2/\beta}$ (β=0.55) power law, and the *dashed-dotted line* the extrapolation of the $S_{\text{chain}}(Q,t)$ times to higher Q-values. The *dashed line* is the prediction from the Sköld ansatz for $\langle \tau_{\text{pair}} \rangle$. The *picture* symbolises a polymer melt where one chain is labelled. *Circles* represent length scales of observation corresponding to different Q-scales

5 Intermediate Length Scales Dynamics

At high Q on the other hand, we note that in Fig. 5.24 the extrapolated single chain relaxation times coincide with the collective relaxation times. This means that for length scales below the mean interchain distance the scattering volumes are such that for labelled single chain and for a coherently scattering multichain system, the single chain dynamics is observed. In the cartoon in Fig. 5.24 such a situation is depicted: the small circle (with diameter smaller than the interchain distance) contains atoms of one single chain; for bigger observation distances, the single chain study addresses atoms in the labelled chain.

We now address the connection between the self-motions of protons and the collective motions. Figure 5.24 shows that the structural relaxation time τ_s is about one order of magnitude slower than the characteristic time for proton self-motion at $Q_{\max}$. However, the power-law increase of τ_{self} towards low Q slows down the self-motion such that the collective times become much faster at intermediate length scales. Is it possible to relate both experimental sets of data with any existing Ansatz?

The relation between collective and self-motion in simple monoatomic liquids was theoretically deduced by de Gennes [233] applying the second sum rule to a simple diffusive process. Phenomenological approaches like those proposed by Vineyard [194] and Sköld [234] also relate pair and single particle motions and may be applied to non-exponential functions. The first clearly fails to describe the PIB results since it considers the same time dependence for both correlators. Taking into account the stretched exponential forms for $S_{\text{pair}}(Q,t)$ (Eq. 4.21) and $S_{\text{self}}(Q,t)$ (Eq. 4.9), the Sköld approximation:

$$S_{\text{pair}}(Q,t) \approx S_{\text{self}}\left(\frac{Q}{\sqrt{S(Q)}}, t\right) \tag{5.26}$$

allows us to relate both characteristic timescales. Assuming τ_{self} to follow the power law of Eq. 4.11, the prediction for the collective time reads as:

$$\tau_{\text{pair}}(Q,T) = a_0(T) S(Q)^{1/\beta} Q^{-2/\beta} \tag{5.27}$$

Figure 5.24 shows that this approach fails not only quantitatively but also qualitatively. Neither is the strong increase of the collective times relative to the self-motion in the peak region of $S_{\text{pair}}(Q)$ explained (this is the quantitative failure) nor is the low Q plateau of $\tau_{\text{pair}}(Q)$ predicted (this is the qualitative shortcoming). We note that for systems like polymers an intrinsic problem arises when comparing the experimentally accessible timescales for self- and collective motions: the pair correlation function involves correlations between all the nuclei in the deuterated sample and the self-correlation function relates only to the self-motion of the protons. As the self-motion of carbons is experimentally inaccessible (their incoherent cross section is 0), the self counterpart of the collective motion can never be measured. For PIB we observe that the self-correlation function from the protonated sample decays much faster than the pair correlation function even close to the minimum of $S_{pair}(Q)$. This difference could not be accounted for by any of the above approaches. The motional amplitudes of protons are apparently larger than those of the carbons, as found in MD-simulations on fully atomistic polymers [235]. Allowing an arbitrary shift factor to account for this effect, the τ_{pair} modulation in the peak region could be described rather well [147]; however, the plateau observed at low Q for collective times cannot be reproduced by any approach based on renormalizations of the self-motion, considering the strong Q-dependence characteristic for the incoherent time in this region.

Finally, we note that the temperature dependencies obtained for both the characteristic times for self-motion (see Fig. 4.14b) and for collective motions at intermediate length scales (see Fig. 5.16c) correspond to the same apparent activation energy of about 0.43 eV. This coincidence is evidenced in Fig. 5.25. Thus the motion of protons in this polymer follows a weaker temperature dependence than the structural relaxation. This implies that the mean square displacement a proton performs during the characteristic time of the structural relaxation increases with decreasing temperature. In the temperature range investigated, it changes from 17 Å^2 at 390 K to about 30 Å^2 at 335 K [147]. Apparently, the average distance a proton has to move during the relaxation of the structure grows considerably with decreasing temperature. We note that such an observation does not seem to be general, since for PI the same thermal evolution is observed for τ_{self} and τ_s [9].

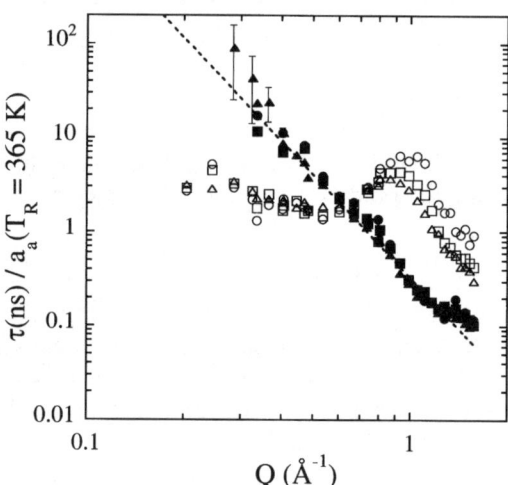

Fig. 5.25 Result of applying shift factors corresponding to an activation energy of 0.43 eV to the relaxation times observed for the collective dynamics (*empty symbol*) and the self-correlation (*full symbol*) of PIB: 335 K (*circle*), 365 K (*square*), and 390 K (*triangles*) (reference temperature 365 K). The *dotted line* through the self-correlation data shows the $Q^{-2/0.55}$ dependence implying Gaussian behaviour (Reprinted with permission from [147]. Copyright 2002 The American Physical Society)

5.4.2
Stressing the Internal Viscosity Approach

In Sect. 5.3. the different nature of the motions within a Gaussian blob and within the Rouse regime has been established. Though this result is a big step ahead, the question still remains how the dynamics can be described in the crossover region between the α-relaxation and the Rouse regime, i.e. to realize how to connect the mean squared displacements in Fig. 5.19 and to understand the collective dynamics in the low Q-plateau. For this a theory is necessary. In the case of PIB, the characterization of the single chain dynamics at intermediate length scales was possible with the approach of Allegra et al. In the evaluation it was found that the coupling effect was at full strength for Rouse mode numbers around 10. For the investigated chain of 69 monomers this defines a length scale of about 7 monomers. With the characteristic ratio of PIB C_∞=6.73 and a C–C bond length of ℓ_0=1.54 Å, 7 monomers span an end-to-end distance of $\sqrt{\langle R_e^2 \rangle} = \sqrt{\ell_0^2 C_\infty 2N} = 15$ Å. On this length scale the dissipative process ought to dominate the single chain structure factor. In the following we discuss such an approach for PIB.

First of all, it is noteworthy that the activation energy for the internal viscosity time τ_a is identical with that for both, τ_{self} and τ_{pair} in the intermediate length scales region. The values of the different timescales are depicted in Fig. 4.20 for comparison. At 390 K the value of τ_a is similar to the average value of the structural relaxation and slightly slower than the average time for collective motion at intermediate length scales. The agreement between the thermal behaviour of τ_a and the latter suggests that the microscopic mechanism reflected at intermediate length scales is the same. According to the single chain study, this process should be due to correlated jumps across the rotational barriers in the polymer. On the other hand, this mechanism should also be responsible for the stress relaxation, as it controls the evolution of τ_{pair} towards the hydrodynamic limit. It seems also to be behind the self-motion of the protons and, finally, can be detected by NMR. Recent molecular dynamics simulations on PIB [236] have shown that NMR results mainly reflect *trans-gauche* transitions while the transitions between the *trans-trans* states occur more rapidly and essentially provide an additional channel for PIB in order to relax local stresses. Thus, considering this interpretation, it would follow that the common microscopic mechanism behind these measurements at intermediate length scales involves such a *trans-gauche* conformational transitions. The structural relaxation would take place additionally by the *trans-trans* changes and therefore it would show a different apparent activation energy.

We now may ask whether the Allegra model can reproduce some of the features experimentally found for the other correlators. The incoherent intermediate scattering function can be calculated from an adaptation of Eq. 3.16 to Fig. 3.18 to the results of Allegra et al. We note that in the Q-range where experimental data are available, the line shape deduced from the model is significantly less stretched than the experimental one [187]. Therefore, the characteristic times have to be compared through the average times $\langle \tau \rangle$ of the KWW functions describing the curves. They are shown as empty circles in Fig. 5.24. The agreement between the data obtained from the two approaches is remarkable. Thus, we have established a relationship between the observed self-correlation function and the intrachain viscosity effect present in the single chain structure factor. Using Eq. 4.15 we may even evaluate the mean square proton displacement related to this Rouse relaxation limiting process in PIB. It results in $\langle r^2(\tau_a) \rangle = 27.7$ Å^2, about 84% of the interchain distance of $d_{\text{chain}} = 6.3$ Å.

The dynamic structure factor cannot be obtained from the Allegra model. However, the behaviour of the single chain dynamic structure factor can be extrapolated towards high Q. The corresponding average times are shown as the dotted line in Fig. 5.24. As discussed above, the coincidence of this extrapolation with $\langle \tau_{\text{pair}} \rangle$ beyond the first peak is meaningful, giving again support to the model.

Finally, it is worthy of remark that, though the comparison between the timescales leads to an almost perfect agreement between the predictions of the Allegra and Ganazzoli model and the collective and self-motion results, it is evident that clear differences appear when comparing the spectral shapes of the respective functions. The model delivers close to exponential decays for both correlators while experimentally one observes significantly stretched relaxation function ($\beta \cong 0.5$).

6
The Dynamics of More Complex Polymer Systems

6.1
Polymer Blends

The study of polymer blends has been a very active field in polymer physics for the past 20 years (see [47, 175, 237–250] and references therein). From the point of view of the polymer blend dynamics, the so-called dynamic miscibility, i.e. how the dynamics of each component is modified in the blend, has been deeply investigated. In an ideal two-component miscible polymer blend one could, in principle, expect a completely homogeneous dynamic behaviour, in the meaning that the time scale as well as the relaxation function of each component becomes similar in the blend. Although this has been observed in some particular cases [243], weakly interacting thermal miscible polymer blends are in general dynamically heterogeneous. This heterogeneity, which manifests itself in a failure of the time-temperature superposition principle [245], shows itself in two ways [47, 175, 237–242]. On the one hand, when the segmental dynamics (α-relaxation) of a single component of a blend is selectively investigated, it is found that in the vicinity of the glass transition temperature T_g the response extends over a very broad time/frequency range. On the other hand, distinctly different local segmental mobilities for the two blend components have been observed, mainly at temperatures well above T_g. Thus, one of these two manifestations seem to dominate the blend dynamics, depending on the temperature range considered. At high temperature the response of each component has the shape corresponding to the pure polymer, and therefore the distinct local segmental mobility of the components is apparent. However, at low temperature the response of each component extends over a very broad time range, and the distinct segmental mobility of the two components is not so evident. Whether these two manifestations have the same microscopic origin or refer to two different aspects of the polymer blend dynamics, is still controversial.

The models or theoretical approaches for polymer blend dynamics proposed up to date can be classified in two separate groups, depending on which of these sources of dynamic heterogeneity is considered to be the most relevant. The so-called thermal concentration fluctuation models [246, 247] are based on the idea that the local concentration fluctuations, inherent to any miscible blend, are quasi-stationary near the glass transition (typically for $T_g<T<T_g+60$ K) because their average relaxation time is much longer than that of the α-relaxation in that temperature range. This leads to a distribution of local concentration throughout the blend and thereby to a distribution of characteristic relaxation times for the α-relaxation of each component. This qualitatively explains the dramatic relaxation broadening observed experimentally when a single component of a blend is selectively investigated. The first formal development of this idea was proposed by Fischer and co-workers [246], and later Kumar et al.

proposed a more refined model [247]. Nevertheless, the thermal concentration fluctuations approach fails to explain the persistence of different segmental mobilities for the two blend components well above the T_g (typically T_g+100 K). As mentioned, in this temperature range the shape of the component response is indistinguishable from that observed in the pure polymer [47, 175, 238], providing evidence that the concentration fluctuations do not play a significant role in determining the shape of the segmental dynamics relaxation. However, the difference between the characteristic times of the α-relaxation of the two components remains.

On the other hand, there are other models that consider that the chain connectivity [237, 248] is the main factor responsible for the dynamic heterogeneity of polymer blends. This approach, first proposed by Chung et al. [237], gives a comprehensive understanding of the dynamic heterogeneity without any particular temperature range restriction. The basic idea is that in a miscible blend of polymers A and B, chain connectivity imposes that the local environment of a segment of polymer A is (on average) necessarily richer in polymer A compared to the bulk composition. This source of local concentration variation, which is in principle independent from thermal concentration fluctuations, implies that the average α-relaxation time of the lowest T_g component segments in the blend will always be significantly smaller than that of the highest T_g component segments. However, this formalism does not account for the strong broadening of the component response observed in the vicinity of T_g. Thereby, the two mechanisms of chain connectivity and thermal concentration fluctuations seem to be highly complementary. A minimal model that combines both mechanisms has recently been proposed and successfully compared with dielectric data corresponding to two different but complementary polymer blend systems [251].

Concerning the general problem of dynamic miscibility in miscible polymer blends, a fundamental question is whether or not there is a particular length scale beyond which the system becomes dynamically homogeneous. For instance, from an intuitive point of view, we should expect that chain connectivity effects are very important at short distances but that they vanish at larger length scales. It is nowadays well known [175, 252, 253] that very localized motions in polymer blends (e.g. methyl-group rotation and the β-relaxation) are totally heterogeneous, in the sense that they are hardly sensitive to blending. Each polymer component retains its individuality at this level. In the approach proposed by Lodge and McLeish [248], the chain connectivity effects ("self-concentration") are calculated at the Kuhn length, i.e. one could consider that these effects would be relevant for spatial scales of the order of, or shorter than, 10 Å. It is noteworthy that, following a similar procedure of calculation, chain connectivity effects are negligible for characteristic lengths three times larger. Beyond these simple qualitative arguments, the existence of a particular length scale for dynamic homogeneity has not been considered until now as an essential ingredient of any model for polymer blend dynamics.

6.1.1
Neutron Scattering and Polymer Blend Dynamics

Quasi-elastic neutron scattering (QENS) in general is a very suitable technique for studying polymer blend dynamics, in particular in the high temperature range, where two distinct local segmental mobilities are usually observed. First of all, QENS is a microscopic technique, which provides space-time information about the geometry and speed of the molecular motions involved through the momentum (Q) and energy ($\hbar\omega$) transfer dependence of experimental magnitudes. It is worthy of remark that the spatial resolution is of utmost importance when dealing with questions such as the length for dynamic homogeneity. Second, QENS is a "selective" technique, which in principle allows us to follow the dynamics of each of the two components of a given blend separately. Due to the very different scattering cross-sections of hydrogen and deuterium, it is possible to highlight the dynamic response of one of the components in the blend by selective deuteration. In the case of the local segmental motions involved either in the β-relaxation or α-relaxation, the usual procedure is to work with samples where one of the components is protonated and the other deuterated. In this case, the neutron scattering intensity is dominated by the incoherent scattering of the hydrogen atoms. Therefore, the dynamics of the protonated polymer in the blend determines neutron scattering spectra and, in this way, the dynamic response of each of the two components of the blend can experimentally be isolated. For this kind of experiment, time-of-flight (TOF) or backscattering (BS) techniques seem to be more appropriate than neutron spin echo (NSE). In the latter case, the incoherent contribution to the scattering will be masked by the coherent contributions from the deuterated component. On the other hand, in order to investigate large scale dynamics (single chain structure factor) the experimental procedure involves samples with a 10% labelled (protonated) chain fraction of one of the components in otherwise deuterated matrices. In this case, NSE turns out to be the most suitable technique.

In spite of the apparent capabilities of neutron scattering techniques, they have not yet been fully exploited in the field of polymer blend dynamics. Most of the results reported until now in this field correspond to relaxation techniques (dielectric spectroscopy, mechanical relaxation, NMR, etc.). It is also noteworthy that the available theoretical approaches are based on relaxation results as well. There are only a few examples where QENS has been used to investigate the effect of blending on either localized motions (e.g. methyl group rotations or β-relaxations) [252, 253] or α-relaxation [175, 253, 254]. All these works involve mainly TOF or BS techniques, in some cases in combination with dielectric spectroscopy [175, 253] or MD-simulations [254]. As already mentioned, local motions are hardly affected by blending. Moreover, heterogeneous behaviour for the α-relaxation is also observed by QENS in good agreement with high temperature dielectric results or with MD-simulations. To our knowledge, there are only two works reported so far [47, 255] concerning large-scale dynamics and NSE. In [47], NSE is combined with BS techniques in an effort to cover a wide Q-range, which allows investigation of both large-scale dynamics and α-relaxation in the same system. The main results obtained are summarized below.

6.1.2
Neutron Spin Echo Results in Polymer Blends

In [47], the crossover from local segmental dynamics (α-relaxation) to large-scale dynamics (Rouse regime) in miscible polymer blends, and the question of the existence of a length scale for dynamic homogeneity, has been investigated by a combination of NSE and BS techniques. In this way, almost two orders of magnitude in momentum transfer were covered. The system investigated was the miscible blend of polyisoprene and polyvinyl ethylene (PI/PVE), which is considered as a canonical polymer blend. The low-Q Rouse regime was covered by means of NSE spectrometers while a BS spectrometer was used in the high-Q regime, where the local segmental motions involved in α-relaxation manifest themselves.

For the experiments, hydrogenated and fully deuterated PI and PVE materials were synthesized anionically. In order to study the single chain structure factor, samples with a 10% labelled (protonated) chain fraction in otherwise deuterated matrices were prepared [hPI/dPI, hPVE/dPVE, hPI/(dPI/dPVE) and hPVE/(dPI/dPVE)]. The number averaged molecular weights were: $M_n^{hPI}=$ 82,000; $M_n^{dPI}=$91,000; $M_n^{hPVE}=$81,000; and $M_n^{dPVE}=$90,000 with $M_W/M_n<1.02$ (M_W: weight averaged molecular weight). The component-wise study of the self-correlation function dwells on the dominant incoherent scattering from a protonated polymer in a blend at higher Q. For these studies the following samples were prepared: hPI, hPVE, hPI/dPVE, and hPVE/dPI.

The investigations of the single chain dynamic structure factors were performed at the NSE spectrometer IN11 at the Institute Laue-Langevin (ILL) in Grenoble. Seven different Q-values in the range $0.05 \leq Q \leq 0.2$ Å^{-1} at three different temperatures (T=418 K, 368 K, and 330 K) were studied. Figure 6.1 presents spectra at 418 K from the two pure polymers as well as from PVE and PI in the blend as a function of the Rouse scaling variable $Q^2\sqrt{t}$. In all cases the data follow the Rouse scaling very well and, at least for smaller values of the scaling variable, the Rouse dynamic structure factor (Eq. 3.24) also provides a satisfying description of the experimental results. Comparing the decay of the relaxation functions, we realize (i) that the time scales of the Rouse motion in PI and PVE are significantly different, and (ii) that the relaxation curves for pure PI and PVE in the blend are practically identical, indicating close to homogeneous dynamics.

Q-dependent Rouse rates were obtained by fitting each spectrum separately. For a comparison with the BS data (see below) the obtained values for $W\ell^4(Q)$ were transformed into average relaxation times for the Rouse self-correlation function by $\langle\tau^R(Q)\rangle=18\pi Q^{-4}/[W\ell^4(Q)]$ (Eq. 3.18).

In order to relate to the BS-results at 330 K, temperature shift factors a_T covering the appropriate T-range were also measured using the very high res-

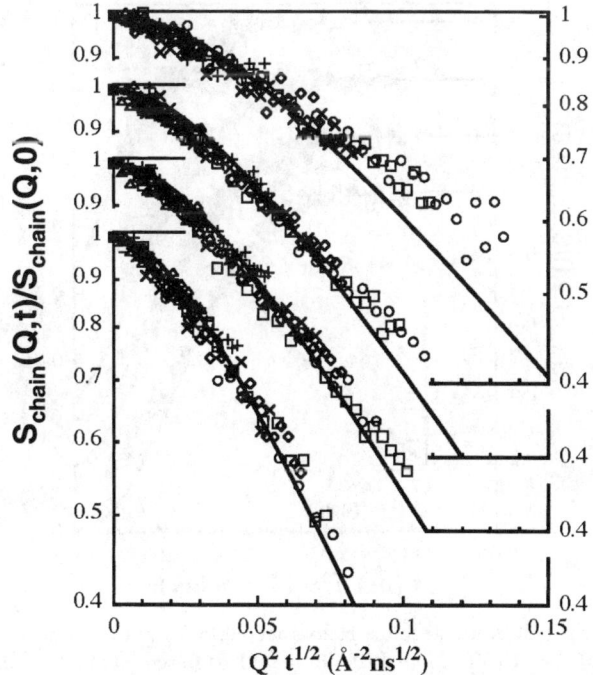

Fig. 6.1 NSE results at 418 K on the single chain dynamic structure factors from PVE, PVE in PI/PVE, PI in PI/PVE and PI (from above). The different symbols correspond to the following Q-values (*dash* 0.05 Å^{-1}, *empty square* 0.077 Å^{-1}, *plus* 0.10 Å^{-1}, *cross* 0.13 Å^{-1}, *empty diamond* 0.15 Å^{-1}, *empty square* 0.18 Å^{-1}, *empty circle* 0.20 Å^{-1}). *Solid lines* Rouse structure factors. (Reprinted with permission from [47]. Copyright 2000 The American Physical Society)

olution NSE-spectrometer IN15 at the ILL. Experiments were performed at $Q=0.15$ Å^{-1} on all four samples for nine different temperatures covering a temperature range $305 \leq T \leq 422$ K. Employing the time-temperature superposition principle Fig. 6.2 presents the spectra as a function of the rescaled time t/a_T (reference temperature 368 K). The respective shift factors for the polymers in the blend agree very well with the corresponding dielectric results for the end-to-end vector relaxation (normal mode) for PI (see insert in Fig. 6.2). Furthermore, the authors emphasized that the shift factors for PVE and PI in the blend follow identical temperature dependence. Thus, the finding that at 418 K, on the level of the Rouse modes, the blend dynamics is homogeneous seems to be not fortuitous but holds over the entire temperature range investigated.

The relaxation spectra for the four polymer systems at higher Q's ($0.2 \leq Q \leq 1.9$ Å^{-1}) were investigated by the BS-spectrometer at the FRJ-2 reactor in Jülich. The temperature range was 270 K $\leq T \leq 340$ K. Figure 6.3 displays spectra for the four polymer systems at a representative Q-value ($Q=1.1$ Å^{-1}) and at 330 K, where BS reveals results covering a maximum Q-range. The data were

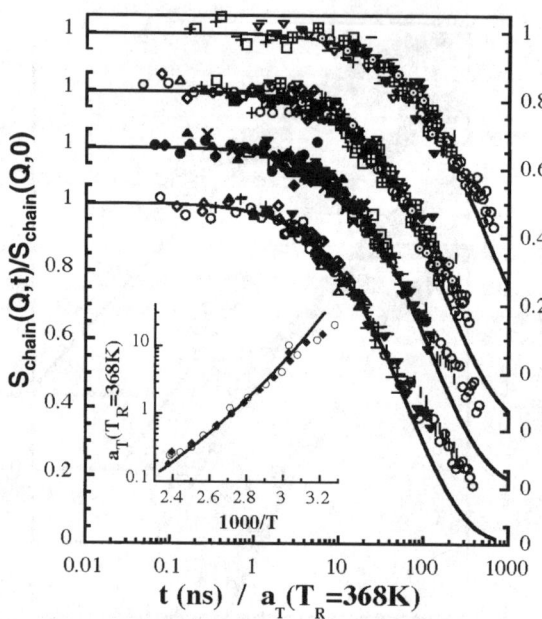

Fig. 6.2 Single chain dynamic structure factors at $Q=0.15$ Å^{-1} vs. the rescaled time t/a_T. From above PVE, PVE in PI/PVE, PI in PI/PVE and PI. Different symbols refer to different temperatures in the interval 305 K$\leq T \leq$422 K. *Solid lines* Rouse structure factors. *Insert* Temperature shift factors a_T for the polymers in the blend (*filled diamond* PI in PI/PVE, *empty circle* PVE in PI/PVE, *solid line* dielectric results). (Reprinted with permission from [47]. Copyright 2000 The American Physical Society)

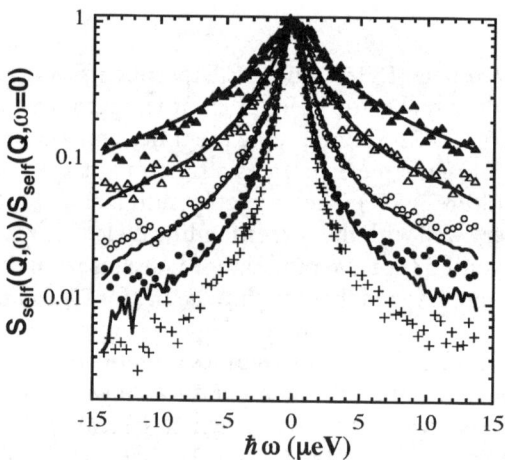

Fig. 6.3 BS-results at 330 K and at $Q=1.1$ Å^{-1}. From above PI, PI in PI/PVE, PVE in PI/PVE, PVE, and instrumental resolution. *Solid lines* fit with the Fourier transform of Eq. 4.9. (Reprinted with permission from [47]. Copyright 2000 The American Physical Society)

compared to the instrumental resolution function determined on the samples at 10 K. Starting from PI, where the spectrum is broadest, the spectra narrow successively in going to hPI/dPVE then to hPVE/dPI and finally to pure PVE. In distinct contrast to the Rouse regime, the spectra from the two components in the blend show a strongly different broadening implying – independently of any analysis – significantly different or heterogeneous relaxation rates.

The data were fitted to a stretched exponential function (Eq. 4.9) setting the stretching parameter β to its dielectric value. The solid lines included in Fig. 6.3 display the resulting curves. These fits lead to the Q-dependent characteristic relaxation times $\tau_{KWW}(Q)$, which are converted to average relaxation times by Eq. 5.25 (see Fig. 6.4).

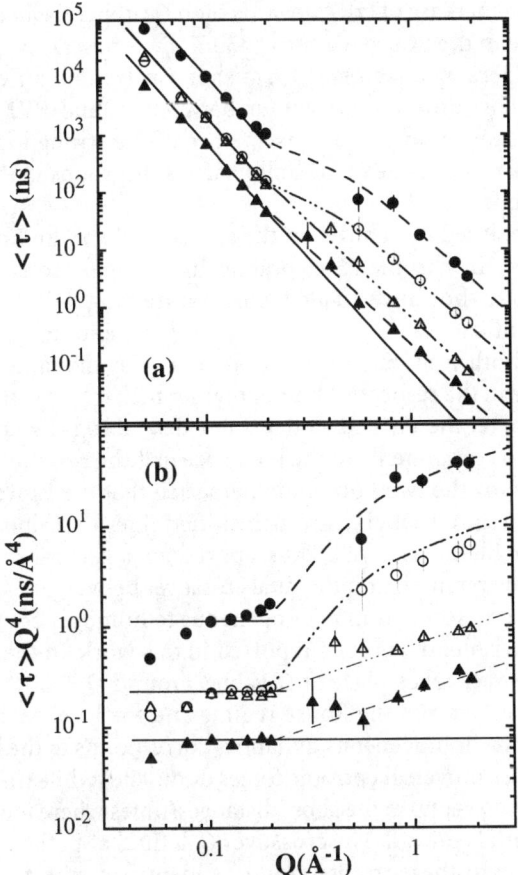

Fig. 6.4 a Characteristic times $\langle \tau \rangle$ at 330 K. PI *filled triangle*, PI in PI/PVE *empty triangle*, PVE in PI/PVE *empty circle*, PVE *filled circle*. b $\langle \tau \rangle Q^4$ vs. Q. *Solid lines* Rouse prediction valid for low Q; *dashed and dashed dotted lines* are guides for the eye. (Reprinted with permission from [47]. Copyright 2000 The American Physical Society)

Figure 6.4a presents the Q-dependent average relaxation times for the four samples at 330 K. They represent the main result of this work. For this presentation the average Rouse relaxation times $\langle\tau\rangle(Q)$ obtained at 418 K were shifted to 330 K applying the shift factors of Fig. 6.6. Figure 6.4b emphasizes the Q-dependent crossovers to local dynamics in plotting $\langle\tau\rangle Q^4$. Thereby, the intrinsic Q-dependence of the Rouse relaxation is taken out. We can first concentrate on the pure homopolymer melts. At low Q (PI $Q\leq0.2$ Å^{-1}, PVE $Q\leq0.15$ Å^{-1}) the dynamics of both polymers is in accordance with the Rouse model (solid lines in Fig. 6.4a,b). Moving to shorter scales (higher Q) for PI $\langle\tau\rangle(Q)$ decreases only slightly less than an extrapolation of the Rouse relaxation would prescribe. In Fig. 6.4b this reveals itself as a slight upturn of $\langle\tau\rangle Q^4$ vs. Q. In contrast to PI, approaching local scales the timescale of motion in PVE is significantly retarded compared to the Rouse-like Q^{-4} behaviour. In Fig. 6.4b this crossover manifests itself in a strong increase of $\langle\tau\rangle Q^4$ towards high Q (upper dashed line). As a consequence, while in the Rouse regime at 330 K the timescale of motion between PI and PVE differs by a factor of 15, at short scales this difference increases to about 150. This different behaviour between PI and PVE is qualitatively understood by the authors as a consequence of the strong local hindrance of conformational motion due to the bulky vinyl side groups in PVE. Now we can turn to the blend.

As already pointed out above, on the spatial scale of the Rouse motion the chain dynamics in the blend is practically homogeneous (low Q-part in Fig. 6.4a,b) in the whole investigated temperature range – both respective shift factors are identical (see Fig. 6.2). On the other hand, towards higher Q or shorter scales, both components in the blend display the same crossover to local dynamics as in the respective homopolymer melts. This can be directly seen in Fig. 6.4b. There, the dashed-dotted lines describing the Q-dependence of $\langle\tau\rangle Q^4$ for the two components in the blend are just the result of shifting the respective curves for the two homopolymers such that they agree in the homogeneous Rouse regime. Obviously, such shifted lines describe the component crossover in the blend very well. Thus, apart from a general shift in timescale, both components retain their individual crossover behaviours also in the blend.

Therefore, a crossover from heterogeneous to homogeneous dynamics in a miscible polymer blend has been reported in this work. In the system investigated this crossover takes place at Q-values around $Q\approx0.2$ Å^{-1}, i.e. just in the crossover region between the Rouse regime and the local segmental α-relaxation. Thereby, the homogeneous dynamics corresponds to the large scales and long times (where universal entropic forces dominate) while the heterogeneous dynamics seems to set up at the short distances/times where local inter- and intra-chain potentials prevail. The crossover Q defines a spatial crossover length which, according to these results, could be identified with the length for dynamic homogeneity discussed above. In the framework of this interpretation, these results would mean the first microscopic probe of the existence of such a length scale. However, more experimental results in other polymer blends are necessary in order to clarify the general character of these results.

On the other hand, in [255] the authors report on NSE measurements carried out in combination with small angle neutron scattering (SANS) on isotopic and binary blends of non-entangled liquid polysiloxanes: polydimethylsiloxane (PDMS) and polyethylmethylsiloxane (PEMS). Two isotopic blends of low molar mass (d-PDMS/h-PDMS; d-PEMS/h-PEMS), and the corresponding binary blends d-PDMS/ h-PEMS and h-PDMS/d-PEMS (d: deuterated, h: protonated) were considered at the critical composition in the homogeneous regime. From the results it seems that both the coil dimensions and the collective dynamics of these two blend systems behave significantly differently in the macroscopically homogeneous regime. Compared to the isotopic mixtures, which exhibit the expected unperturbed chain dimensions and typical Rouse relaxation, in the binary blends considerable coil expansion and spatially restricted Rouse dynamics occur. Both effects become visible far above the critical temperature (Fig. 6.5).

Although these results need to be understood – and confirmed in other similar systems – the authors claim that they are in qualitative agreement with a model of droplet formation and chain localization resulting from the existence of microscopic heterogeneities within the spinodal boundaries of the phase diagram [256].

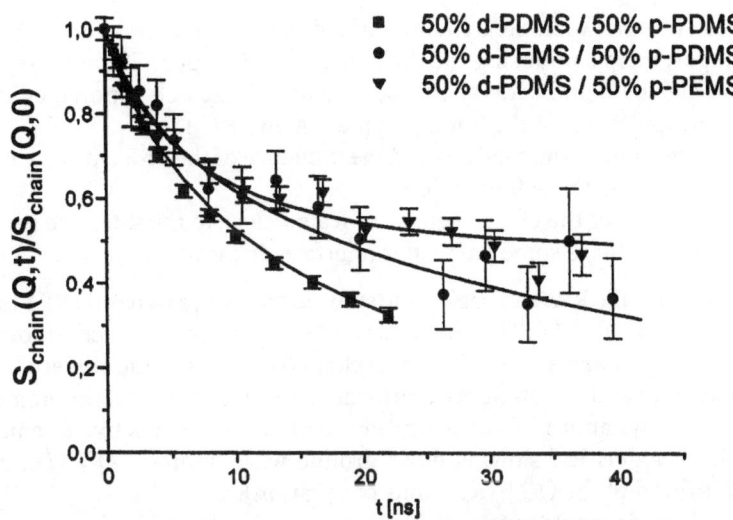

Fig. 6.5 Comparison between NSE spectra at $Q=0.1$ Å and $T=473$ K of both binary blends (*filled circle* and *filled triangle*), and those of the isotopic PDMS blends (*filled square*). The *solid lines* result from a fit t with the dynamic structure factor for the Rouse model (*filled square*) and of a spatially limited Rouse dynamics, as derived in [256] (*filled square* and *filled triangle*). (Reprinted with permission from [255]. Copyright 2003)

6.2
Diblock Copolymers

Diblock copolymers represent an important and interesting class of polymeric materials, and are being studied at present by quite a large number of research groups. Most of the scientific interest has been devoted to static properties and to the identification of the relevant parameters controlling thermodynamic properties and thus morphologies [257–260]. All these studies have allowed for improvements to the random phase approximation (RPA) theory first developed by Leibler [261]. In particular, the role of the *concentration fluctuations*, which occur and accompany the order-disorder transition, is studied [262, 263].

These concentration fluctuations are pivotal to the phase transitions in block copolymer melts and are dynamic in nature. They lead to a renormalization of the relevant interaction parameters and are thought to be responsible for the induction of the first-order nature of the phase transition [264, 265]. Such fluctuations are better studied in dynamic experiments. Thus, one can observe an increasing interest in diblock copolymer dynamics. These dynamic properties are being analysed through experimental, theoretical [266, 267] and computer simulation approaches [268, 269] with the aim of determining the main features of diblock copolymer dynamics in comparison to homopolymer dynamics. There are three main issues:

1. The relation between the dynamics of a diblock copolymer and that of the homopolymers composing the diblock chains. Is it possible to understand the single chain and collective dynamics of A-B diblock copolymer chains from the dynamics of the homopolymers A and B?
2. The concentration fluctuations and the influence of the order-disorder transition (ODT) on these fluctuations.
3. The dynamics of the chain segments located close to the interface between the blocks. Is there some particular interface dynamics?

By analogy with the RPA describing the static properties, a dynamic RPA theory has been developed [270]. This theory describes the response functions of composite polymer systems using the single chain dynamics of the corresponding homopolymers and the static structure factors as input. The main prediction concerns the dynamics of concentration fluctuations (collective dynamics), which display a critical slowing down around wave numbers $Q^* \sim 1/R_g$ in the neighbourhood of the ODT (R_g radius of gyration).

Most of experiments made on block copolymer dynamics have been designed to test the validity and the limits of this dynamic RPA approach [271, 272]. They have investigated the diffusive motions of the block copolymer chains, which occur at length scales larger than the radius of gyration ($Q<1/R_g$). These experiments were performed using light scattering and photon correlation spectroscopy [273–279]; most of them were made on block copolymers in solution since melts are generally beyond the scope of dynamic light scattering experiments. Furthermore, such techniques do not allow the observation of

6 The Dynamics of More Complex Polymer Systems

the single chain dynamics in a melt. There are also experiments dealing with the reptation motions in block copolymers [280, 281].

6.2.1
Dynamic Random Phase Approximation (RPA)

A multicomponent dynamic RPA for incompressible diblock copolymer mixtures was developed by Akcasu [270]. The details of the approach may be found in the above reference. Here we recall some basic ideas dealing with the calculation of the dynamic structure factor $S(Q,t)$ for Rouse chains.

We consider an incompressible $(m+1)$ multicomponent mixture of polymers consisting of m different types of polymer chains within a matrix referred to as "0". Components may be either homopolymers of a given chemical species or, e.g. homopolymer sections in block copolymers. Hydrogenated and deuterated species of the same homopolymer are considered as different components. In this context a diblock copolymer is a two-component polymer system. A mixture of partially protonated diblock chains hA-dB with deuterated diblock chains is consequently regarded as a four-component system.

We denote the fluctuations of the number density of the monomers of component j at a point $\underline{r}$ and at a time t as $\rho_j(\underline{r},t)$. With this definition we have $\langle \rho_j(\underline{r},t) \rangle = 0$. In linear response theory, the Fourier-Laplace transform of the time-dependent mean density response to an external time dependent potential $\underline{U}(\underline{r},t)$ is expressed as:

$$\underline{\rho}(Q,s) = -\underline{\underline{\chi}}(Q,s)\,\underline{U}(Q,s) \tag{6.1}$$

where $\underline{\underline{\chi}}(Q,s)$ is the $(m+1)\times(m+1)$ dynamic response matrix, which is related to the $\underline{\underline{S}}(Q,t)$ by:

$$\underline{\underline{\chi}}(Q,t) = \frac{1}{Vk_BT}\frac{\partial \underline{\underline{S}}(Q,t)}{\partial t} \tag{6.2}$$

with V the volume of the system.

In order to proceed further a reference system is introduced (bare system) where the interaction between the monomers is removed but the chain connectivity is preserved. The response matrix in the bare system is denoted as $\underline{\underline{\chi}}^0(Q,t)$. Then in the bare system the density response is written as:

$$\underline{\rho}(Q,s) = -\underline{\underline{\chi}}^0(Q,s)\,\underline{U}(Q,s) \tag{6.3}$$

The relation of $\underline{\underline{\chi}}^0$ with $\underline{\underline{S}}^0$ is analogous to Eq. 6.2. $\underline{\underline{\chi}}^0$ is related to $\underline{\underline{\chi}}$ in introducing an interaction potential $W_{ab}(r)$ between monomers "a" and "b". In the Laplace domain Eq. 6.1 then becomes:

$$\underline{\rho}(Q,s) = -\underline{\underline{\chi}}^0(Q,s)\,(\underline{U}(Q,s) + \underline{\underline{W}}(Q)\,\underline{\rho}(Q,s)) \tag{6.4}$$

with $\underline{W}(Q)$ being the Fourier transform of the matrix $W_{ij}(r)$. Finally, the incompressibility constraints are taken into account by a Lagrange multiplier $L_i(r,t)$ which may be considered as an external potential, which couples to all monomers equally such that the perturbed average of the total density is zero.

With that Eq. 6.4 becomes:

$$\underline{\rho}(Q,s) = -\underline{\chi}^0(Q,s)\,(\underline{U}(Q,s) + \underline{L}(Q,s)\,\underline{E} + \underline{W}(Q)\,\underline{\rho}(Q,s)) \tag{6.5}$$

where $\underline{E}$ is a column vector with elements "1". The incompressibility enforces the boundary condition $\underline{E}^T\,\underline{\rho}(\underline{r},s)=0$. Now one may either eliminate $L_i(Q,s)$ directly using the boundary condition $\underline{E}^T\underline{\rho}(\underline{r},s)=0$, or one may use the matrix component "0" in order to fulfil the incompressibility constraint. For that purpose we have to assume that "0" is not coupled to other components in the bare system ($\chi^0_{0i}=0$ for $i\neq 0$). Then $L_i(Q,s)$ can be expressed in terms of Eq. 6.5 using the line for the "0" component.

Re-substituting the result for $\underline{L}(Q,s)$ into the rest of Eq. 6.5 we obtain:

$$\underline{\rho}(Q,s) = -\underline{\chi}^0(Q,s)\,(\underline{U}(Q,s) + k_B T\,\underline{v}(Q,s)\,\underline{\rho}(Q,s)) \tag{6.6}$$

where the excluded volume matrix $\underline{v}$ has the following elements:

$$v_{ii}(Q,s) = \frac{1}{k_B T\,\chi^0_{00}(Q,s)} - 2\kappa_{i0}(Q)$$

$$v_{ij}(Q,s) = \frac{1}{k_B T\,\chi^0_{00}(Q,s)} + \kappa_{ij}(Q) - \kappa_{i0}(Q) - \kappa_{j0}(Q) \tag{6.7}$$

where:

$$\kappa_{ij}(Q) = \frac{1}{k_B T}\left[W_{ij}(Q) - \frac{1}{2}[W_{ii}(Q) + W_{jj}(Q)]\right] \tag{6.7a}$$

κ_{ij} is the Flory-Huggins interaction parameter between the i and j monomers. In Eq. 6.6, the matrices have a dimension $(m)*(m)$. We note that the s-dependence of the excluded volume matrix is solely determined by the contribution of the bare susceptibility $\chi^0_{00}(Q,s)$ from the "invisible" matrix component "0". Finally, combining Eq. 6.6 with Eq. 6.1 the response function in the interacting system is given by:

$$\underline{\chi}(Q,s) = [\,\underline{I} + k_B T\,\underline{\chi}^0(Q,s)\,\underline{v}(Q,s)]^{-1}\,\underline{\chi}^0(Q,s) \tag{6.8}$$

where $\underline{I}$ is the unit matrix. For $s=0$, we obtain the static response function $\underline{\chi}(Q)$:

$$\underline{\chi}(Q) = [\,\underline{I} + k_B T\,\underline{\chi}^0(Q)\,\underline{v}(Q)]^{-1}\,\underline{\chi}^0(Q) \tag{6.9}$$

or

$$\frac{1}{\underline{\chi}(Q)} = \frac{1}{\underline{\chi}_0(Q)} + k_B T \underline{v}(Q) \tag{6.9a}$$

Equation 6.8 and Eq. 6.9 are the basic results needed in order to calculate the static and dynamic structure factors $\underline{S}(Q)$ and $\underline{S}(Q,t)$, respectively.

6.2.2
Static Scattering Function

In terms of the linear response theory the static scattering function $\underline{S}(Q)$ relates to the static response function $\underline{\chi}(Q)$ by:

$$\underline{\chi}(Q) = \frac{1}{k_B T V} \underline{S}(Q) \tag{6.10}$$

This relation also holds between $\underline{\chi}^0(Q)$ and $\underline{S}^0(Q)$. From Eq. 6.9 the basic result of RPA for the static structure factor matrix immediately follows:

$$\frac{1}{\underline{S}(Q)} = \frac{1}{\underline{S}^0(Q)} + \underline{v}(Q) \tag{6.11}$$

For an A-B diblock copolymer system, Eq. 6.11 yields the well known Leibler equation [261] for the partial structure factors $S_{ij}(Q)$:

$$S_{aa}(Q) = S_{bb} = -S_{ab} = \frac{(S_{aa}^0 S_{bb}^0 - S_{ab}^{02})}{S_{aa}^0 + S_{bb}^0 + 2S_{ab}^0 - 2\kappa(S_{aa}^0 S_{bb}^0 - S_{ab}^{02})} \tag{6.12}$$

6.2.3
Dynamic Structure Factor

In terms of the Zwanzig-Mori [282, 283] projection operator formalism the equation of motion for the dynamic structure factor is given by:

$$\frac{\partial \underline{S}(Q,t)}{\partial t} = -\underline{\Omega}(Q)\,\underline{S}(Q,t) + \int_0^t \underline{M}(Q,t-u)\,\underline{S}(Q,u)\,du \tag{6.13}$$

concentrating on the short time dynamics and thus neglecting the memory function $\underline{M}(Q,t)$ the Laplace transform reads:

$$\underline{S}(Q,s) = (s\underline{I} + \underline{\Omega}(Q))^{-1}\,\underline{S}(Q) \tag{6.14}$$

where $\underline{I}$ is the unity matrix and $\underline{\Omega}(Q)$ the first cumulant matrix:

$$\underline{\Omega}(Q) = \lim_{t \to 0} \left[\frac{\partial \underline{S}(Q,t)}{\partial t} \right] \underline{S}^{-1}(Q) \tag{6.15}$$

contains the initial relaxation frequencies of the system. The details of the calculations to arrive at $\underline{\underline{S}}(Q,s)$ and $\underline{\underline{S}}(Q,t)$ may be found in [270]. The first cumulant matrix is related to a mobility matrix, which is expressed in terms of bare mobilities of the non-interacting system:

$$\underline{\underline{\mu}}(Q) = \underline{\underline{\mu}}^0(Q) - \frac{\underline{\underline{\mu}}^0(Q)\,\underline{E}\underline{E}^T\,\underline{\underline{\mu}}^0(Q)}{\mu_{00}^0(Q) + \underline{E}^T\underline{\underline{\mu}}^0(Q)\underline{E}} \tag{6.16}$$

and

$$\underline{\underline{\mu}}(Q) = \frac{1}{k_B T}\frac{1}{Q^2}\underline{\underline{\Omega}}(Q)\,\underline{\underline{S}}(Q) \tag{6.17}$$

Since $\underline{\underline{\mu}}(Q)$ only depends on bare mobilities, the interaction expressed by the Flory-Huggins parameters κ_{ij} do not influence the mobility. For the case of Rouse dynamics, which only depends on the local friction, the bare mobility matrix:

$$\mu_{ij}^0(Q) = m_{ij}(Q) = \delta_{ij}\left(\frac{\varphi_i}{\zeta_i}\right) \tag{6.18}$$

is diagonal, ζ_i being the monomeric friction coefficient of component i; and φ_i the volume fraction of component i. Equation 6.18 evolves if we normalize to the total volume of all monomers. For a time-independent mobility matrix Eq. 6.13 may be solved easily with the result:

$$\underline{\underline{S}}(Q,t) = \sum_i \underline{\alpha}_i(Q)\,\underline{\beta}_i^T(Q)\,e^{-\lambda_i(Q)t}\,\underline{\underline{S}}(Q) \tag{6.19}$$

where $\underline{\alpha}_i(Q)$ and $\underline{\beta}_i(Q)$ are the right- and left-hand eigenvectors of the matrix $\underline{\underline{\Omega}}$ with respect to the eigenvalues $\lambda_i(Q)$.

The measured intensity in a scattering experiment depends on the scattering length density contrasts of the different components "i" with respect to the matrix:

$$K_i = \frac{\sum b_i}{v_i} - \frac{\sum b_i}{v_0} \tag{6.20}$$

where b_i are the scattering lengths of the atoms within a monomer of component "i" and v_i is the monomer volume; b_0 and v_0 denote the same quantities for the matrix. The measured intensity follows as:

$$I(Q,t) = \underline{K}^T\,\underline{\underline{S}}(Q,t)\,\underline{K} \tag{6.21}$$

where K is the vector of the component contrast factors K_i. Finally, we note that the dynamic RPA equations allow us to describe the initial dynamic response of polymer mixtures solely on the basis of the static data (static structure factors) and knowledge of the bare mobility of the constituents.

6.2.4
Neutron Scattering Results

A detailed test of dynamic RPA requires a combination of static and dynamic neutron experiments involving investigations of the respective component dynamics. All three issues addressed in the introduction were investigated. We will start with the collective dynamics of a diblock copolymer and then address the single component dynamics. Finally, we will discuss aspects of the interface dynamics. The experiments were performed on diblock copolymers of the PE-PEE and PE-PEP type. In order to access the different dynamics a series of materials with different h-d labelling was employed (see Table 6.1).

6.2.5
Static Structure Factor

For both polymer systems the static structure factors were investigated using small angle neutron scattering and the results interpreted in terms of RPA theory. Figure 6.6 displays the temperature-dependent static structure factor obtained from a PE-PEE melt (sample IV).

This diblock copolymer does not undergo the order-disorder microphase separation; the peak stays broad above 373 K, where the system is in the mean

Table 6.1 Samples used in order to test the dynamic RPA

Polymer	M_w (g/mol)	M_w/M_n	N_{PEE} (PEE monomer number)	f_{PEE} (volume fraction in PEE)
hPEE	21,550	1.02	385	...
dPEE	24,530	1.02	383	...
Sample I: mixture of 20% hPEE in dPEE				
hPEE-dPEE	16,900	1.02	127	0.5
dPE-hPEE	16,500	1.02	136	0.5
dPE-hPEE	16,400	1.02	130	0.5
Sample II: 20% hPE-hPEE in dPE-dPEE				
Sample III: 20% dPE-hPEE in dPE-dPEE				
Sample IV: hPE-dPEE				
hPEP-hPEE	68,000	1.03	553	0.45
dPEP-hPEE	68,000	1.03	533	0.45
dPEP-dPEE	68,000	1.03	480	0.45
Sample V: 10% of hPEP-hPEE in dPEP-dPEE				
Sample VI: hPEP-dPEE				
Sample VII: dPEP-hPEP-dPEE				
hPEP: 24 monomers	68,000	1.04	480	0.45

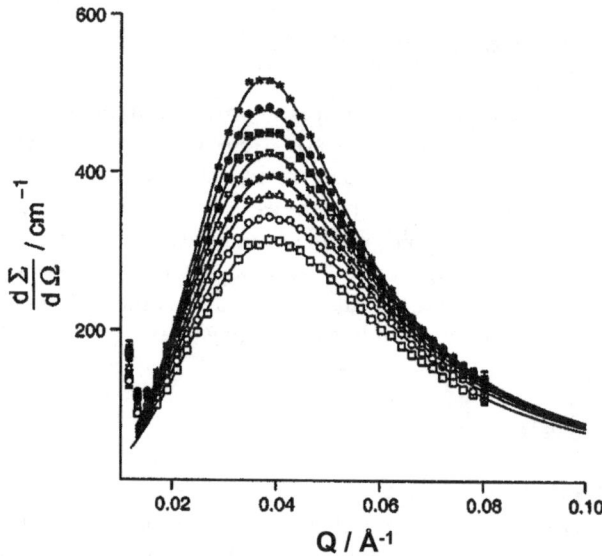

Fig. 6.6 Variation of the static structure factor $S(Q)$ measured on hPE-dPEE diblock copolymer chains (sample IV) as a function of the wave number Q. Temperature: *closed star* 393 K, *closed circles* 403 K, *closed square* 413 K, *inverted triangle* 423 K, *closed star* 433 K, *open triangle* 433 K, *open circle* 453 K, *open square* 463 K. *Solid lines* represent the fit with a two-component static RPA approach (Eq. 6.12). (Reprinted with permission from [44]. Copyright 1999 American Institute of Physics)

field regime and concentration fluctuations are low. The data were analysed in terms of RPA structure factor (Eq. 6.12), where expressions for the partial structure factors allowing for the different segment lengths and thus different radii of gyration of the two blocks were used. From the analysis, temperature-dependent values for the radii of gyration φ of the arm and the temperature-dependent Flory-Huggins parameter were obtained:

$$\kappa_{PE-PEE} = \frac{14.5}{T[K]} - 0.0106 \tag{6.22}$$

The general agreement between mean-field prediction and experimental results was excellent. Other than PE-PEE, PEP-PEE (sample VI) undergoes an ODT which is observed at 473 K. An analysis by the modified RPA theory due to Fredrickson [265] yields a strong stretching of the radii of gyration around the ODT and a temperature-dependent Flory-Huggins parameter:

$$\kappa_{PE-PEE}^{eff} = \frac{4.27}{T[K]} + 1.8 \cdot 10^{-3} \tag{6.23}$$

Figure 6.7 displays the SANS data obtained from the PEP-PEE triblock, where a label was placed at the polymer–polymer connection point (sample VII)

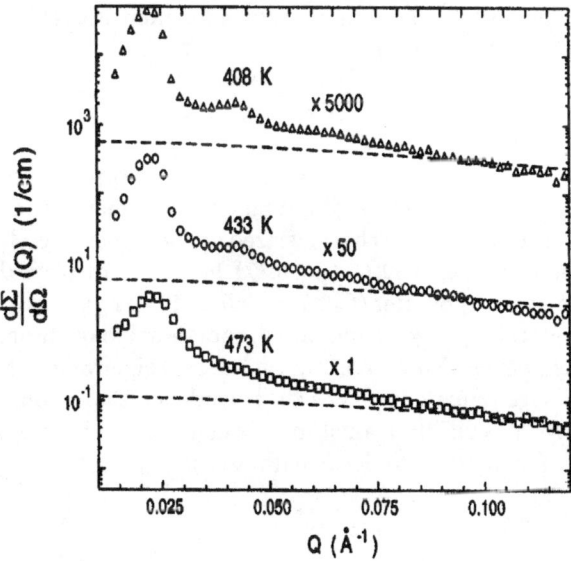

Fig. 6.7 SANS pattern from a junction-labelled PEP-PEE triblock copolymer at different temperatures. *Dashed lines* cross-section from the PEP label. (Reprinted with permission from [284]. Copyright 2002 EDP Sciences)

[284]. Below the ODT such a label highlights the polymer–polymer interface. A main peak around $Q^*=0.02$ Å^{-1} corresponding to a lamellar periodocity $2\pi/d_{lam}$ with $d_{lam}=315$ Å is observed. Its visibility results from the asymmetric nature of the diblock. We note the existence of a second order peak, which is well visible at $T_{ODT}=433$ K. At large $Q \gg Q^*$ the scattering is dominated by the form factor of the PEP-label in the environment of the deuterated monomers at the interface. This form factor may be described by a Debye function $f_{Debye}(x)$ (Eq. 3.23). The absolute cross-section for these labels is given by:

$$\frac{d\Sigma}{d\Omega} = (\rho_d - \rho_h)^2 \, \varphi(1-\varphi) \, \varphi_{int} \, f_{Debye}(x) \qquad (6.24)$$

where ρ_d and ρ_h are the scattering length densities of the protonated and deuterated chains, $\varphi_{int}=d_{int}/d_{lam}$ is the volume fraction of interface and d_{int} is the interface thickness; $\varphi=\Phi_{label} d_{lam}/d_{int}$ is the volume fraction of labels in the interface and Φ_{label} is the overall volume fraction of labelled chains (2.5%). This contribution is dominant at high Q values. The solid lines in Fig. 6.7 display this single label scattering contribution taking for φ_{int} values close to those predicted by Matsen and Bates [285] ($\varphi_{int}=0.16$ and $\kappa N=13.4$ at $T=433$ K; $\varphi_{int}=0.25$ at $\kappa N=12.6$ at $T=473$ K). These values correspond to a value of $d_{int} \approx 50$ Å close to the prediction given by Helfand and Wassermann [286] for the strong-segregation limit ($d_{int}=2\sqrt{b_A^2+b_B^2/12\kappa}$; $b_{A,B}$ being the Kuhn segment lengths of polymer A or B, respectively). The observed absolute intensities at high Q are quantitatively described by Eq. 6.24.

6.2.6
Collective Dynamics

The collective relaxation motion of a diblock copolymer may be expressed in terms of two relaxation modes characterized by the eigenvalues $\lambda_1(Q)$ and $\lambda_2(Q)$ of the first cumulant matrix $\underline{\Omega}(Q)$ (Eq. 6.17). Figure 6.8 displays these eigenvalues for a symmetric copolymer ($f=0.5$, $R_{ga}=R_{gb}=40$ Å, $\kappa_{12}=0$) as a function of Q.

While at high Q both eigenvalues are proportional to the momentum transfer to the fourth power (Rouse regime), at lower Q eigenvalue 1 becomes Q-independent and eigenvalue 2 displays a Q^2-dependence. Under contrast conditions where the block "a" is visible against "b", the eigenvalue 2 has no weight and the total structure factor starts to decay with $\lambda_1(Q)$:

$$I(Q,t) = I(Q)\, e^{-\lambda_1(Q)t} \tag{6.25}$$

In our example $S(Q)$ (see Eq. 6.21) would be the Leibler structure factor (Eq. 6.12) and $\lambda_1(Q)$ describes the collective dynamics of the diblock copolymer melt. For small momentum transfers $\lambda_1(Q)$ is given by:

$$\lambda_1(Q) = \frac{Q^2 k_B T f(1-f)}{N[(1-f)\zeta_a + f\zeta_b]}$$
$$\times \left[\frac{2}{2Q^2(R_{ga}^2 + R_{gb}^2)f^2(1-f)^2} - 2\kappa_{12}N \right] \tag{6.26}$$

where ζ_a and ζ_b are the friction coefficients of the bare components, f the volume fraction of "a" monomer in the diblock, and R_{ga} and R_{gb} are the block radii of gyration. For $\kappa_{12}=0$ it describes the breathing mode of the diblock, where the two arms move with respect to a common centre of mass. At Q^* for finite κ_{12}, $\lambda_1(Q)$ reaches a minimum displaying the critical slowing down of the concentration fluctuations. At $QR_g \gg 1$ in the Rouse limit we have:

$$\lambda_1(Q) = \frac{Q^4 R_g^2 k_B T}{2N\bar{\zeta}} \left[1 - \frac{4\kappa_{12}N}{Q^2 R_g^2} f(1-f) \right] \tag{6.27}$$

$\bar{\zeta}$ is an average friction coefficient that describes the collective dynamics and is given by $\bar{\zeta}=(1-f)\zeta_a+f\zeta_b$, where $N=N_a+N_b$ and $R_g^2=R_{ga}^2+R_{gb}^2$. The second mode stands for the centre of mass diffusion of the diblock, which is invisible for the discussed contrast. Experimentally the collective dynamics was observed on samples IV and VI (hPE-dPEE and hPEP-dPEE). As an example, Fig. 6.9 presents spectra taken on the hPE-dPEE sample at 473 K. The relaxation rate was determined in the initial decay regime ($t<10$ ns) by fitting the Rouse dynamic

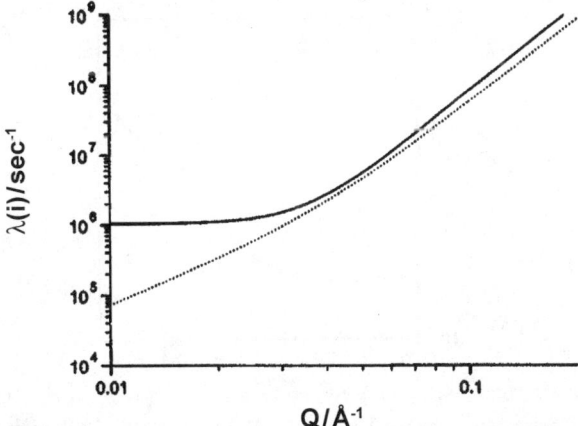

Fig. 6.8 Q dependence of the two eigenvalues $\lambda_1(Q)$ (*solid line*) and $\lambda_2(Q)$ (*dotted line*) predicted by a two-component dynamic RPA approach for the case of an hA-dB labelled diblock copolymer melt. Calculations were performed with $f=0.5$, $R_{ga}=R_{gb}=40$ Å, $N_a=N_b=200$, $\kappa_{12}=0$. $\lambda_1(Q)$ describes the collective mode of the diblock copolymer chains. The Rouse rates were taken from PE and PEE at 473 K (see Table 6.2). (Reprinted with permission from [44]. Copyright 1999 American Institute of Physics)

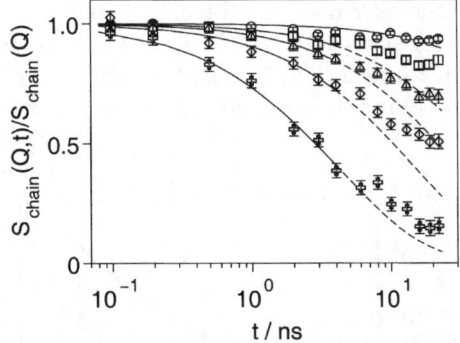

Fig. 6.9 NSE spectra from the hPE-dPEE diblock copolymer melt (sample IV) at 473 K. Q values in each case from above are $Q/\text{Å}^{-1}=0.05, 0.08, 0.10, 0.121, 0.187$. The *solid lines* are the result of a fit with the Rouse dynamic structure factor. Their extent marks the fitting range; the *dashed lines* extrapolate the Rouse structure factor. (Reprinted with permission from [44]. Copyright 1999 American Institute of Physics)

structure factor to the data. Figure 6.10 compares the initial slopes thus obtained with the RPA predictions for the two-component system. They were obtained solely on the basis of the structural data and friction coefficients found for the respective homopolymers.

At large Q quantitative agreement between experiment and RPA prediction is found. Here the first cumulant is proportional to Q^4 – we are in the Rouse

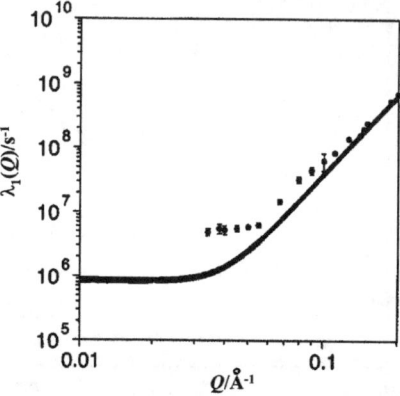

Fig. 6.10 Characteristic frequencies $\lambda_1(Q)$ at 473 K from the hPE-dPEE melt. *Solid line* RPA prediction on the basis of the structural data and the respective polymer friction coefficients. (Reprinted with permission from [44]. Copyright 1999 American Institute of Physics)

regime. However, at lower Q values ($Q<0.06$ Å^{-1}), significant deviations from the RPA results are obvious. The relaxation rates bend into a nearly Q-independent plateau regime around $Q=0.06$ Å^{-1} and they are three to four times faster than predicted. This fast process is not accounted for by the breathing mode of RPA.

Similar observations were made on the higher molecular weight PEP-PEE diblock, which undergoes an ODT at 473 K. Figure 6.11 presents the initial slopes obtained at 473 and 533 K. The RPA predicted collective mode was calculated on the basis of the effective Flory-Huggins parameter and structural

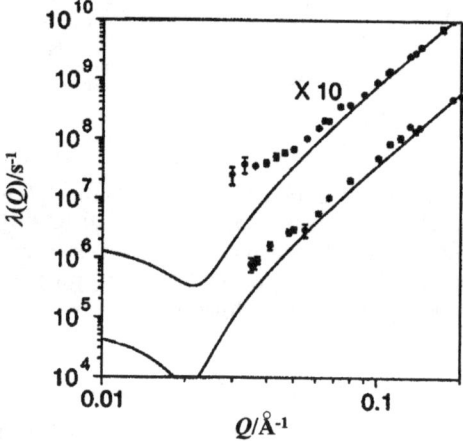

Fig. 6.11 Characteristic frequencies from the long chain diblock (sample IV dPEP-hPEE) at 533 K (*upper data*) and 473 K (*lower data*). The *solid lines* represent the RPA predictions for the collective mode based on structural and homopolymers data. (Reprinted with permission from [44]. Copyright 1999 American Institute of Physics)

parameters taken from the SANS experiment. The friction coefficients were taken from the experiments on the corresponding homopolymers. At high Q there is again quantitative agreement between the measurements and the RPA predictions. But, in the low Q-regime severe deviations from the RPA values are found, again revealing a much faster relaxation of concentration-like fluctuations than obtained theoretically. The observed faster relaxations slow down in the neighbourhood of the order-disorder transition but remain faster than those predicted by the theory.

At high momentum transfers, RPA predicts the observation of an average friction coefficient $\bar{\zeta}=f\zeta_a+(1-f)\zeta_b$ (see Eq. 6.27), which is very well supported by the experiment. We note that this mixing rule differs from the rule obtained for polymer blends by viscosity measurements [287]. From these measurements it was inferred that the inverse friction coefficients should average. If this rule held, a Rouse factor of $W\ell^4=5\times10^4$ $Å^{-4}s^{-1}$ for hPE-dPEE (533 K) instead of 3.45×10^4 from Eq. 6.27 and 3.5×10^4 from the experiment would have been expected. Moreover, the collective Rouse factor of the PEP-PEE copolymer is not very different from that for the PE-PEE system in spite of the dynamics of the PE homopolymer being twice as fast as that of the PEP homopolymer. This is well described by RPA, while the viscosity mixing rule fails.

In the low Q-regime RPA describes well the static structure factor for the short chain melt, where the ODT is sufficiently far away ($\kappa N\approx7$). In the dynamics we would expect the diblock breathing mode to take over around $QR_g\approx2$ ($Q\approx0.04$ $Å^{-1}$). Instead, deviations from Rouse dynamics are already observed at Q values as high as $QR_g\cong5$. At $QR_g=3$ a crossover to a virtually Q-independent relaxation rate about four to five times faster than the predicted breathing mode is found. This phenomenon is only visible under h-d labelling. Under single chain contrast (see below) these deviations from RPA are not seen. Thus, the observed fast relaxation mode must be associated with the block contrast.

Similarly, the results on dPEP-hPEE already exhibit deviations from the RPA-Rouse regime around $QR_g\cong7$ ($T=533$ K). Thereby, the experimentally observed rates differ from the predicted rates by up to more than one order of magnitude (see Fig. 6.11). At the phase transition temperature ($T_{ODT}=473$ K), however, the effect is greatly reduced and a general slowing down of the fluctuations is found, though some deviations persist.

6.2.7
Single Chain Dynamics

The single chain dynamics of one given block or of one chain in a diblock copolymer melt is observed if a matched deuterated diblock is mixed with a small amount of labelled diblocks, where the label could be a protonated "a" or "b" block or a protonated chain. In terms of the dynamic RPA such a system is a four-component polymer mixture. It is characterized by four different relaxation modes $\lambda_1-\lambda_4$ which – depending on the contrast conditions – appear with weights $w_1(Q)-w_4(Q)$ in the dynamic structure factor. These weights depend both on the volume fraction and on the contrast factors K_i of the components in the mixture. Figure 6.7 displays the Q-dependent eigenvalues obtained for symmetric A-B diblock copolymer chains with $N_a=N_b$, $f=0.5$ and $R_{ga}=R_{gb}=40$ Å. Due to the incompressibility constraint the weight $w_4(Q)=0$, i.e. the λ_4 mode is invisible. For the example of Fig. 6.12 it is assumed that the component A is the fastest $\zeta_b/\zeta_a=4$. The eigenvalues depend on friction coefficients, the volume fraction of the components A and B and the radii of gyration of the blocks. However, they do not depend on the labelling. The observable dynamics of this system results from a weighted (weights w_1, w_2, w_3, w_4) average of these four

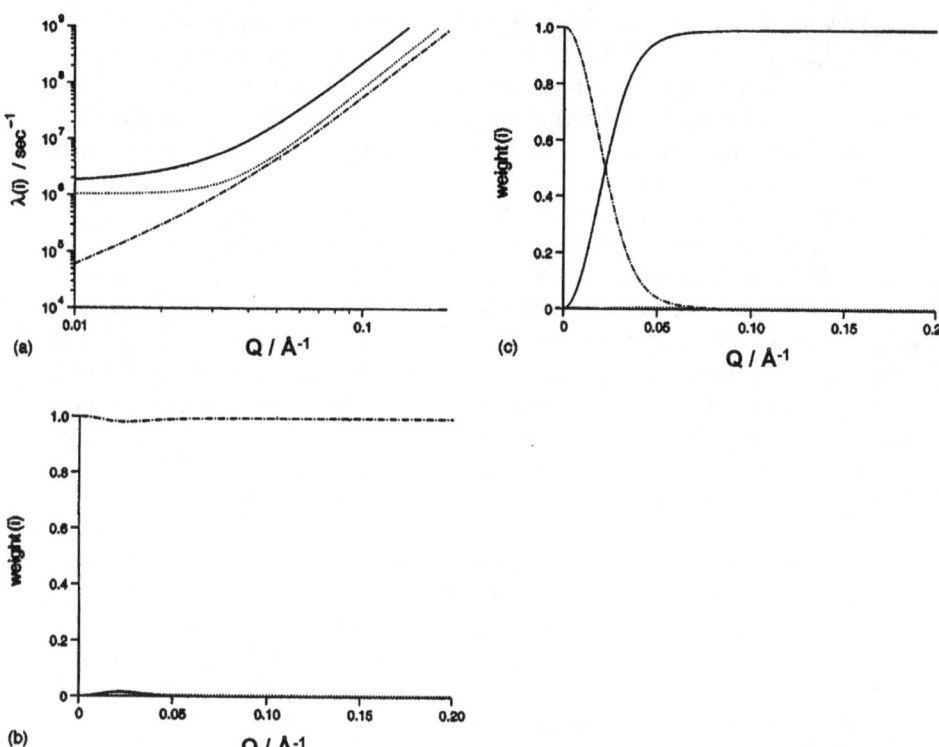

Fig. 6.12 a Q dependence of the eigenvalues predicted by a four-component RPA approach for a mixture of two kinds of symmetric diblock copolymer chains differing by their labelling. λ_1 (*solid line*), λ_2 (*dash-dotted line*), λ_3 (*dotted line*). Calculations were performed with $f=0.5$, $R_{ga}=R_{gb}=40$ Å, $N_a=N_b=200$, $\kappa_{12}=0$. **b** and **c** present the corresponding weights weight$_1(Q)$ (*solid line*) and weight$_2(Q)$ (*dash-dotted line*), weight$_3(Q)$ (*dotted line*) of the eigenvalues plotted in **a**. They have been calculated for different labelling conditions. **b** Mixture of 1% of dA–hB diblock copolymer chains in a dA–dB matrix. **c** Mixture of 1% of hA–dB diblock copolymer chains in a dA–dB matrix. The time scales are set by the friction coefficients of PEE and PE homopolymers at 473 K. (Reprinted with permission from [44]. Copyright 1999 American Institute of Physics)

eigenvalues representing relaxation rates. Eigenvalue λ_3 is identical to the collective mode of the pure dA-hB diblock copolymer. From Fig. 6.12b, giving the weights for an admixture of 1% volume fraction of dA-hB block copolymer to the deuterated matrix it is evident that λ_2 describes the dynamics of the slow B block. In the limits of small ($QR_g \ll 1$) and high ($QR_g \gg 1$) Q the following expressions hold:

$$\lambda_2(Q) = \frac{Q^2 k_B T}{N(1-f)\zeta_b + f\zeta_a} \quad QR_g \ll 1$$

$$\lambda_2(Q) = k_B T\, Q^4 \frac{\ell_b^2}{12\zeta_b} \quad QR_g \gg 1$$

(6.28)

l_b being the segment length of polymer B.

At low momentum transfer λ_2 describes the translational Rouse diffusion coefficient of the whole diblock, considering $N(1-f)$ segments exerting the friction ζ_b and Nf segments exerting the friction ζ_a. In the high Q limit, RPA predicts a strange result: The block B is predicted to undergo the same Rouse dynamics as a B homopolymer in the melt of B polymers. A proper average over the mixed surrounding is not predicted.

Similarly from Fig. 6.12c it is evident that at high Q, $\lambda_1(Q)$ describes the motion of the A arm in the diblock copolymer melt. At lower Q in the region where the static structure factor under hA-dB labelling shows its maximum (Q^*) the weight crosses over to mode (λ_2), the mode describing the translational diffusion of the whole molecule. Thus, no matter which arm we label, at low Q the translational diffusion of the total diblock is always seen. Again also for mode $\lambda_1(Q)$ we give the asymptotic values for small and large Q:

$$\lambda_1(Q) = \frac{3}{2}\frac{[(1-f)\zeta_b + f\zeta_a]\, k_B T}{N\zeta_a\zeta_b(R_{ga}^2 + R_{gb}^2)f(1-f)} \quad (QR_g \ll 1)$$

$$\lambda_1(Q) = \frac{k_B T\, Q^4 \ell_a^2}{12\zeta_a} \quad (QR_g \gg 1)$$

(6.29)

Again, at high Q the RPA predicts that the dynamics of arm A is identical to the Rouse motion of an A polymer in an A homopolymer melt. At low Q, $\lambda_1(Q)$ turns into a breathing mode with a non-vanishing relaxation rate at $Q=0$, as the collective mode $\lambda_3(Q)$.

Finally, we note that while the eigenvalues of $\underline{\underline{\Omega}}$ describe the relaxation properties inherent to the polymer system under consideration, the initial slopes of the measured dynamic structure factor:

$$\Gamma(Q) = -\lim_{t\to 0}\frac{\partial I(Q,t)}{\partial t}\frac{1}{I(Q)} = \frac{\underline{K}^T\underline{\underline{\Omega}}\,\underline{\underline{S}}(Q)\,\underline{K}}{\underline{K}^T\underline{\underline{S}}(Q)\,\underline{K}}$$

(6.30)

depend on the observation conditions and thus on the chosen labelling. In general the eigenvalues combined with the weights may be transferred directly into slopes that are meaningful from an experimental point of view.

Caution is advisable (exemplified for the case of labelling a small fraction of A blocks in the system) looking at the weights displayed in Fig. 6.12c, that at low Q the weight crosses over from eigenvalue λ_1 to eigenvalue λ_2. At the same time the values λ_1 and λ_2 become grossly different. In an actual experiment a relaxation corresponding to λ_2 would be clearly observed. However, calculating the initial slope following Eq. 6.30 leads to $\Gamma(Q) = \dfrac{k_B T Q^4}{N_a \zeta_a}$.

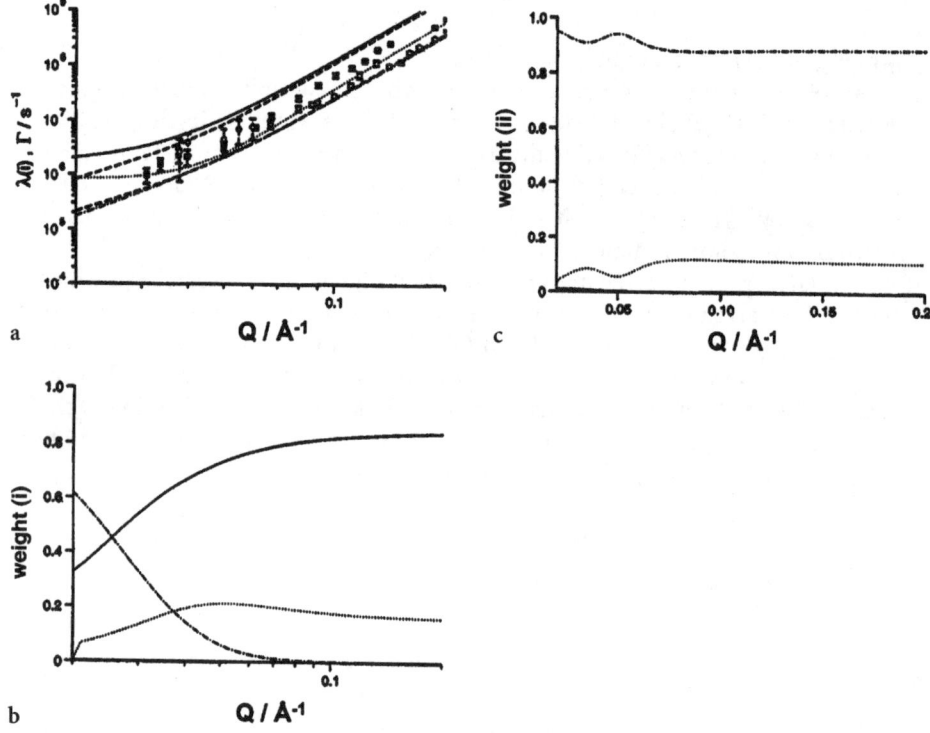

Fig. 6.13 a Rouse relaxation rates from the labelled PE arm (*closed square*) and the labelled PEE arm (*open circle*) of a PE-PEE diblock copolymer in a deuterated diblock copolymer matrix at 473 K. The various lines represent eigenvalues derived from the four-component dynamic RPA approach: λ_1 (*solid line*), λ_2 (*dash-dotted line*), λ_3 (*dotted line*). **b** and **c** The relative weights weight$_1$ (*solid line*), weight$_2$ (*dash-dotted line*), weight$_3$ (*dotted line*) of the eigenvalues plotted in **a** for several labelling conditions. **b** Relative weights for a mixture of a 20% of hPE-dPEE diblock chains in a dPE-dPEE matrix (Sample II). **c** Relative weights for a mixture of a 20% of dPE-hPEE diblock chains in a dPE-dPEE matrix (Sample III). The first cumulant, which is the weighted average of the eigenvalues, is represented in **a**: *upper dashed line* first cumulant of Sample II, *lower dashed line* first cumulant of Sample III. (Reprinted with permission from [44]. Copyright 1999 American Institute of Physics)

6 The Dynamics of More Complex Polymer Systems

Thus, the initial slope corresponds to that of an A arm performing translational diffusion without knowing about the slower B arm. Similarly the prediction for a small fraction of labelled B arms leads to the unphysical result $\Gamma(Q)=k_B TQ^2/N_b\zeta_b$, again a single arm performing centre of mass diffusion.

Experimentally, the component dynamics of the two arms was investigated on the PE-PEE system (samples II and III). The results are displayed in Fig. 6.13 and compared with the respective RPA predictions. Figure 6.13a displays the Q-dependent eigenvalues as well as RPA initial slopes together with the experimental initial slopes obtained from samples II (PE label) and sample III (PEE label). Figures 6.13b and c show the weights of the different modes in the dynamic structure factor, which are determined by the chain labelling. In the case of the labelled PE at high Q the fast mode dominates; for the labelled PEE the opposite occurs. At lower Q under both labelling conditions the weight stays with the slower mode, signifying the translational diffusion of the whole molecule. Finally, the RPA first cumulant, which is the respective proper average for the different labelling conditions, is indicated by a dashed and a point dashed line.

As pointed out above, the RPA theory predicts that the dynamics of the respective homopolymers should be observed at high Q in the Rouse regime. While the experiment shows that the predicted Q dependencies are reproduced well by the data, the absolute values for the observed relaxation rates disagree with the predictions (see Table 6.2). In particular the observed Rouse factors for PE are considerably smaller than predicted, $(W\ell^4)_{expt}=2\times10^4$ Å^4 s^{-1} compared to $W\ell^4_{RPA}=3.8\times10^4$ Å^4 s^{-1} at $T=473$ K. At low Q values, the two blocks display the same single chain dynamics.

In order to understand why the RPA approach predicts the wrong result we go back to Eq. 6.16, insert the expressions for the bare mobilities (Eq. 6.18) and arrive at:

$$\mu_{ij}(Q) = \delta_{ij}\left(\frac{\varphi_i}{\zeta_i}\right) + \frac{1}{\sum_\kappa \frac{\varphi_\kappa}{\zeta_\kappa}+\frac{\varphi_0}{\zeta_0}}\left(\frac{\varphi_i\varphi_j}{\zeta_i\zeta_j}\right) \tag{6.31}$$

Table 6.2 Experimental Rouse rates ($W\ell^4$) for the pure homopolymers and for the block-copolymer melts obtained from the high Q behaviour of $\Gamma(Q)$ compared to the predictions of the RPA theory. Values of $W\ell^4$ are given in 10^{-13} Å^4 s^{-1}

T (K)	PEE	PE[a]	20% hPE-dPEE in dPE-dPEE		dPE-dPEE		20% dPE-hPEE in dPE-dPEE	
			Experimental	Theoretical[b]	Experimental	Theoretical[b]	Experimental	Theoretical[b]
473	0.8±0.15	4.2±0.15	2±0.2	3.8	1.8±0.15	1.7	0.97±0.15	0.87
533	2.2±0.2	8±0.1	4.1±0.4	7.4	3.5±0.2	3.6	2.5±0.2	2.2

[a] Reference [15].
[b] Rouse factor given by the RPA theory.

With the aid of Eq. 6.30 the relaxation rates of Eq. 6.28 and Eq. 6.29 evolve. For a vanishing concentration φ_ε of the labelled component, the leading term in the appropriate diagonal element of μ_{ij} is given by $\varphi_\varepsilon/\zeta_{i\varepsilon}$ all admixtures from the second term are proportional to φ_ε^2 and thus do not contribute to first order. Then the first cumulant becomes equal to that of the labelled homopolymers. Thus, even for a system in which two components A and B exhibit strongly different friction coefficients, the RPA theory will predict a relaxation rate close to that of the labelled homopolymer for the single chain dynamics. This result is counterintuitive and obviously spurious but it is related deeply to the basic assumption of dynamic RPA, namely to the fact that the bare mobility matrix is diagonal and does not allow for off-diagonal elements describing, e.g. mutual friction in the bare system. Within the RPA framework the bare systems are assumed to be mixed already; however, without any mutual interaction as expressed by the κ parameters, the changes that occur in the real system are a consequence of a linear response to the switching on of the κ parameters. However, nothing equivalent is provided for the friction terms, which are inherently difficult because friction cannot be cast into an interaction potential. The bare mobility matrix $\underline{\underline{\mu}}^0$ completely ignores that the friction – even of a mixture with all κ parameters equal to 0 – may be changed to a different embedding. The unusually fast relaxation of the collective modes at low momentum transfers, even in a regime where a critical slowing-down due to the closeness of the phase transition would have been expected, suggests the existence of additional forces possibly due to surface tension at surfaces even near to the ODT and above.

This conjecture was investigated on the PEP-PEE triblock copolymers with labels at the junction point (sample VII) which showed an ODT at T_{ODT}=433 K. Figure 6.14 presents the dynamic structure factors obtained at 433 and 473 K in the short-time regime (t<10 ns). We observe a strong Q-dependent decay of the structure factor. In a first qualitative evaluation the NSE data have been

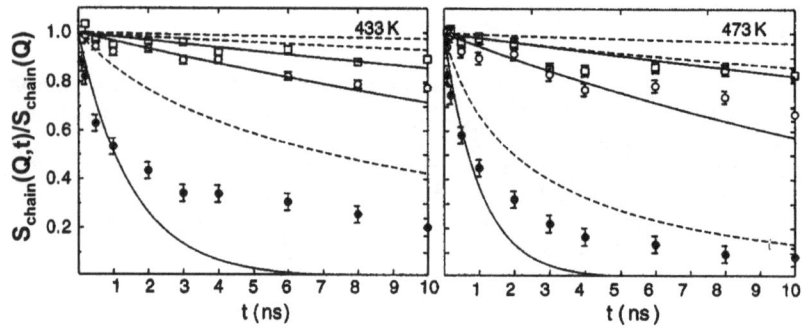

Fig. 6.14 NSE spectra from the junction-labelled PEP-PEE triblock copolymer (sample VII) at different Q-values (*filled circles* Q=0.2 Å^{-1}, *open circles* Q=0.08 Å^{-1}, *squares* Q=0.05 Å^{-1}). The *solid lines* indicate the initial slope fits. The *dashed lines* are the expectations for Rouse relaxation in the system (see text). (Reprinted with permission from [284]. Copyright 2002 EDP Sciences)

Table 6.3 Relaxation times from an initial slope evaluation of the NSE spectra compared to the corresponding predictions from the Rouse model

Q (Å^{-1})	T=433 K			T=473 K		
	τ_{Rouse} (ns)	τ_{exp} (ns)	τ_{Rouse}/τ_{exp} (ns)	τ_{Rouse} (ns)	τ_{exp} (ns)	τ_{Rouse}/τ_{exp} (ns)
0.05	1440	61±10	23	384	52±10	7.4
0.08	220	31±7	7	58	18±3	3
0.1	–	–	–	24	4±0.5	6
0.114	53	10±2	5.3	–	–	–
0.14	23	6±1	3.8	–	–	–
0.18	8.6	2.2±0.6	3.9	–	–	–
0.2	5.6	1.7±0.5	3.3	1.5	1±0.5	1.5

parameterized in terms of initial slopes. The corresponding fits are indicated by solid lines.

The resulting relaxation times are given in Table 6.3. In order to compare with the Rouse dynamics, the broken lines in each figure display the predictions of this theory for our sample, taking the Rouse rate from an earlier experiment on the same materials. The corresponding Rouse times are also displayed in Table 6.3. Obviously, the chain dynamics close to the junction point are greatly accelerated. This strong enhancement of the relaxation is evidence for the existence of additional forces beyond entropic forces acting on the junction points.

If the PEP-PEE triblock chains are arranged in microdomains separated by well-defined interfaces, then the additional fast relaxation process observed by NSE would result from the undulation motions of the interface. The thermal motions of the interface are driven by the surface tension and overdamped. With an effective viscosity η_{eff} in the interfacial region, the relaxation rate $\tau(Q)^{-1}$ follows the linear dispersion $\tau(Q)^{-1} = Q \cdot \frac{\gamma}{4\eta_{eff}}$ [288]. In the Q-range of the NSE experiment the scattering is dominated by the single-chain scattering of the labels, which tag the local interface motions in a incoherent way. The corresponding single-chain motion is effected by two processes: (i) the hPEP section undergoes the entropy-driven Rouse motion active in polymer melts and (ii) in addition, it follows the undulations of the interface. The two processes are approximated as independent and are modelled by a Rouse chain with an anchor in the interface that follows the undulations. The time correlation function for the displacements due to undulations assumes:

$$f(t) = \langle u^2 \rangle - \langle u(t)\, u(0) \rangle = \frac{k_B T}{4\pi\gamma} \left\{ \ln\left(\frac{\xi}{\ell}\right) + Ei\left(1, t/\tau\left(\frac{2\pi}{\xi}\right)\right) - Ei\left(1, t/\tau\left(\frac{2\pi}{\ell}\right)\right) \right\} \quad (6.32)$$

$\langle u^2 \rangle = \frac{1}{(2\pi)^2} \int d^2q \langle u^2(q) \rangle$ denotes the mean-squared amplitude of the surface fluctuations with $\langle |u^2(q)| \rangle = \frac{k_B T}{q^2 \gamma}$. Since the integral is logarithmically divergent, the integration has to be restricted to the interval between a minimal and a maximal Q, which are related to the finite domain size ξ and a molecular size ℓ [289]. With that we arrive at $\langle u^2 \rangle = \frac{k_B T}{2\pi\gamma} \ln\left(\frac{\xi}{\ell}\right)$. The time-dependent contributions are given by the exponential integral function $Ei(1, x) = \int_1^\infty \frac{e^{-xt}}{t} dt$. Applying the Gaussian approximation and performing an angular average of the one-dimensional undulation motion perpendicular to the local interface plane, the self-correlation function is obtained:

$$S_{\text{int}}^{\text{self}}(Q, t) = \frac{1}{2} \sqrt{\pi} \frac{\text{erf } Q\sqrt{f(t)}}{Q\sqrt{f(t)}} \qquad (6.33)$$

The total structure factor, finally, is given by a product of $S_{\text{int}}^{\text{self}}(Q,t) \cdot S_{\text{Rouse}}^{\text{self}}(Q,t)$. ($S_{\text{Rouse}}^{\text{self}}$ has to be calculated for a chain with one end fixed to a surface). Figure 6.15 displays a fit of the measured spectra at both temperatures with the complete dynamic structure factor where the Rouse relaxation rate was taken from an earlier experiment. Fit parameters were the surface tension and the effective local viscosity of the short labelled PEP segments. The data are well

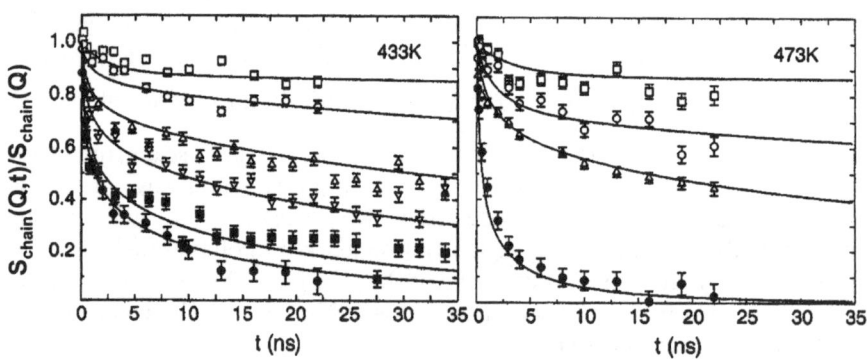

Fig. 6.15 Dynamic structure factor from the junction-labelled triblock copolymer for different Q-values. T=433 K: *filled circles* Q=0.20 Å^{-1}, *filled squares* Q=0.18 Å^{-1}, *open triangles down* Q=0.14 Å^{-1}, *open triangles up* Q=0.114 Å^{-1}, *open circles* Q=0.08 Å^{-1}, *open squares* Q=0.05 Å^{-1}. T=473 K: *filled circles* Q=0.20 Å^{-1}, *open triangles up* Q=0.10 Å^{-1}, *open circles* Q=0.08 Å^{-1}, *open squares* Q=0.05 Å^{-1}. The *solid lines* are result of the fit with the complete structure factor for surface undulations and Rouse motion. (Reprinted with permission from [284]. Copyright 2002 EDP Sciences)

described by this approach. For the surface tension, values of $\gamma=2.3\times10^{-3}$ N/m at 433 K and $\gamma=2.4\times10^{-3}$ N/m at 473 K are revealed. These values are well comparable with the theoretical values ($\gamma=1.5\times10^{-3}$ N/m at 433 K and $\gamma=1.4\times10^{-3}$ N/m at 473 K) deduced from the relation developed by Helfand [286] for the strong segregation regime. For the local viscosity the first reveals $\eta_{\text{eff}}=1.4\times10^{-3}$ Ns/m^2 at 433 K. This value needs to be compared to the local viscosity that one would expect, e.g. for a monomeric unit in the frame of the Rouse

$$\eta_{\text{eff}} = \frac{3k_B T}{W\ell^4} \frac{\rho\ell N_a}{36 M_0} = 2.7\times10^{-3} \text{ Ns/m}^2 \text{ (}N_a \text{ Avogadro number, } M_0 \text{ model monomeric weight, } \rho \text{ polymer density)}.$$

We note that via $\langle u^2 \rangle$ (which depends only on γ) and $\tau(q)$ (which depends on η_{eff}/γ) both quantities are separately evaluated by the fitting procedure. The mean-square amplitudes of the undulations $\langle u^2 \rangle$ are 15 Å at 433 K and 15.5 Å at 473 K. The contribution of the interface undulations to the scattering intensity decreases with increasing Q and temperature. This feature is particularly obvious at the short-time limit through the variation of the ratio $\tau_{\text{Rouse}}/\tau_{\text{exp}}$.

The combination of careful chemical synthesis with NSE and SANS experiments sheds some light on the fast relaxation processes observed in the collective dynamics of block copolymers melts. The results reveal the existence of an important driving force acting on the junction points at and even well above the ODT. Modelling the surface forces by an expression for the surface tension, it was possible to describe the NSE spectra consistently. The experimental surface tension agrees reasonably well with the Helfand predictions, which are strictly valid only in the strong-segregation limit. Beyond that, these data are a first example for NSE experiments on the interface dynamics in a bulk polymer system.

6.3
Gels

Until now there have only been very few investigations that utilize NSE spectroscopy to address the dynamics in polymer gels. Three regimes have been observed [290–293]. At very low Q, frozen fluctuations lead to a dominating scattering intensity that displays no dynamics. At intermediate Q, the scattering intensity results from fluctuations due to the compressibility of the chain network. The dynamic signature is a collective diffusion, i.e. $S(Q,t)/S(Q) = \exp(-D_c Q^2 t)$. Finally, as $Q\xi>1$ the intensity behaves as $1/Q^\nu$ and the observed dynamics represent the local chain motions and have the signatures of Zimm dynamics, as observed in a corresponding polymer solution without cross-links. The latter was observed on poly(dimethylsiloxane)(PDMS)-toluene gel [290] and on poly(N-isopropyl-acrylamide)(PNIPA)-water gel [292, 293].

According to [291] the intensity scaled to the scattering contrast factor $\Delta\rho^2$ may be written as:

$$I(Q)/\Delta\rho^2 = S_s(Q) + S_f(Q,t) \tag{6.34}$$

with a contribution from the frozen inhomogenieties:

$$S_s(Q) = \frac{8\pi\, \Xi^3 \langle \Delta\varphi^2\rangle}{(1+Q^2\Xi^2)^2} \tag{6.35}$$

where Ξ and $\langle\Delta\varphi^2\rangle$ are the length scale and amplitude of the inhomogeneities. The dynamic intensity is:

$$S_f(Q,t) = \frac{k_B T \varphi^2}{M} \frac{1}{1+(Q\xi)^2} f(Q,t) \tag{6.36}$$

where ξ is the correlation length and M the longitudinal elastic modulus, the form of which discriminates between a polymer solution $M=\varphi\partial\Pi/\partial\varphi$ and a gel network $M=\varphi\partial\omega/\partial\varphi+4/3G$ with $\omega=\Pi_{mix}-G$, where G is the shear modulus of the gel and Π the osmotic pressure. The dynamic factor $f(Q,t)=\exp(-D_c Q^2 t)$ for $(Q\xi)\ll 1$ and $\approx\exp(-(\Gamma Q^3 t)^{2/3})$ for $(Q\xi)\gg 1$.[1] The collective diffusion constant depends on the ratio of the modulus M and a friction coefficient ξ, $D_c=M/\varphi\xi$.

Figure 6.16 [291] displays data from a poly(fluorosilicone)(PFS)-acetone gel and illustrates how the large Fourier time range of IN15 enables a clear sepa-

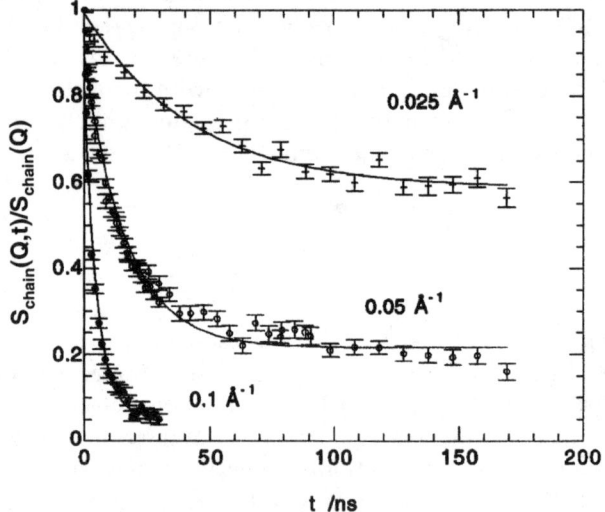

Fig. 6.16 NSE relaxation curves obtained from a 16% volume fraction poly(fluorosilicone) gel in acetone using the high resolution NSE spectrometer IN15 at the ILL, Grenoble. The existence of plateaus that represent the level of excess scattering from static inhomogenieties at low wave vector Q is clearly visible. The decay rates times of the dynamic parts yield the collective diffusion coefficient of the gel. (Reprinted with permission from [291]. Copyright 2002 American Chemical Society)

---

[1] This formulation is used in the literature; however, see also Eq. 6.48 which yields a better approximation of the Zimm dynamics.

ration of the frozen static component that corresponds to the plateau values visible at large t and low Q. Figure 6.17 contains data points indicating the amount of the dynamic scattering intensity in the total small angle scattering signal as deduced from a fit of $S_f(Q,t)$ to the data in Fig. 6.16 and a limiting value from light scattering. The corresponding intensity of the non-cross-linked solution was found to be 40% lower and the derived diffusion constant about 30% lower than that of this gel. Dividing out the modulus that enters both $S_f(0)$ and D_c leads to the conclusion that the friction coefficient ξ in the gel is significantly lower than in solution.

An NSE investigation on PNIPA microgels [294] was focussed on the dependence of the collective diffusion D_c on the cross-linker density. Here the centre of mass diffusion of the microgel particles was suppressed by choosing a density where they assemble to a colloidal crystal. The NSE data obtained at the IN11 in Grenoble covered the Q-range between 0.04 Å^{-1} and 0.19 Å^{-1} on a time window 0.4–25 ns. Only single exponential decay according to collective diffusion was observed. The diffusion constant D_c varies from 4.6×10^{-11} m^2 s^{-1} at 1% cross-linker N,N'-methylenebisacrylamide (BIS) to 3.1×10^{-11} m^2 s^{-1} at 5% BIS concentration. PNIPA shows a volume phase transition that leads to a collapse of the gel at about 34 °C, the above experiment was done in the swollen phase at 20 °C. In [293] the changes in the scattering contributions upon approaching the volume phase transition were also addressed.

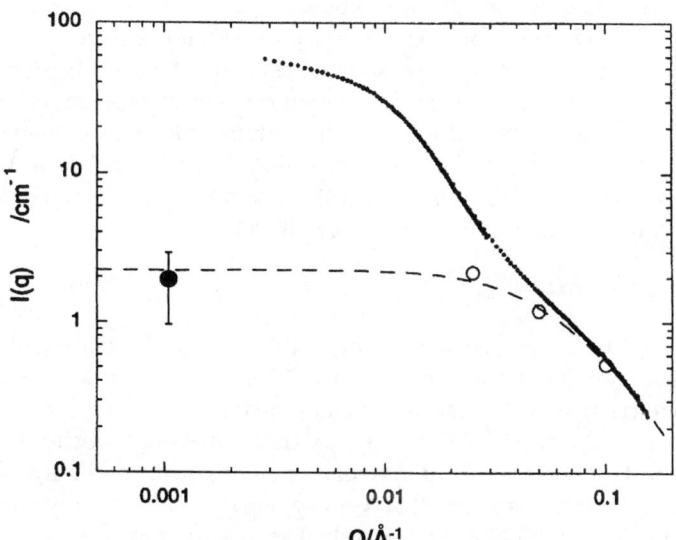

Fig. 6.17 SANS of the poly(fluorosilicone) gel showing the excess scattering from frozen inhomogenieties at low Q. The *dashed line* corresponds to the dynamic part of the scattering, *open circles* are derived from NSE experiments, the *solid circle* was obtained by dynamic light scattering [291]. (Reprinted with permission from [291]. Copyright 2002 American Chemical Society)

The only attempt so far to measure an anisotropy of D_c in a stretched gel (PDMS-toluene, stretching ratio 1.25 and 1.4) found, if any, only marginal anisotropy ratios (1.02±0.04 and 1.11±0.06) [295].

6.4
Micelles, Stars and Dendrimers

Polymeric micelles, stars and dendrimers in solution consist of a number of polymer chains that form relatively compact aggregates that exhibit internal dynamics and overall diffusion. Whereas the association of polymer in a micelle is usually driven by physical interactions, the star and dendrimer architecture is generally achieved by chemical bonds.

6.4.1
Micelles

Amphiphilic molecules as low molecular surfactants form micelles in water or oil (organic solvents). Analogously, block copolymers may form aggregates in selective solvents. The less soluble blocks aggregate in a compact core surrounded by a corona of the soluble chains that extend into the solvent region. In the simplest case the micelle has spherical symmetry and low polydispersity. The corona may be considered as a polymeric brush. Due to its segmental mobility and compressibility it bears some resemblance with a dense or semi-dilute solution and therefore exhibits sizeable dynamic density fluctuations. NSE experiments on micelles also contain the centre of mass diffusion term as a factor. Only where the scattering intensity, due to the average scattering length density distribution of the micelle, becomes comparable or lower than the fluctuation or "blob" scattering from the brush, may the additional dynamic of the brush be found in the NSE signal. The NSE relaxation signal obtained therefore contains two components and can be described by:

$$S(Q,t)/S(Q) = \exp(-\Gamma_{slow}t)[A_{slow} + (1 - A_{slow})\exp(-\Gamma_{fast}t)] \qquad (6.37)$$

where $\Gamma_{slow}=DQ^2$ is associated with the centre of mass diffusion and the fast process with the fluctuations in the brush [296, 297]. The ratio $(1-A_{slow})/A_{slow}$ reflects the relative intensities of blob and medium scale density fluctuation scattering from the brush to the average particle scattering intensity. In both investigations [296, 297], aqueous systems containing hydrophilic-hydrophobic diblock copolymers were used that form aggregates with hydrodynamic radii R_H of about 70 Å and 140 Å, respectively. The slow component yields a diffusion constant well in agreement with particle radii from other experiments, i.e. $R_H=k_BT/6\pi\eta D$. The fast relaxation rate Γ_{fast} is about one order of magnitude larger than the slow rate and shows a Q-dependence that deviates from that of a simple diffusion. Castelletto et al. [297] investigated the micellar system at different concentrations and temperatures that yield liquid, hard and soft gel phases. Besides the effect of the temperature-dependent viscosity $\eta(T)$ the nature of the phase did not reflect in the observed dynamics.

The non-aqueous system of spherical micelles of poly(styrene)(PS)-poly-(isoprene)(PI) in decane has been investigated by Farago et al. and Kanaya et al. [298, 299]. The data were interpreted in terms of corona brush fluctuations that are described by a differential equation formulated by de Gennes for the breathing mode of tethered polymer chains on a surface [300]. A fair description of $S(Q,t)$ with a minimum number of parameters could be achieved. Kanaya et al. [299] extended the investigation to a concentrated (30%, PI volume fraction) PS-PI micelle system and found a significant slowing down of the relaxation. The latter is explained by a reduction of osmotic compressibility in the corona due to chain overlap.

Non-spherical micelles of poly(ethylene)(PE)-poly(ethylene-propylene)(PEP) in decane are self-assembling in the form of extended platelets that have a crystalline PE-core and a planar PEP brush on both sides. Due to the large size of the platelets the centre of mass diffusion is extremely slow and allows a clear separation of the density fluctuation in the brush. NSE experiments [301] have been analysed in terms of the model of de Gennes [300]. The friction coefficient and modulus of the brush were found to be similar to those of a typical gel.

6.4.2
Stars

Earlier NSE investigation of star polymers addressed the chain dynamics inside the individual star molecules [302, 303] (see [5] for a comprehensive review). A recent study using the ultrahigh resolution instrument IN15 on 18-arm poly(isoprene)[PI] stars in d-methylcyclohexane was rather focused on the interaction of stars [304]. This was achieved by using centre-labelled stars (only 1/4 of the arm length starting from the centre was protonated). In concentrated solutions above the overlap concentration, a plateau of the relaxation in $S(Q,t)/S(Q)$ was observed that indicates a caging or confinement of the star centres in a region with a radius that is closely related to the blob size of the surrounding star solution. Various labelling schemes achieved by combination of partially labelled and fully deuterated (i.e. matched) stars addressed the self-correlation of the stars as well as the collective dynamics. Figure 6.18 illustrates the difference between both scattering functions that is only observed at low Q (beyond the correlation peak in $S(Q)$). On small length scales (larger Q) both correlation functions have the same behaviour. The relaxation plateaus due to confinement are also clearly visible. The collective diffusion, as obtained from the initial decays of $S(Q,t)/S(Q)$ (lines in Fig. 6.18), scales as $D=D_0/S(Q)$, and extra influence of a hydrodynamic factor $H(Q)$ could not be identified. The self-correlations in contrast exhibit Zimm scaling, i.e. $\Gamma \sim Q^3$. The separation of collective and single-star correlation depends crucially on the labelling and matching only available for neutron scattering.

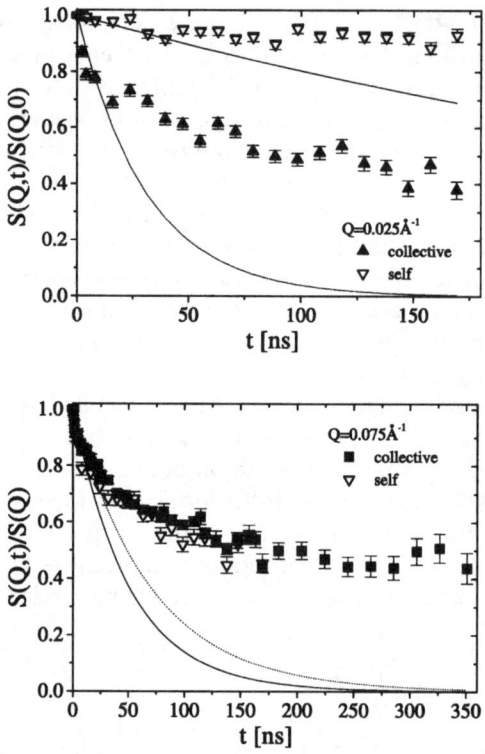

Fig. 6.18 Normalized intermediate scattering function from centre-labelled 18-arm PI solutions. "Collective" corresponds to a 4.85% solution of labelled stars, whereas the "self" data stem from a solution of 1% labelled and 16.6% non-labelled stars. Note the maximum Fourier time of 350 ns (λ=1.9 nm), which was obtained at the IN15 in the case of these strong scattering samples. (Reprinted with permission from [304]. Copyright 2002 Springer)

6.4.3
Dendrimers

Dendrimers have a star-like centre (functionality e.g. 4) in contrast to a star; however, the ends of the polymer chains emerging from the centre again carry multifunctional centres that allow for a bifurcation into a new generation of chains. Multiple repetition of this sequence describes dendrimers of increasing generation number g. The dynamics of such objects has been addressed by Chen and Cai [305] using a semi-analytical treatment. They treat diffusion coefficients, intrinsic viscosities and the spectrum of internal modes. However, no expression for $S(Q,t)$ was given, therefore, up to now the analysis of NSE data has stayed on a more elementary level.

Star-burst polyamidoamine (PAMAM) dendrimers with primary amine end groups in d-methanol [306] and with hydroxyl or glucopeptide end groups in D_2O [307] have been investigated by NSE in order to identify internal segment

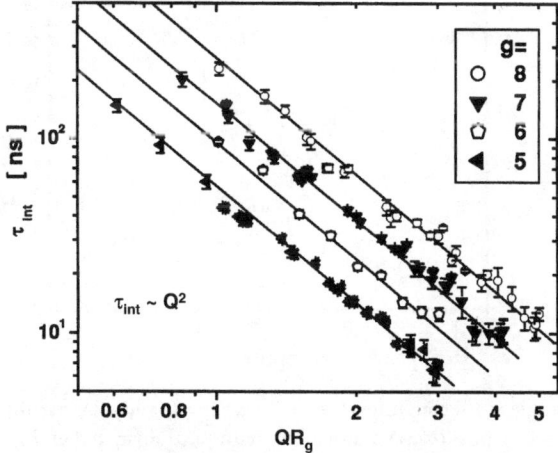

Fig. 6.19 Relaxation rates from single exponential fits to the NSE data from PAMAM dendrimers of generation g=5–8 (5%) in d-methanol. The *solid lines* are derived from NSE data from the FRJ2-NSE (Jülich) and MESS (Saclay) spectrometers and show the prediction for simple Stokes-Einstein diffusion of hard spheres at finite concentration. (Reprinted with permission from [306]. Copyright 2002 American Institute of Physics)

motions. The denrimers in methanol showed only simple centre of mass diffusion with no indication for internal dynamics for generations g=5–8 (see Fig. 6.19) even in the Q-regime where blob scattering has significant contributions to the scattering intensity. The internal structure of these dendrimers seems to be very compact such that segmental motions are suppressed and slowed down.

In contrast, Funayama et al. [307] see indications for a fast component in a double exponential fit for the PAMAM dendrimers (g=5) with –OH or sugar end groups in D_2O at low concentration (1%), whereas they observe only diffusion $D_{eff} \approx D_0/S(Q)$ at 10% concentration, (i.e. a single exponential decay). A fast extra signal has been fitted with amplitudes between 4 and 17% to the data from 1% solutions and is attributed to motions of the amido-amine segments inside the dendrimer. However, the data coverage and statistics for a clear identification of the fast signal is marginal.

Carbosilane dendrimers with perfluorinated end groups in perfluorohexane were studied by Stark et al. [308]. The significant deviations from simple diffusion that are observed in the NSE data in this case are attributed to shape fluctuations following a procedure that had been developed for the analysis of micro-emulsion droplet fluctuations [309].

Since the scattering signal from shape fluctuations is only significant compared to the average form factor, where the latter has minima or zeroes, the normalized intermediate scattering is mainly sensitive to the fluctuations at these minima, as can be seen in Fig. 6.20. Only for generations $g>3$ do these dendrimers exhibit a spherical shape with the corresponding minima; lower generations were found to be elliptical, i.e. without zeroes in the angular aver-

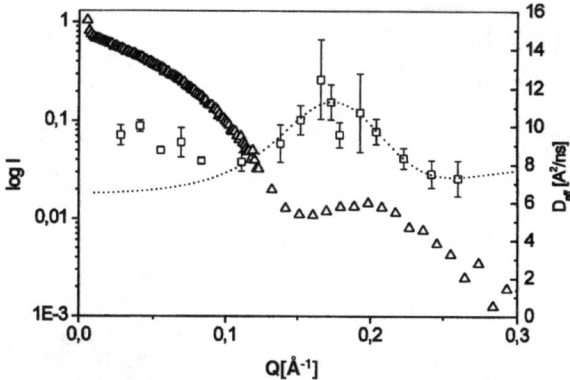

Fig. 6.20 Small angle scattering intensity (*triangles* log I) and effective diffusion $D_{eff}(Q)$ as obtained from $g=4$ carbosiloxane dendrimers with perfluorinated end groups in perfluorohexane. The *dashed line* is a fit to the prediction of a model for shape fluctuations of micro-emulsion droplets, the resulting bending modulus was $0.5\ k_BT$. (Reprinted with permission from [308]. Copyright 2003 Springer Berlin Heidelberg New York)

aged scattering. Therefore, fluctuation scattering is largely masked and the NSE data ($Q=0.028-0.278$ Å^{-1}, up to 25 ns) correspond to simple diffusion with an apparent diffusion constant that matches the value for D_{eff} obtained by photon correlation spectroscopy. The applied model for the explanation of the NSE data for the $g=4$ dendrimer implies a low viscosity liquid-like interior of the dendrimers and an incompressible interface formed by the end groups.

The question remains whether the inelastic intensity that becomes visible at the minima of the average form factor is really due to shape fluctuations or rather stems from density fluctuations (blob segmental motions) inside the dendrimer.

6.5
Rubbery Electrolytes

Polymeric electrolytes can possibly be used to build safe, non-toxic modern battery systems, e.g. Li-batteries. In this context the understanding of the ionic conduction mechanism of dissolved alkali salts is of major importance. Besides macroscopic measurements of transport coefficients, the investigation of mobilities on a molecular level is essential to identify the relevant conduction mechanisms.

Corresponding inelastic or quasi-elastic neutron scattering investigations on polymeric electrolytes have been focussed on polyethyleneoxide PEO and polypropyleneoxide PPO containinig Li-salts [310–315] or NaI [316–318]. These investigations cover the momentum transfer range 0.4 Å$^{-1}<Q<0.16$ Å^{-1} and do not extend into the SANS regime, i.e. they are most sensitive to local dynamics of segments. The first (interchain) peak of the structure factor $S(Q)$ for the pure (deuterated) systems are located at $Q=1.2$ Å^{-1} for PPO [319] and at

$Q=1.5$ Å^{-1} for PEO [320]. Here the relevant time scales range from picoseconds to a few nanoseconds, where an overlap between Fourier transformed spectra from backscattering or inverted TOF instruments and NSE data exists and is used to extend the time range. In all these investigations a significant (more than one order of magnitude) reduction of the segmental mobility upon salt addition has been observed. In addition the stretching parameter β, as obtained from stretched exponential fits to the intermediate scattering function, decreases.

The experiments on alkali iodides, PEO$_x$-NaI or PEO$_x$-LiI [316–318] were performed on PEO chains of 23 or 182 (-CH$_2$-CH$_2$-O-) monomers and O:ion ratios between 15 and 50. The incoherent scattering from protonated polymers was measured using IN11C, which yields the intermediate scattering function of the self-correlation. The experiments were performed in the homogeneous liquid phase where the added salt is completely dissolved and no crystalline aggregates coexist with the solution, i.e. at temperatures around 70 °C.

The results were compared to MD-simulations [317]. Whereas the scattering function of pure PEO could be well described, the dynamics of the salt-loaded samples deviates from the predictions obtained with various electrostatic interaction models. The best but still not perfect and – at least for longer times – unphysical model assumes Hookean springs between chains to simulate the Na-ion mediated transient cross-links [317].

Mao et al. [310] investigated PEO with LiClO$_4$ or Li-TFSI=Li-N(CF$_3$SO$_2$)$_2$ as salts using QENS. First data were obtained via inverse TOF(IRIS) spectroscopy with a resolution of $\Delta E=15$ µeV. However, to get a better data quality for the slow processes observed, additional NSE (IN11C) experiments up to $t=1.7$ ns ($\sim\Delta E=0.4$ µs) proved necessary both for PEO-LiClO$_4$ [313] and PEO-Li-TFSI [311]. Experiments were performed on deuterated PEO samples with molecular weights of $M_w=40$ kg/mol and 52 kg/mol, respectively, in the Q-range around the first structure factor peak at 1.5 Å^{-1} [320], starting from 0.5 Å^{-1} and extending to 1.6 Å^{-1}. Data treatment procedures were analogous to those used for the classification of dynamics in glass-forming polymers:

$$\frac{S_{\text{pair}}(Q,t)}{S_{\text{pair}}(Q,0)} = A(Q) \exp[-\{t/\tau(Q)\}^{\beta(Q)}] \qquad (6.38)$$

with $A(Q)$ an amplitude factor due to fast motions (vibrations and relaxations) that occur at times shorter than resolved in the experiment:

$$\tau(Q) = \tau_0 \left(\frac{Q_0}{Q}\right)^\nu \qquad (6.39)$$

for incoherent scattering and

$$\tau(Q) = \tau_0 \sqrt{S(Q)} \left(\frac{Q_0}{Q}\right)^\nu \qquad (6.40)$$

for coherent scattering.

Gaussian behaviour of the segmental displacement is indicated by $\beta\nu=2$. Analysis of the data obtained at $T=348$ K and 373 K by fitting the parameters $\beta(Q)$ and $\tau(Q)$ revealed a strong slowing down of the dynamics upon salt addition by about one order of magnitude for a ratio O:Li=7.5:1 and 2.5 decades for O:Li=3:1 in PEO. The value of β decreases from $\beta(Q=1.5\ \text{Å}^{-1})=0.73$ (0.426) to 0.324 (0.390) at 348 K (323 K), which was attributed to an additional degree of "randomness" due to the Li-mediated transient cross-linking. Also a "positive correlation" of the values of $\tau(Q)$ and $\beta(Q)$ with $S(Q)$ was found. To explain the observations, the data were complemented with medium resolution incoherent scattering results from hPEO [310] and compared to results from MD simulations [313] conducted to validate a quantum chemistry based potential Ansatz for the PEO interactions. The data trends of the NSE data were well reproduced by the MD calculations whereas the TOF data showed less agreement, possibly due to the difficulty in extracting unique values for $\tau(Q)$ and $\beta(Q)$. However, they reveal the presence of a faster process of rotational character. The product $\beta\nu$ was found to be 1.3 rather than 2 for the medium resolution incoherent scattering from hPEO. The slower relaxation has translational (diffusion) character and closely couples to Li-ion dynamics, as concluded from the combination of neutron experiments and MD calculations.

An extensive investigation on the system PPO-LiClO$_4$ is reported by Carlsson et al. [314]. Using deuterated PPO (dPPO) with a molecular weight M_n=2 kg/mol

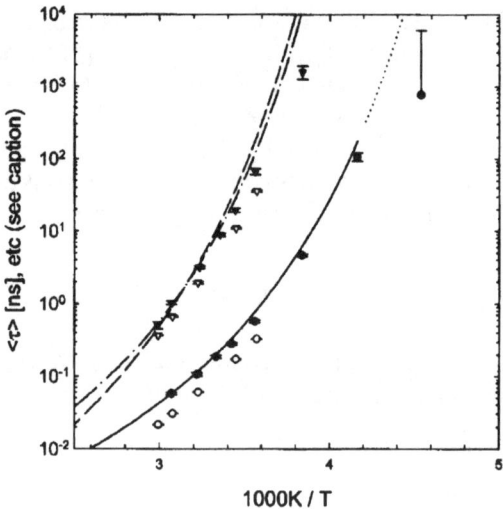

Fig. 6.21 Arrhenius plot of the relaxation time $\langle \tau(Q=14.5\ \text{nm}^{-1})\rangle$ for dPPO and dPPO-LiClO$_4$ (Li:O=1:16) (*upper curves*). *Solid symbols* were derived from NSE (IN11) results, *open symbols* from inverted TOF (IRIS). The *solid and dashed lines* represent the temperature dependence of the scaled viscosity for the two systems. The *dash-dotted line* is the scaled resistivity. (Reprinted with permission from [314]. Copyright 2002 American Institute of Physics)

the study was focussed on the structure factor peak at $Q=1.45$ Å^{-1} and on the temperature dependence of the relaxation functions and the effect of the Li-salt on it. It was expected that the results would reveal the influence of the ions on the segmental mobility and that the latter would correlate with the conductivity. The NSE and TOF data were analysed by fitting with stretched exponentials, $\exp[-(t/\tau_K)^\beta]$ yielding a stretching exponent $\beta=0.67$ (0.56) for pure PPO and $\beta=0.45$ (0.37) for PPO-LiClO$_4$. The possible explanation that heterogeneities cause the lower value of β was excluded by comparing the temperature dependence of β and the relaxation function shape itself to a heterogeneity model. The broadening of the relaxation time distribution needed to fit the $S_{\mathrm{pair}}(Q,t)$ data would be far too large if compared with expected and observed DSC widths of the glass transition. The average relaxation times $\langle\tau\rangle=\tau_K\beta^{-1}\Gamma(\beta^{-1})$ as a function of temperature are shown in Fig. 6.21. Except for the lowest temperature they exhibit virtually the same T-dependence of the bulk viscosity, and for the salt-containing system, as the resistivity. Deviations of $\langle\tau\rangle$ and β between NSE and TOF results are explained by the different importance of the admixture of incoherent scattering due to residual protons and the weaker incoherent scattering from deuterons.

In Fig. 6.22 the results of a viscosity scaling by $t \to t \times T/\eta\,(T)$ of the relaxation data are shown. Such a scaling is motivated by the Rouse model and should hold for the α-relaxation. The pure PPO data (right) behave according to this expectation; in contrast the PPO-LiClO$_4$ curves deviate considerably. This indicates that the coupling factor between microscopic friction and viscosity depends on temperature, possibly due to transient cross-linking via Li-ions.

Breaking and reforming of such cross-links relate to the "renewal time" τ_R of the DDH model [321] of the conductivity. Here the mean squared displacement without renewal events saturates after a short time to a value of $\langle r^2(\infty)\rangle$ until a restructuring establishes the start of a new diffusion step. Then $D(0)=\langle r^2(\infty)\rangle/(6\tau_R)$ and via the Nernst-Einstein relation:

$$\langle r^2(\infty)\rangle = \frac{6\tilde{\sigma}\tau_R k_B T}{n_{\mathrm{ions}}q_e^2} \quad (6.41)$$

where $\tilde{\sigma}$ is the conductivity, n_{ions} the carrier density and q_e their charge. Now τ_R is identified with $\langle\tau\rangle$, which yields reasonable $\langle r^2(\infty)\rangle^{1/2}$ values between 0.1 and 0.5 Å. The trend to increase more than $\langle r^2(\infty)\rangle \propto T$ is expected from harmonic vibrations with increasing T. The latter fact leaves the suspicion that the exposed DDH model still contains some oversimplification or wrong assumptions.

A number of investigations on polymeric electrolytes that are out of the scope of this review rely solely on the TOF/backscattering methods, e.g. [322, 323].

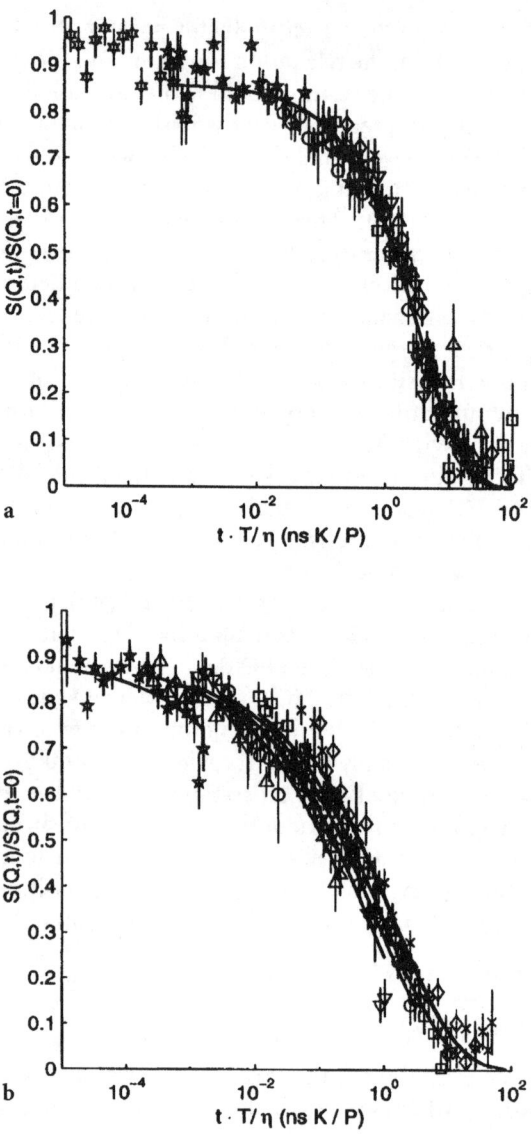

Fig. 6.22 Viscosity scale relaxation functions $S(Q,t\times T/\eta(T))/S(Q,0)$ obtained from NSE experiments at different temperatures between 220 K and 310 K for pure PPO (**a**) and between 280 K and 334 K for PPO-LiClO$_4$ (**b**). The scaling only holds for the pure PPO. (Reprinted with permission from [314]. Copyright 2002 American Institute of Physics)

6.6
Polymer Solutions

Dilute polymer solutions exhibit a universal regime, where Zimm dynamics [218, 220] are observed. Here the local friction of the Rouse model is supplemented by a dominating hydrodynamic coupling of polymer segments. In Chap. 5 we have already introduced the Zimm model. There we discussed deviations occurring at small length scales as a consequence of dissipation through rotational jumps. Here we will address results from the universal regime, where mode-dependent characteristic ratios and dissipation effects may be neglected. Starting from Eq. 5.17 and Eq. 5.20 the Fourier transformed Zimm equation reads:

$$\frac{\partial \tilde{x}(p,t)}{\partial t} = \sum_q H_{pq}(-k_q \tilde{x}(Q,t) + f_q) \tag{6.42}$$

With $k_q = q^2 6\pi^2 k_B T/(N\ell^2)$ for $q=0, 1, 2, \ldots N$ and

$$H_{pq} = \frac{1}{N^2} \int_0^N \int_0^N \cos\left(\frac{p\pi n}{N}\right) \cos\left(\frac{q\pi m}{N}\right) H_{nm} \, dn \, dm \tag{6.43}$$

which after some approximations [6] that leave only the diagonal terms leads to:

$$H_{pq} \cong \frac{\delta_{pq}}{2\eta_s b \sqrt{N 3\pi^3 p}} + \frac{\delta_{pq}}{2\zeta N} = \frac{\delta_{pq}}{2\zeta N}\left(B\sqrt{\frac{2N}{p}} + 1\right) \tag{6.44}$$

Thus, the remaining difference from the Rouse model is a mode-dependent friction coefficient $\zeta_p = (H_{pp})^{-1}$ for $(p>0)$, which leads to a relaxation mode spectrum with a different mode number p-dependence. The second term in H_{pq} is the bead friction with the surrounding medium (solvent), which is the only term present in the Rouse model. The ratio of the diagonal (Rouse-like) friction and the solvent-mediated interaction strength may be expressed by the draining parameter $B = (\zeta/\eta_s)/(6\pi^3 \ell^2)^{1/2}$. The Rouse model has $B=0$, whereas the assumption (segment=sphere) of $\ell=2a$ and $\zeta=6\pi\eta_s a$ leads to $B=0.69$. In the classic derivation of the Zimm scattering function [220] only the first hydrodynamic interaction term of H_{pq} is observed to yield results for large N and dominate the low modes (small p):

$$\tau_p = \frac{\zeta_p}{k_p} = \frac{\eta(N\ell^2)^{3/2}}{k_B T \sqrt{3\pi}} p^{-2/3} \tag{6.45}$$

The centre of mass diffusion $D = k_B T/\zeta_0$ follows from:

$$\zeta_0^{-1} = H_{00} = \frac{1}{N^2} \int_0^N \int_0^N H_{nm} \, dn \, dm = \frac{3}{8} \eta_s \ell \sqrt{6\pi^3 N} \tag{6.46}$$

As pointed out in [6] the extended coil conformation in good solvents leads to different exponents for $\zeta_p = \alpha_1 \eta_s \ell N^\nu p^{1-\nu}$, and $\tau_p = \alpha_2 p^{-3\nu} \eta (N^\nu \ell)^3 / k_B T$ and $D = \alpha_3 k_B T / (\eta^a N^\nu \ell)$ with the Flory exponent ν and the numerical prefactors α_i also dependent on the conformation.

Using the same Ansatz as for the Rouse model the scattering function $S(Q,t)$ for the Zimm model simply emerges by replacing the above expressions for D and τ_p (for θ-solvents) into the summation, analogous to (Eq. 3.19). In the limit

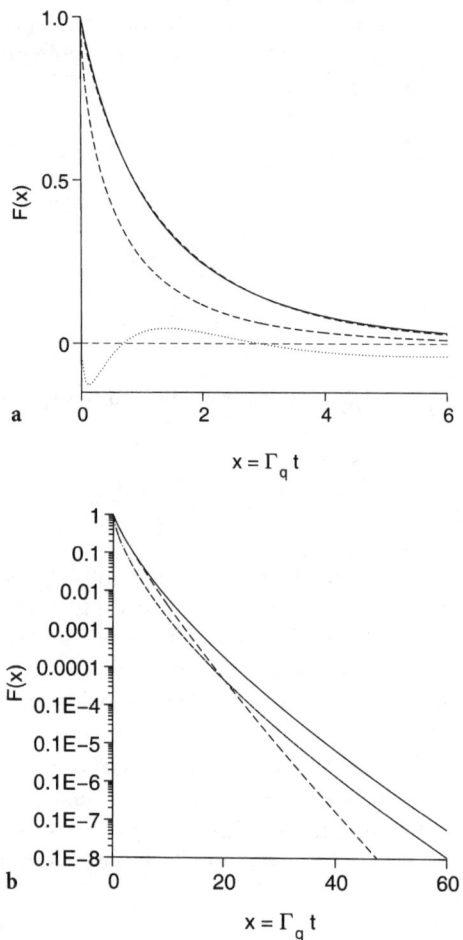

Fig. 6.23 a Comparison of the universal Zimm relaxation function $F(x)$ to stretched exponentials. The *dotted line* is the residual error ×10 to the best fit ($a=1.354$, $\beta=0.852$). The β value of 2/3 only applies at large values of x with $F(x)<10^{-3}$, which are irrelevant for NSE data. The *dashed blue line* in the right part of the figure corresponds to the asymptotic form given in [6]: $\exp(-1.35 x^{2/3})$. b Comparison of the integration result (*solid line*), the approximation Eq. 6.48 (*dashed line*) and the asymptotic form (*dashed-dotted line*). Only for very large values of $x>30$ does the asymptotic value of β emerge

of large QR_g the centre of mass diffusion can be neglected and the well-known scaling form for the Zimm dynamic structure factor emerges [6, 220]. At intermediate Q-values, with a typical Q-dependence of the relaxation rate, $\sim Q^3$ also expressed in the scaling property $S(Q,t)/S(Q)=F(k_B T/6\pi\, Q^3 t/\eta_s)$ with a universal function $F(x)$. As pointed out above, details depend on the chain conformation, i.e. the quality of the solvent. The observed relaxation rate is predominantly determined by the (effective) solvent viscosity η_s. At low Q ($QR_g<1$), the centre of mass diffusion (rate $\sim Q^2$) becomes the dominant process and limits the validity of the universal behaviour [5]. The dynamics is dominated by the universal behaviour in a large Q-range, particularly for long Gaussian chains:

$$F(x) = \int_0^\infty \exp\left[-u - x^{2/3}\frac{2}{\pi}\int_0^\infty \frac{\cos(yux^{-2/3})}{y^2}\{1 - \exp(-y^{2/3}/\sqrt{2})\}\,dy\right] du \quad (6.47)$$

which can be approximated over the range where most of the decay happens by a stretched exponential:

$$F(x) \approx \exp[-(x/a)^\beta] \quad (6.48)$$

with $a\approx1.354$ and $\beta\approx0.85$ as illustrated in Fig. 6.23. Note that these coefficients, which are valid in the practical relevant range of $x=0-6$, are different from the asymptotic form ($\beta=2/3$) given in [6]. The derivative $dF(x)/dx\approx\exp[-(x/b)^B]$ in the relevant regime of x is also well represented by a stretched exponential with $b\approx0.827$ and $B\approx0.75$.

6.6.1
Semidilute Solutions

In the semidilute regime ($c>c^*$) the scattering intensity reflects the segment concentration fluctuations on the screening length scale (or blob size) $\xi\approx R_E(c/c^*)^{-\nu/(3\nu-1)}$ with the overlap concentration $c^*\approx N/R_E^3$ [37] with the Flory exponent $\nu=1/2$ in θ-solvents and close to 5/3 in good solvents. For $Q<1/\xi$ collective concentration fluctuations with diffusive behaviour determine the scattering intensity while for large Q the Zimm dynamics of segments prevails. By employing the technique of zero average [324] the single chain structure factor can also be measured [325]. In that case a mixture (1:1) of h-polymer and d-polymer in a d/h-solvent mixture with a scattering length density equal to the average scattering length density of the polymer mixture is prepared, i.e. the contrast for the h-polymer is equal to $\Delta\rho_h=-\Delta\rho_d$ the negative value of the contrast of the d-polymer. In [325] it is shown that in the absence of isotope dependence of interactions (which in many cases is at least approximately given) the resulting scattering intensity is that of an ensemble of single chains. The difference between simple d-solvent vs h-polymer contrast and zero average contrast is illustrated in Fig. 6.24.

Figure 6.25 shows the dispersion of the effective diffusion Γ_Q/Q^2 with Γ_Q the first cumulant of the relaxation function. Three branches are visible. A: collec-

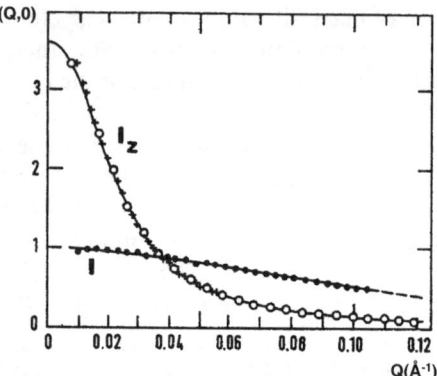

Fig. 6.24 Comparison of the scattering from a semidilute PDMS solution under normal polymer contrast (I) revealing the correlation length with the single chain scattering (I_z) as obtained by a zero average contrast preparation (see text). The line through I_z represents a Debye function with R_g=7 nm whereas the line through I corresponds to a Lorentzian (Ornstein-Zernike) with a correlation length ξ=1 nm. (Reprinted with permission from [325]. Copyright 1991 EDP Sciences)

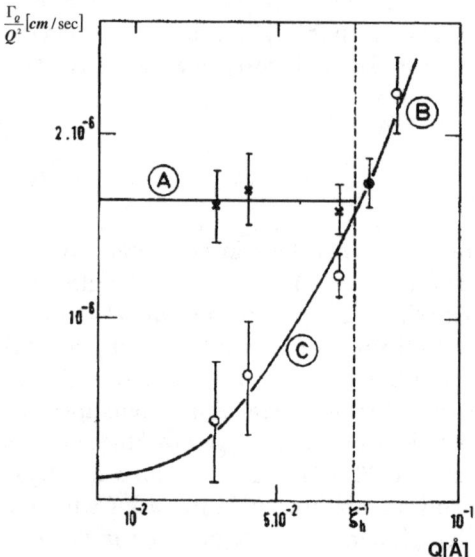

Fig. 6.25 Relaxation rates as function of Q in semidilute PDMS/toluene solutions. A collective concentration fluctuation seen in normal contrast, C single chain motion as seen in zero average contrast. B Zimm regime of local chain relaxations, equal in both contrasts. (Reprinted with permission from [325]. Copyright 1991 EDP Sciences)

6 The Dynamics of More Complex Polymer Systems

tive concentration fluctuations (normal contrast). B: local segment motions, Zimm dynamics, (both contrasts). C: single chain fluctuations and diffusion (zero average contrast). In regime $A(Q\xi \ll 1)$ the characteristic rate parameter is Eq. 5.113 in [6]:

$$\Gamma_Q^A = \frac{k_B T}{6\pi \eta_s \xi} Q^2 = D_c Q^2 \tag{6.49}$$

i.e. it depends on the solvent viscosity η_s and the correlation length ξ. On the other hand the same derivation for dense solutions yields for $Q\xi \gg 1$:

$$\Gamma_Q^B = \frac{k_B T}{16\eta_s} Q^3 \tag{6.50}$$

for the normal contrast, which may be compared to the single chain Zimm dynamics with the first cumulant:

$$\Gamma_q^Z = \frac{k_B T}{6\pi \eta} Q^3 \tag{6.51}$$

The observation that branches A and B in Fig. 6.25 merge at large Q is consistent with the predictions for Γ_Q^B and Γ_q^Z since 6π and 18.84 deviate from 16 by less than 15% and statistical errors of the experiment and systematic uncertainties in methods to extract the cumulant exceed this difference. In [325] for both the collective concentration fluctuations and the local Zimm modes the observed rates are too slow by a factor of 2 if compared to the predictions with η (the solvent viscosity) and ξ (the correlation length) as obtained from the SANS data. It is suggested that this discrepancy may be removed by the introduction of an effective viscosity η_s^{eff} that replaces the plain solvent viscosity η_s.

Finally at very low Q, i.e. $QR_g \ll 1$, branch C should level at the centre of mass respectively self-diffusion D_{self} of the polymer chains that in the semidilute and dense regime is strongly reduced by the factor $(c/c^*)^{7/4}$ (=6.6 here) due to interchain interaction, compared to its limiting value at infinite dilution [6] (5.144).

6.6.2
Theta Solution/Two Length Scales

In a θ-solvent the effective binary interactions between segments are zero. Thus, coil dimensions are independent of concentration and $\nu=1/2$ applies. In semidilute solutions there are two length scales, the molecular weight independent correlation length $\xi \approx R_E(c/c^*)$ and the tube diameter d measuring the distance needed to encounter a certain number of binary contacts [326]. These two lengths scale differently with concentration c in good and in θ-solvents: $d \sim \xi \sim c^{-3/4}$ and $d \sim c^{-2/3}$, $\xi \sim c^{-1}$, respectively. Adam et al. have investigated the differences in the dynamic properties of high molecular weight polymers between

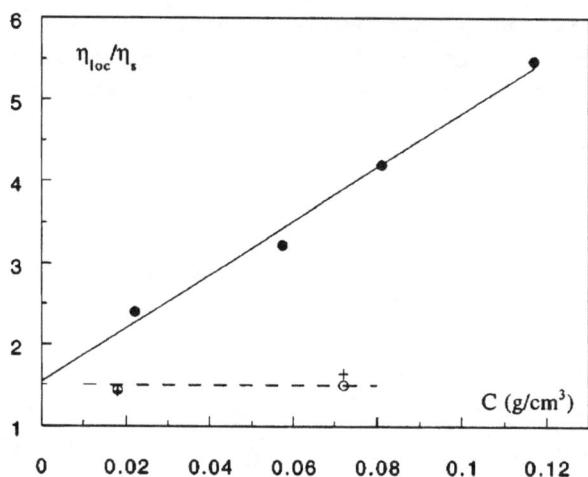

Fig. 6.26 Local viscosity as function of the polymer concentration C scaled to the solvent viscosity η_s for PS in the good solvent d-benzene at two temperatures (*open circles* T=30.6 °C, *plus* 65 °C). And in cyclohexane at the θ-temperature (*full circles*). (Reprinted with permission from [327]. Copyright 1996 The American Physical Society)

semidilute solutions at θ- and good solvent conditions [327, 328]. Peculiar effects were found in the NSE results from θ-solutions. In a good solution of polystyrene (PS) with a molecular weight of M=6,777 kg/mole in d-benzene the local viscosity, as deduced from the high Q-behaviour of Γ_Q^z is consistent with the solvent viscosity. On the other hand, solutions of the same polymer in cyclohexane at the θ-temperature yields a strong concentration dependence of the effective local viscosity, as illustrated by Fig. 6.26.

When the temperature is increased above the θ-temperature the enhanced local viscosity at higher concentration drops. The variation is the same as observed for the macroscopic viscosity [329]. In [328] polyisoprene (PI) and PS in different solvents have also been investigated and the authors observe that the slopes of the concentration dependence of the scaled local viscosities for PS and PI have a ratio of 1.7, which matches the value of the concentration ratio either on the Kuhn length (1.6) or the persistence length (1.7) for the two polymers.

The final conclusion from the observed concentration dependence is that under θ-conditions friction can occur between monomers, not only in the same blob but also between monomers belonging to any polymer in the solution. However, the underlying mechanism for this property has not yet been unravelled.

A rescaling of D(Q) with respect to a length scale $ls=(d_k\xi_\theta)^{1/2}$ the mean distance between two successive self-knots where d_k=20 Å is one adjustable parameter leads to the good superposition shown in Fig. 6.27.

An analogous PS solution (M_w=1,030 kg/mol) in cyclohexane at one concentration (0.052 g/cm³) has been investigated by Brulet et al. [330]. However, the NSE experiments were performed under zero average contrast, i.e. the

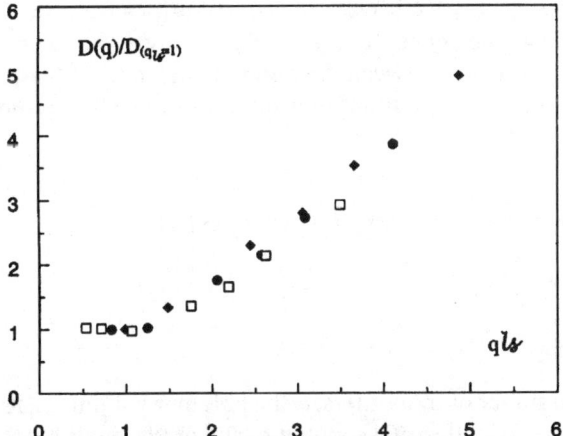

Fig. 6.27 Reduced diffusion coefficient versus reduced wave vector by scaling it to the length scale $ls=(a\xi_\theta)^{1/2}$. Data were obtained near the θ-temperature $T=39.1$ °C. Different symbols *diamonds*, *circles* and *squares* correspond to concentrations $c=0.057, 0.081$ and 0.177 g/cm^3, respectively. (Reprinted with permission from [327]. Copyright 1996 The American Physical Society)

single chain structure factor has been observed. As a consequence the data show a different behaviour, especially at low Q. In particular a huge effect on $D_{\text{eff}}(Q<1/ls; ls=24$ Å$)$ at the θ-temperature is observed. ls is close to the hydrodynamics screening length ξ_h [325], which separates the regimes related to non-draining ($Q>1/\xi_h$) from free draining at small Q. Here the effective diffusion is enhanced by a factor of ~5 compared to good solvent conditions obtained at $T=\theta+24$ °C. The value of ls deduced from the location of the knee or minimum in $D_{\text{eff}}(1/ls)$, which deviates by a factor of ~2 from the 46 Å that result from the parameters used in [327]. The enhanced relaxation at θ-condition is interpreted as a consequence of the "self-entanglement rigidity". In contrast to the results of Adam et al. an increase of the local viscosity at $Q>1/ls$ upon approaching the θ-temperature has not been detected.

6.6.3
Multicomponent Solutions

A variant of the zero average contrast method has been applied on a solution of a symmetric diblock copolymer of dPS and hPS in benzene [331]. The dynamic scattering of multicomponent solutions in the framework of the RPA approximation [324] yields the sum of two decay modes, which are represented by exponentials valid in the short time limit. For a symmetric diblock the results for the observable scattering intensity yields conditions for the cancellation of either of these modes. In particular the zero average contrast condition, i.e. a solvent scattering length density ρ_S that equals the average of both polymer blocks $(\rho_H+\rho_D)/2$, leaves only contributions from the "interdiffusion mode" Γ_I whereas the concentration mode Γ_C prevails as the only component when there is a contrast between the polymer and solvent scattering and both blocks are equal. The latter situation would result in a homopolymer solution of $2M_w$ in the present case:

$$\frac{S(Q,t)}{N\varphi} = \left[\frac{\rho_H - \rho_D}{2}\right]^2 (P_{1/2}(Q) - P_1(Q)) \exp(-\Gamma_I t)$$
$$+ \left[\frac{\rho_H - \rho_D}{2} - \rho_S\right]^2 \frac{P_1(Q)}{1 + v\varphi N P_1(Q)} \exp(-\Gamma_C t) \qquad (6.52)$$

N is the total number of monomers, φ the polymer volume fraction and P_1 and $P_{1/2}$ the form factors of the total copolymer and of the single blocks respectively. $V=v_D=v_H$ is the excluded volume interaction parameter which relates to the second virial coefficient $A_2=vN/(2Mc)$.

Under the simplifying assumption that hydrodynamic interactions may be neglected, the only new parameter that controls the dynamics is a monomeric friction coefficient ζ (Rouse model). Then the prediction for the rate Γ_I is given by:

$$\Gamma_I(Q) = \frac{k_B T}{N\zeta(c)} \frac{Q^2}{P_{1/2}(Q) - P_1(Q)} \qquad (6.53)$$

where any concentration dependence is lumped into the friction coefficient $\zeta(c)$ in terms of the generalized mobility $\mu=1/(N\zeta)$. In the Rouse model the mobility is independent of Q; however, since the scattering intensity allows measurement of $P_{1/2}(Q)-P_1(Q)$ and the NSE relaxation data $S(Q,t)/S(Q)$ yield $\Gamma_I(Q)$ the experimental results may be expressed in terms of $\mu(c,Q)$. Any Q-dependence then indicates deviations from the Rouse model. The increase of $\mu(c,Q)$ found for small Q is interpreted as an influence of the hydrodynamic interactions. Besides that dependence a strong reduction of the mobility with increasing concentration (above c^*) is observed while the Q-dependence of μ becomes weaker due to increased screening of the hydrodynamic interactions.

6.7
Biological Macromolecules

The application of neutron spin-echo spectroscopy to the analysis of the slow dynamics of biomolecules is still in its infancy, but developing fast. The few published investigations either pertain to the diffusion of globular proteins in solution [332–334] or focus on the internal subnanosecond dynamics on the length scale, <10 Å as measured on wet powders [335, 336]. The latter regime overlaps with other quasi-elastic neutron scattering methods as backscattering and TOF spectrometry [337–339].

Motivated by the fact that inside biological cells the protein concentration reaches volume fractions φ up to 0.3, the transport properties of myoglobin – a globular protein with $R_g \approx 16$ Å – under such "crowded" conditions has been investigated in [332] using NSE. The properties are derived from the effective diffusion as obtained from the NSE signal from salt-free solutions in D_2O (c=5.2–35 mM; φ=0.07–0.44) at 37 °C. At such concentrations the static structure factor $S(Q)$ obtained from SANS exhibits a correlation peak. Furthermore, the forward scattering is suppressed indicating a repulsive interaction. An analysis of $S(Q)$ using a rescaled [340] mean spherical approximation (MSA) [341] revealed an effective charge of $|Z_p|$=1.5–2 per protein. NSE experiments were performed covering a Fourier time range from 30 ps to 200 ns. The observed relaxations did not show stretching, i.e. were single exponential and thus:

$$S(Q,t)/S(Q) = \exp(-D_{\text{eff}}Q^2 t) \tag{6.54}$$

At low Q the experiments measure the collective diffusion coefficient D_c of concentration fluctuations. Due to the repulsive interaction the effective diffusion increases $\sim 1/S(Q)$. Well beyond the interaction peak at high Q, where $S(Q) \equiv 1$, the measured diffusion tends to become equal to the self-diffusion D_s. A hydrodynamics factor $H(Q)$ describes the additional effects on $D_{\text{eff}} = D_0 H(Q)/S(Q)$ due to hydrodynamics interactions (see e.g. [342]). Variations of $D(Q)S(Q)$ with Q (Fig. 6.28) may be attributed to the modulation with $H(Q)$ displaying a peak, where $S(Q)$ also has its maximum. For the transport in a crowded solution inside a cell the self-diffusion coefficient D_s is the relevant parameter. It is strongly (approximately exponentially) concentration dependent and decreases by more than one order of magnitude in the investigated concentration regime [343]; at the highest concentration it is more than two orders of magnitude below the Einstein-Stokes value for D_0.

Another investigation – motivated by the observation of an extra slow relaxation observed by DLS [344] – on interacting protein spheres in solution resulting in $D_{\text{eff}}(Q)$ obtained from NSE experiments is reported in [333]. Larger aggregates consisting of 24 units of the protein apoferritin, in the form of a spherical shell with an outer diameter of 120 Å in D_2O, were investigated. The electrostatic interaction was modified by adding various amounts of NaCl to the solutions with different protein volume fractions φ=0.05–0.2. At low salt concentration and higher volume fraction a strongly peaked structure factor is observed, which is explained in terms of paracrystalline ordering. For weak interaction (high salt, low φ) a weakly modulated $S(Q)$ close to the Perkus-Yevick prediction emerges. Even in the highly ordered state the aggregates are not confined as is seen in Fig. 6.29 from the nearly full decay of the relaxation.

Figure 6.30 displays the relaxation rate $\Gamma(Q)=D_{\text{eff}}(Q)Q^2$. The line corresponds to the Einstein-Stokes diffusion of a sphere with 69 Å radius (from SANS a radius of 60 Å was obtained). The dip at $Q \approx 0.035$ Å^{-1} and the enhancement at lower Q for the high concentrations with low salt corresponds to a modulation that follows the inverse of the (paracrystalline) structure factor. Unlike the case

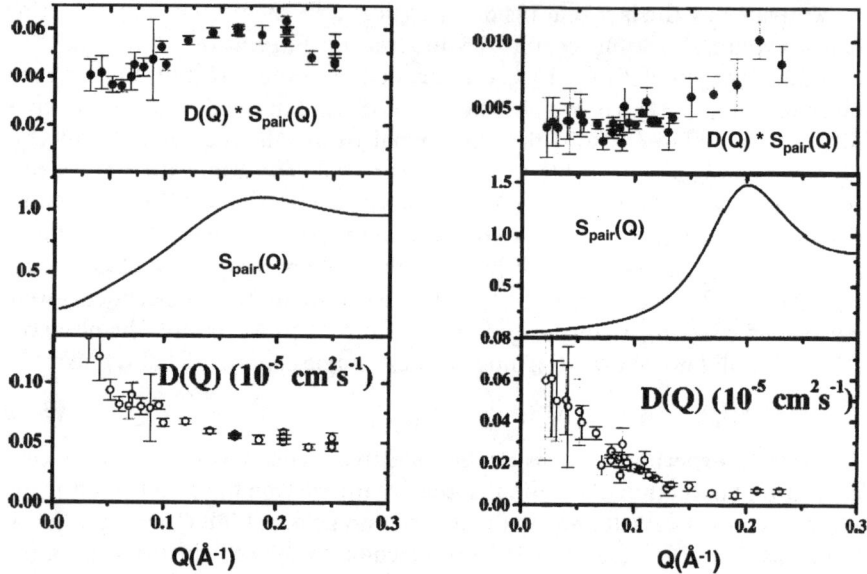

Fig. 6.28 From *bottom to top*: effective diffusion constant from NSE experiments, $S(Q)$ deduced from RMSA fits, and $D_0H(Q)=D(Q)S(Q)$ with $H(Q)$ the hydrodynamic factor. The *left* side corresponds to a myoglobin solution of concentration $c=14.7$ mM and the *right* side $c=30$ mM. Note the strong reduction of the value of D upon concentration increase. (Reprinted with permission from [332]. Copyright 2003 Elsevier)

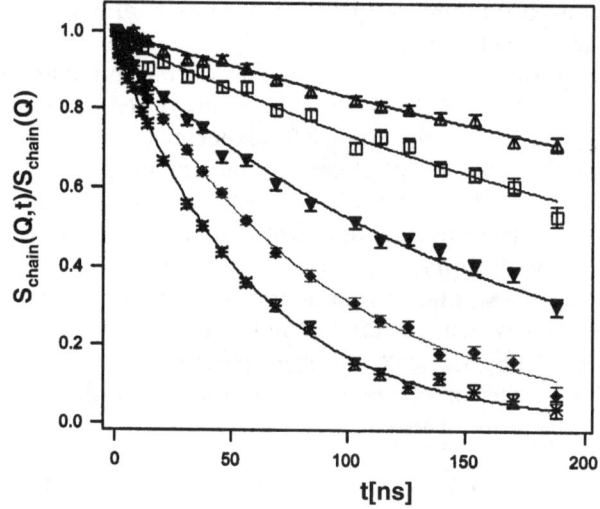

Fig. 6.29 $S(Q,t)/S(Q)$ measured at the IN15 on a 156 mg/mL apoferritin solution with 0.01 mM salt. Despite the paracrystalline order no permanent restriction of motion is present as indicated by the virtually full decay of the relaxation curve ($Q=0.09$ Å^{-1}). (Reprinted with permission from [333]. Copyright 2003 Elsevier)

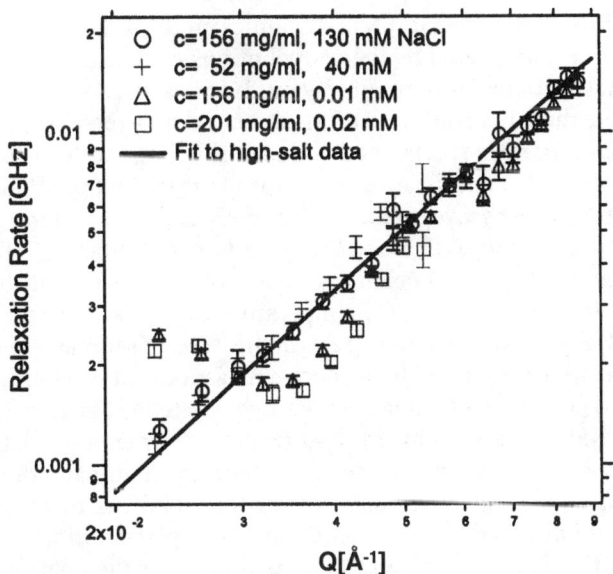

Fig. 6.30 Relaxation rate $\Gamma(Q)=D_{eff}(Q)Q^2$ from apoferritin solutions. The *straight line* corresponds to simple diffusion and is fitted to the high-Q asymptote of the data. (Reprinted with permission from [333]. Copyright 2003 Elsevier)

of crowded myoglobin solutions, obviously no concentration effect on the limiting value at high Q ($D_{eff} \rightarrow D_{self}$) is visible.

In neutron scattering work aiming at the internal protein dynamics, NSE experiments at low Q have been used in order to measure the centre of mass diffusion. This, quantity is needed as input for the interpretation of TOF spectra taken from the same solution. In [339] the internal dynamics of the photochemical reaction centre from the photosynthetic bacterium *Rhodobacter sphaeroides* has been investigated by QENS. In this experiment the protein was solubilized with surfactants. Both the solvent as well as the surfactant and all exchangeable hydrogens of the protein were deuterated such that the incoherent scattering mainly resulted from the photoactive part. At low Q the contrast between the protonated inner part of the protein and the surrounding creates coherent scattering. At $Q \leq 0.15$ Å^{-1} the centre-of-mass diffusion coefficient D_{cm} was measured by NSE. The value obtained was compatible with that obtained by DLS at a 100 times lower concentration. Difference TOF spectra of the solution and the solvent (buffer solution) were interpreted in terms of translational diffusion following the NSE-determined D_{cm} and an internal degree of freedom, which was modelled by a Lorentzian. The corresponding width extrapolates towards low Q to a relaxation $\tau(0)=7$ ps. This time stays constant up to $Q \cong 0.8$ Å^{-1} and then rises by 30% in the region $Q=0.8-1$ Å^{-1}. The proton motion behind this observation appears to be restricted to length scales of about 3 Å.

The interpretation of intraprotein dynamics on the basis of experiments in solution is always hampered by the contributions of centre of mass diffusion and rotational motions. A common way out is to investigate "wet" powders of proteins where there is already a water shell around the protein such that it may function, but the sample still is a soft solid without large scale diffusion of the proteins. A careful comparison between results from hydrated powders and solutions of the proteins myoglobin and lysozyme has been performed in [345]; however, only relying on TOF spectroscopy. A larger number of TOF and high resolution backscattering experiments – many of them analysing the incoherent elastic intensity in terms of a temperature-dependent mean square displacement (Debye-Waller-factor) – rely on this type of sample (see e.g. [346]).

A first attempt to extend such incoherent measurements on fully protonated hydrogenated powder of glutamate dehydrogenase to the enhanced resolution available by NSE is reported in [347]. At $Q=0.3$ Å^{-1}, where the relative contribution of incoherent scattering to the total scattering dominates the NSE spectrum, data up to $t=26$ ns have been collected. Beyond 1–2 ns the data reveal a plateau at about $S_{self}(Q,t>2 \text{ ns})/S_{self}(Q) \approx 0.6$. The plateau indicates a spatial limitation of the motional process. Its value is given by the Fourier transform of the non-decaying part of the self-correlation function and is also called the elastic incoherent structure factor (EISF). It measures the asymptotic ($t \rightarrow \infty$) distribution of a moving particle. This technique has also been applied to hydrated parvalbumin [338], where incoherent NSE data are compared to the EISF derived from medium resolution backscattering (BS). Data over a larger Q-range are compatible with proton diffusion on a sphere with an average radius 1.2–1.3 Å. At $t=150$ ps – corresponding to the 10 μeV BS resolution – both methods coincide. However, the NSE data that extend to 0.8 ns indicate a further drop of the EISF at longer times. This result shows that apparent motional restrictions depend on the time scale of observation.

A fully deuterated powder of the photosynthetic protein C-phycocyanin (CPC) has been investigated by NSE in a similar (Q,t)-range [335]. Here the coherent scattering conveying the pair correlation $S_{pair}(Q,t)/S_{pair}(Q)$ in terms of the intermediate scattering function was obtained and compared to molecular dynamics simulations. As illustrated in Fig. 6.31 the salient features of the NSE data were quantitatively and qualitatively reproduced by the MD results [348].

Comparisons as in Fig. 6.31 serve as tool to improve and validate MD simulation results and methods and will help to develop more efficient simulation methods. The interplay between validating experiments and successively improved simulations is a very promising approach for arriving at a very detailed picture of the internal motion within biopolymers.

Finally, Fig. 6.32 compares the internal dynamics of hydrated C-phycocyanin with that of the same system plus trehalose, a well known cryoprotecting disaccharide. Fig. 6.32 shows that the dynamics is slowed down by about 1.5 decades. The observation is interpreted as a slowing down of the protein dynamics due to the viscosity increase of the water shell by the added trehalose molecules. This finding is corrobated by the changes of the mean-squared-dis-

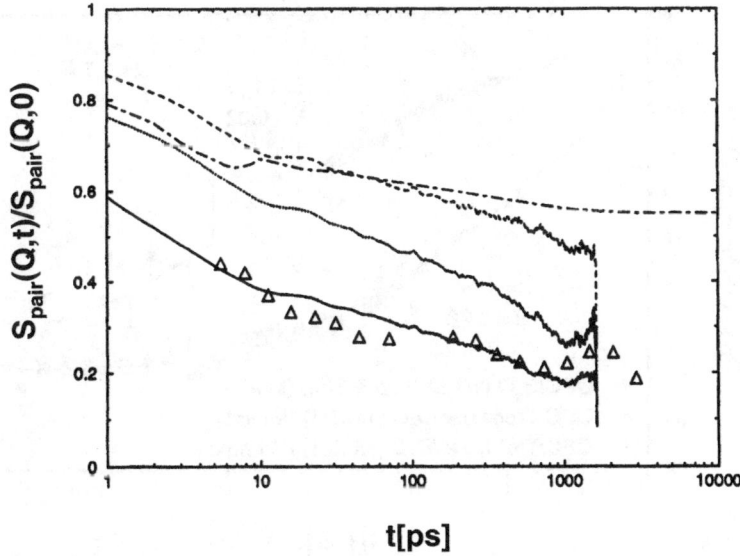

Fig. 6.31 Normalised intermediate scattering function from C-phycocyanin (CPC) obtained by spin-echo [335] compared to a full MD simulation (*solid line*) exhibiting a good quantitative matching. In contrast the MD results from simplified treatments as from protein without solvent (*long dash-short dash line*), with point-like residues (C_α-atoms) (*dashed line*) or coarse grained harmonic model (*dash-dotted line*) show similar slopes but deviate in particular in terms of the amplitude of initial decay. The latter deviation are (partly) explained by the employed technique of Fourier transformation. (Reprinted with permission from [348]. Copyright 2002 Elsevier)

placements $\langle u^2 \rangle$ derived from incoherent scattering experiments. They show a "dynamic transition" (i.e. a kink in the slope of $\langle u^2(T) \rangle$) occurring at 220 K in myoglobin/D_2O. This transition is not visible with added trehalose – possibly due to a slowing down of the dynamics (see e.g. [349]).

The observation of large scale motions of protein substructures, which are related to their function, by inelastic neutron spectroscopy (particularly NSE) has not yet been accomplished successfully and still waits for a working example. The associated motion must have a large amplitude such that it significantly changes the form factor of the protein. Effects of centre of mass diffusion and rotation must be separable or suppressed. Furthermore, most importantly, the large motions must occur as thermally exited fluctuations around equilibrium. Many of the large scale motions pertaining to function that are described in the literature (see e.g. [350]) occur as a response to an external stimulus like change of ion concentration, pH, etc. Such motions are only accessible by comparing the end states and eventually the kinetics between them with methods like SANS, SAXS or spectroscopic labels. NSE and QENS have to rely on genuine fluctuation dynamics. A close feedback to MD simulation seems to be the most promising way to achieve progress in this field.

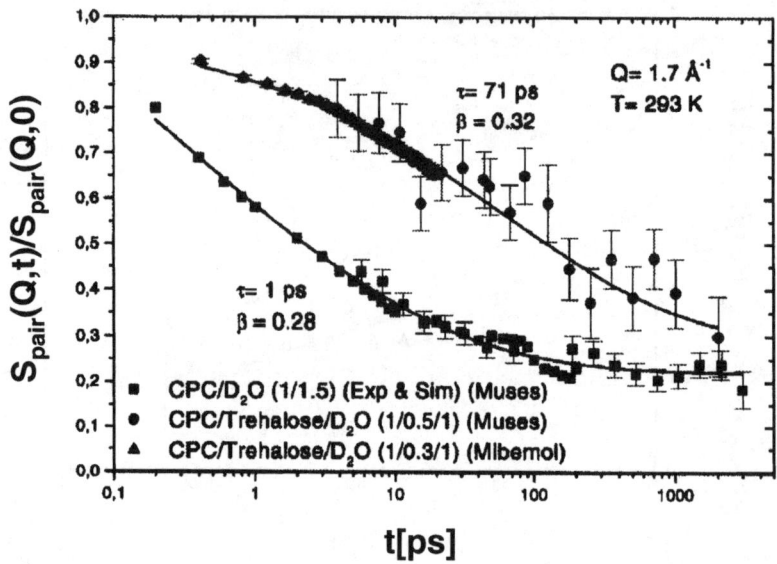

Fig. 6.32 NSE and Fourier transformed TOF (Mibemol) data. The *lower* curve corresponds to the data shown in Fig. 6.31, the *upper* curve stems from a sample with added trehalose (1 g protein/0.3–0.5 g trehalose/1 g D$_2$O). (Reprinted with permission from [336]. Copyright 2002 Springer, Berlin Heidelberg New York)

7
Conclusions and Outlook

In this review we have presented an overview on the state of the art of neutron spin echo spectroscopy in polymer dynamics. Up to now, the dynamics of homopolymers has been mainly addressed covering a very broad range of length and time scales. For glassy relaxation, both the secondary β-relaxation and the primary α-relaxation have been addressed. A key experiment on polybutadiene showed that the α-relaxation relates to the relative motion of adjacent chains, while the β-process is a more local process within a given chain. Measurements of the self-motion in the α-relaxation regime revealed highly Q-dispersive relaxation times, which are consistent with an essentially sublinear diffusion process underlying the α-relaxation. This sublinear α-regime is well separated from the universal regime of entropy-driven dynamics, which has been confirmed in great detail both from measurements of the single chain dynamic structure factor and from the self-motion. Addressing the leading process which limits the universal Rouse dynamics, NSE investigations clearly showed that for flexible polymers this leading mechanism is the dissipation process due to rotational jumps between different isomeric states.

At larger scales, confinement effects determine the dynamics of long-chain polymer melts. The experimental dynamic structure factors from such systems rule out all competing models to reptation, which so far have produced predictions for $S(Q,t)$. From these results it is clear that any more fundamental model of polymer dynamics in the melt must contain features of the tube confinement as phenomenologically introduced by the reptation concept. The NSE data have also quantitatively confirmed contour length fluctuations as the leading process limiting a tube confinement at early times. Though the picture of the dynamics of homopolymer melts now appears to be rather complete there are still a number of outstanding problems to be solved in the future:

i. At large scales the mechanism of constraint release, where the tube confinement is released by laterally moving chains, still needs to be addressed and understood microscopically.
ii. The mechanism of the glass transition is still not clear and further experiments, in particular in the merging regime between the α- and the β-relaxation, may lead to deeper insights.
iii. The detailed molecular nature of the secondary β-relaxation and its relation to the α-process needs further scrutiny.

In the future we will witness a drive towards more complexity. In this review, we have discussed a number of preliminary experiments pointing in this direction. In polymer blends, the question of dynamic mixing on a local scale was addressed and the Rouse dynamics in miscible polymer blends was studied. How the tube confinement evolves in blends where the two components have different tube diameters is a completely open question. Also, the question of how the monomeric friction coefficient in a blend arises from those of the two components has not yet been answered microscopically. Similarly, the dynamics of diblock copolymers has been barely touched. There exists one set of experiments addressing the dynamic RPA theory, which reveals major discrepancies. This includes some fast relaxation processes that appear to be related to the motion of the junction points within a diblock copolymer melt. In the microphase separated state they create the interface but, even in the weak segregation regime and above, such fast motional processes are present. Further systematic experiments to address more complex multicomponent mixtures would clearly be desirable.

Going to more complex systems such as aggregates, micelles, polymer brushes or polymers with architectures like stars, dendrimers, combs, etc. or gels, the scientific arena is wide open for new investigations revealing new phenomena and new insights. This is even more true for the dynamics of proteins and biomaterials in general, where at present basically only diffusion processes or very local dynamics have been addressed.

Composite soft matter systems combining polymers with amphiphiles and/or colloids and possibly biomaterials, where each of the components plays a specific role, will also be a challenge for the future. Such complexity will cover a wide range of length and time scales posing challenging problems to basic science. Desirable systems show complex interaction potentials with several minima, generating different structures according to mechanical and thermal history. The understanding of the interplay of geometry and topology and the characterization of interfacial features will be of utmost importance for future developments and the design of novel materials.

Neutron scattering in combination with advanced chemistry is the necessary tool for facing these new challenges. The focus is on linking chemical architecture to microscopic and macroscopic properties. The interplay between computer simulations and neutron scattering also promises to become particularly effective because of the common ability of neutron scattering and computer simulation to focus in on key structural units. Future trends will require a wide variety of experiments including investigations of dilute components or on very small amounts of matter, such as particular topological points or interfaces. All these experiments will require improved instrumentation, such as the upcoming neutron spin echo instrument at the American Spallation Source SNS, which will increase the dynamic range of the NSE technique to six orders of magnitude in time and move the resolution limit to the microsecond regime.

Acknowledgements We are indebted to our collaborators in Jülich and San Sebastián, who helped to shape many of the experiments as they are presented in this review. In particular we would like to thank Ms. Saida Oubenkhir for the technical support and patience in preparing this manuscript.

References

1. Flory PJ (1953) Principles of polymer chemistry. Cornell University Press, London
2. Kirste RG, Kruse WA, Schelten J (1973) Makromol Chem 162:299
3. de Gennes PG (1971) J Chem Phys 55:572
4. Schleger P, Farago B, Lartigue C, Kollmar A, Richter D (1998) Phys Rev Lett 81:124
5. Ewen B, Richter D (1997) Adv Polym Sci 134:1
6. Doi M, Edwards SF (1986) The theory of polymer dynamics. Clarendon, Oxford
7. Gotro JT, Graessley WW (1984) Macromolecules 17:2767
8. Zorn R, Richter D, Farago B, Frick B, Kremer F, Kirst U, Fetters LJ (1992) Physica B 180 & 181:534
9. Arbe A, Colmenero J, Alvarez F, Monkenbusch M, Richter D, Farago B, Frick B (2003) Phys Rev E 67:051802
10. Zorn R, Frick B, Fetters LJ (2002) J Chem Phys 116:845
11. Zorn R, Arbe A, Colmenero J, Frick B, Richter D, Buchenau U (1995) Phys Rev E 52:781
12. Fetters LJ, Lohse DJ, Colby RH (1996) In: Mark JE (ed) Physical properties of polymers handbook. AIP, New York
13. Marshall W, Lovesey SW (1971) Theory of thermal neutron scattering. Oxford University Press
14. Mezei F (1980) (ed) Neutron spin echo lecture notes in physics, vol 128. Springer, Berlin Heidelberg New York
15. Monkenbusch M (1990) Nucl Inst and Methods A 287:465
16. Williams WG (1988) Polarized neutrons. Claredon, Oxford
17. Golub R, Gähler R (1987) Phys Lett A 123:43
18. Monkenbusch M, Schätzler R, Richter D (1997) Nucl Inst Methods A 399:301
19. Köppe M, Bleuel M, Gähler R, Golub R, Hank P, Keller T, Longeville S, Rauch U, Wuttke J (1999) Physica B 266:75
20. Monkenbusch M (1999) Nucl Inst Methods A 437:455
21. Farago B (1999) Physica B 268:270
22. Schleger P, Alefeld B, Barthelemy JF, Ehlers G, Farago B, Giraud P, Hayes C, Kollmar A, Lartigue C, Mezei F, Richter D (1998) Physica B 241–243:164
23. Schleger P, Ehlers G, Kollmar A, Alefeld B, Barthelomy JF, Casatta H, Farago B, Giraud P, Hayes C, Lartigue C, Mezei F, Richter D (1999) Physica B 266:49
24. Rosov N, Rathgeber S, Monkenbusch M (1998) Abstracts of papers of the ACS 216 172-PMSE Part 2
25. Takeda T, Komura S, Seto H, Nagai M, Kobayashi H, Yokoi E, Zeyen CME, Ebisawa T, Tasaki S, Ito Yj, Takahashi S, Yoshizawa H (1995) Nucl Inst Methods A364:186
26. Ohl M, Monkenbusch M, Richter D (2003) Physica B 335:153
27. Longeville S (2000) J Physique IV 10:59
28. Stejskal EO, Tanner JE (1965) J Chem Phys 42:288
29. Fleischer G, Fujara F (1994) In: NMR basic principles and progress, vol 30. Springer, Berlin Heidelberg New York
30. Brown W (1993) Dynamic light scattering. Claredon, Oxford
31. Lumma D, Borthwick MA, Falus P, Lurio LB, Mochrie GSJ (2001) Phys Rev Lett 86:2042
32. Lumma D, Lurio LB, Mochrie SGJ, Sutton M (2000) Rev Sci Inst 71:3274
33. Sikharulidze I, Dolbnya IP, Fera A, Madsen A, Ostrovskii BI, de Jeu W (2002) Phys Rev Lett 88:115503
34. Ferry JD (1970) Viscoelastic properites of polymers. Wiley, New York
35. Bird RB, Armstrong RC, Hassager O (1977) Dynamics of polymeric liquids, vol 1. Wiley, New York

36. Bird RB, Hassager O, Armstrong RC, Curtis ChF (1977) Dynamics of polymeric liquids, vol 2. Wiley, New York
37. de Gennes PG (1979) Scaling concepts in polymer physics. Cornell University Press, Ithaca
38. Rouse PR (1953) J Chem Phys 21:1272
39. Richter D, Monkenbusch M, Willner L, Arbe A, Colmenero J, Farago B (2004) Europhys Lett 66:239–245
40. Monkenbusch M (2003) In: Mezei F, Pappas C, Gutberlet T (eds) Neutron spin echo spectroscopy. Basics, trends and applications, vol 601. Springer, Berlin Heidelberg New York
41. de Gennes PG (1967) Physics (USA) 3:37
42. Richter D, Ewen B, Farago B, Wagner T (1989) Phys Rev Lett 62:2140
43. Wischnewski A, Monkenbusch M, Willner L, Richter D, Farago B, Kali G (2003) Phys Rev Lett 90:058302
44. Montes H, Monkenbusch M, Willner L, Rathgeber S, Fetters LJ, Richter D (1999) J Chem Phys 110:10188
45. Haley JC, Lodge TP, He Y, Ediger MD, von Meerwall ED, Mijovic J (2003) Macromolecules 36:6142
46. Fetters LJ, Lohse DJ, Richter D, Witten TA, Zirkel A (1994) Macromolecules 27:4639
47. Hoffmann S, Willner L, Richter D, Arbe A, Colmenero, C, Farago B (2000) Phys Rev Lett 85:772
48. Smith GD, Paul W, Monkenbusch M, Willner L, Richter D, Qin XH, Ediger MD (1999) Macromolecules 32:8857
49. Richter D, Butera R, Fetters LJ, Huang JS, Farago B, Ewen B (1992) Macromolecules 25:6156
50. Richter D, Farago B, Butera R, Fetters LJ, Huang JS, Ewen B (1993) Macromolecules 26:795
51. Luettmer-Strathmann J (2001) Int J Thermophys 22:1507
52. Paul W, Smith GD, Yoon DY, Farago B, Rathgeber S, Zirkel A, Willner L, Richter D (1998) Phys Rev Lett 80:2346
53. Harmandaris VA, Mavrantzas VG, Theodorou DN, Kroger M, Ramirez J, Ottinger, HC, Vlassopoulos D (2003) Macromolecules 36:1376
54. Padding JT, Briels WJ (2002) J Chem Phys 117:925
55. Smith GD, Paul W, Monkenbusch M, Richter D (2000) Chem Phys 261:61
56. Guenza M (1999) J Chem Phys 110:7574
57. Edwards SF, Grant JMV (1973) J Phys A 6:1169
58. Doi M, Edwards SF (1978) J Chem Soc Farad Trans 274:1789; 274:1802; 275:38
59. Fatkullin N, Kimmich R (1995) Phys Rev E 59:3273
60. Richter D, Baumgärtner A, Binder K, Ewen B, Hayter JB (1981) Phys Rev Lett 47:109
61. de Gennes PG (1981) J Phys (Paris) 42:735
62. McLeish TCB (2002) Adv Phys 51:1
63. Ronca GJ (1983) J Chem Phys 79:79
64. Hess W (1988) Macromolecules 21:2620
65. Chaterjee A, Loring R (1994) J Chem Phys 101:1595
66. des Cloiseaux J (1990) Macromolecules 23:3992
67. Schweizer KS (1989) J Chem Phys 91:5802
68. Schweizer KS, Gregorz S (1995) J Chem Phys 103:1934
69. Richter D, Willner L, Zirkel A, Farago B, Fetters LJ, Huang JS (1993) Phys Rev Lett 71:4158
70. Richter D, Willner L, Zirkel A, Farago B, Fetters LJ, Huang JS (1994) Macromolecules 27:7437
71. Wischnewski A, Monkenbusch M, Willner L, Richter D, Likhtmann AE, McLeish TCB, Farago B (2002) Phys Rev Lett 88:058301

References

72. Richter D, Farago B, Fetters LJ, Huang JS, Ewen B, Lartigue C (1990) Phys Rev Lett 64:1389
73. Butera R, Fetters LJ, Huang JS, Richter D, Pyckhout-Hintzen W, Zirkel A, Farago B, Ewen B (1991) Phys Rev Lett 66:2088
74. Wischnewski A, Richter D (2000) Europhys Lett 52:719
75. Graft R, Heuer A, Spiess HW (1998) Phys Rev Lett 80:5738
76. Pyckhout-Hinzen W (2003) (unpublished results)
77. Carella JH, Graessley WW, Fetters LJ (1984) Macromolecules 17:2767
78. Kremer K, Grest GS (1990) J Chem Phys 92:5057
79. Pütz M, Kremer K, Grest GS (2000) Europhys Lett 49:735
80. Baschnagel J, Binder K, Doruker P, Guser AA, Hahn O, Kremer K, Mattice WL, Müller-Plathe F, Murat M, Paul W, Santos S, Suter UW, Tres V (2000) Adv Polym Sci 152:41
81. Krecr T, Baschnagel J, Müller M, Binder K (2001) Macromolecules 34:1105
82. Lodge T (1999) Phys Rev Lett 83:3218
83. Pearson DS, Fetters LJ, Graesseley WW, Strate GV, von Meerwall E (1994) Macromolecules 21:711
84. Bartels CR, Crist B, Graessly WW (1984) Macromolecules 17:2702
85. Crist B (1989) Macromolecules 22:2857
86. von Seggern J, Klotz S, Cantour HJ (1991) Macromolecules 24:3300
87. Tao H, Lodge TP, von Meerwall ED (2000) Macromolecules 33:1747
88. Colby RH, Fetters LJ, Graesseley WW (1987) Macromolecules 20:2226
89. Abdul Goad M, Pyckhout-Hintzen W, Kahle S, Allgeier J, Richter D, Fetters LJ (2004) Macromolecules 37(21):8135–8144
90. Likhtmann AE, McLeish TCB (2002) Macromolecules 35:6332
91. Clark N, McLeish TCB (1993) Macromolecules 26:5264
92. Millner ST, McLeish TCB (1998) Phys Rev Lett 81:725
93. Fetters LJ, Lohse DJ, Millner ST (1999) Macromolecules 32:6847
94. Harward RN (ed) (1973) The physics of glassy polymers. Applied Science, London
95. Götze W (1991) In: Hansen JP, Levesque D, Zinn-Justin J (eds) Liquids, freezing and the glass transition. North-Holland, Amsterdam
96. Götze W, Sjögren L (1992) Rep Prog Phys 55:241
97. Dianoux AJ, Petry W, Richter D (ed) (1993) Dynamics on disordered materials II. North-Holland, Amsterdam
98. Special issue (1995) Science vol 265
99. Ngai KL, Riande E, Ingram MD (eds) (1998) Proceedings of the third international discussion meeting on relaxations in complex systems. J Non-Cryst Solids vols 235–237
100. Giordano M, Leporini D, Tosi M (eds) (1999) Special issue: Second workshop on non-equilibrium phenomena in supercooled fluids, glasses and amorphous materials. J Phys: Condens Matter 11(10A)
101. Angell CA, Ngai KL, McKenna GB, McMillan PF, Martin SW (2000) J Appl Phys 88:3113
102. Ngai KL, Floudas G, Rizos AK, Riande E (eds) (2002) Proceedings of the fourth international discussion meeting on relaxations in complex systems. J Non-Cryst Solids vols 307–310
103. Andreozzi L, Giordano M, Leporini D, Tosi M (eds) (2003) Special issue: Third workshop on non-equilibrium phenomena in supercooled fluids, glasses and amorphous materials. J Phys: Condens Matter 15(11)
104. Anderson PW (1995) Science 267:1615
105. Colmenero J, Alvarez F, Arbe A (2002) Phys Rev E 65:041804
106. Cummins H Z (1999) J Phys: Condens Matter 11:A95
107. Goldstein M (1969) J Chem Phys 51:3728
108. Stillinger FH, Weber TA (1984) Science 228:983

109. Stillinger FH (1995) Science 267:1935
110. Sastry S, Debenedetti PG, Stillinger FH (1998) Nature 393:554
111. McCrum NG, Read BE, Williams G (1967) Anelastic, dielectric effects in polymeric solids. Wiley, London
112. Kohlrausch F (1863) Pogg Ann Phys 119:352
113. Williams G, Watts DC (1970) Trans Faraday Soc 66:80
114. Vogel H (1921) Phys Z 22:645
115. Fulcher GS (1925) J Am Chem Soc 8:339; 8:789
116. Johari GP, Goldstein M (1970) J Chem Phys 53:2372
117. Cole KS, Cole RH (1941) J Chem Phys 9:341
118. Wu L, Nagel SR (1992) Phys Rev B 46:11198
119. Frick B, Richter D (1995) Science 267:1931
120. Colmenero J, Mukhopadhyay R, Alegría A, Frick B (1998) Phys Rev Lett 80:2350
121. Chahid A, Alegría A, Colmenero J (1994) Macromolecules 27:3282
122. Alvarez F, Colmenero J, Zorn R, Willner L, Richter D (2003) Macromolecules 36:238
123. Frick B, D. Richter D, Ritter Cl (1989) Europhys Lett 9:557
124. Richter D, Frick B, Farago B (1988) Phys Rev Lett 61:2465
125. Richter D, Arbe A, Colmenero J, Monkenbusch M, Farago B, Faust R (1998) Macromolecules 31:1133
126. Arrighi V, Pappas C, Triolo A, Pouget S (2001) Physica B 301:157
127. Faivre A, Levelut C, Durand D, Longeville S, Ehlers G (2002) J Non-Cryst Solids 307–310:712
128. Arbe A, Moral A, Alegría A, Colmenero J, Pyckhout-Hintzen W, Richter D, Farago B, Frick B (2002) J Chem Phys 117:1336
129. Colmenero J, Arbe A, Richter D, Farago B, Monkenbusch M (2003) In: Mezei F, Pappas C, Gutberlet T (eds) Neutron spin echo spectroscopy. Basics, trends and applications, vol 601. Springer, Berlin Heidelberg New York
130. Berry GC, Fox TG (1968) Adv Polym Sci 5:261
131. Plazek DJ, Plazek DL (1983) Macromolecules 16:1469
132. Sanders JF, Valentine RH, Ferry JD (1968) J Polym Sci A2 6:967
133. Arbe A, Richter D, Colmenero J, Farago B (1996) Phys Rev E 54:3853
134. Tölle A, Wuttke J, Schober H, Randall O G, Fujara F (1998) Eur Phys J B 5:231
135. Törmälä PJ (1979) Macromol Sci-Rev Macromol Chem 17:297
136. Dejean de la Batie J, Lauprêtre F, Monnerie L (1989) Macromolecules 22:2617
137. Baillif PY (1998) Thesis, Le Mans
138. Moe NE, Qiu Xiao Hua, Ediger MD (2000) Macromolecules 33:2145
139. Schaefer D, Spiess HW, Suter UW, Fleming WW (1990) Macromolecules 23:3431
140. Santangelo P, Roland CM (unpublished data)
141. Pschorn U, Rössler E, Sillescu H, Kaufmann S, Schaefer D, Spiess HW (1991) Macromolecules 24:398
142. Fytas G, Ngai KL (1988) Macromolecules 21:804
143. Pearson DS, Fetters LJ, Younghouse CB, Mays JW (1988) Macromolecules 21:478
144. Lovesey SW (1984) Theory of neutron scattering from condensed matter. Clarendon, Oxford
145. Arbe A, Colmenero J, Richter D (2002) In: Kremer F, Schönhals A (eds) Polymer dynamics by dielectric spectroscopy and neutron scattering – a comparison in: broadband dielectric spectroscopy
146. Arbe A, Colmenero J, Monkenbusch M, Richter D (1998) Phys Rev Lett 81:590
147. Farago B, Arbe A, Colmenero J, Faust R, Buchenau U, Richter D (2002) Phys Rev E 65:051803

148. Colmenero J, Alegría A, Arbe A, Frick B (1992) Phys Rev Lett 69:478
149. Colmenero J, Arbe A, Alvarez F, Monkenbusch M, Richter D, Farago B, Frick B (2003) J Physics: Condens Matter 15:S1127
150. Springer T (1972) Neutron scattering for the investigation of diffusive motions in solids, liquids. Springer tracts in modern physics, vol 64. Springer, Berlin Heidelberg New York
151. Zorn R (1997) Phys Rev B 55:6249
152. Colmenero J, Arbe A, Alegría A, Ngai KL (1994) J Non-Cryst Solids 172–174:229
153. Colmenero J, Arbe A, Alegría A, Monkenbusch M, Richter D (1999) J Phys: Condens Matter 11:A363
154. Arbe A, Colmenero J, Alvarez J, Monkenbusch M, Richter D, Farago B, Frick B (2002) Phys Rev Lett 89:245701
155. Narros A, Alvarez F, Arbe A, Colmenero J, Richter D, Farago B (2004) Physica B 350: 1091; J Chem Phys 121:7
156. Squires GL (1996) Introducction to the theory of thermal neutron scattering. Dover, New York
157. Sillescu H (1999) J Non-Cryst Solids 243:81
158. Alvarez F, Alegría A, Colmenero J (1991) Phys Rev B 44:7306
159. Adam G, Gibbs JH (1965) J Chem Phys 43:139
160. Cohen MH, Grest GS (1981) Phys Rev B 24:4091
161. Alegría A, Barandiarán JM, Colmenero J (1984) Phys Stat Sol (b) 125:409
162. Fischer EW, Donth E, Steffen W (1992) Phys Rev Lett 68:2344
163. Schmidt-Rohr K, Spiess HW (1991) Phys Rev Lett 66:3020
164. Cicerone MT, Blackburn FR, Ediger MD (1995) J Chem Phys 102:471
165. Williams G, Fournier J (1996) J Chem Phys 104:5690
166. Kob W, Donati C, Plimpton SJ, Poole PH, Glotzer SC (1997) Phys Rev Lett 79:2827
167. Doliwa B, Heuer A (1998) Phys Rev Lett 80:4915
168. Hall DB, Dhinojwala A, Torkelson JM (1997) Phys Rev Lett 79:107
169. Cicerone MT, Ediger MD (1996) J Chem Phys 104:7210
170. Havlin S, Ben-Avraham D (1987) Adv Phys 36:695
171. Rahman A, Singwi KS, Sjölander A (1962) Phys Rev 126:986
172. Sciortino F, Gallo P, Tartaglia P, Chen S-H (1996) Phys Rev E 54:6331
173. Caprion D, Matsui J, Schober HR (2000) Phys Rev B 62:3709
174. Mossa S, Di Leonardo R, Ruocco G, Sampoli M (2000) Phys Rev E 62:612
175. Cendoya I, Alegría A, Alberdi JM, Colmenero J, Grimm H, Richter D, Frick B (1999) Macromolecules 32:4065
176. Gilbert M (1994) J M S-Rev Macromol Chem Phys C 34:77
177. Butters G (ed) Particulate nature of PVC (1982) Applied Science, London
178. Wenig W (1978) J Polym Sci, Polym Phys Edn 16:1635
179. Blundell DJ (1979) Polymer 20:934
180. Walsh DJ, Higgins JS, Druke CP, McKeown JS (1981) Polymer 22:168
181. Ballard DGH, Burgess AN, Dekoninck JM, Roberts EA (1987) Polymer 28:3
182. Scherrenberg RL, Reynaers H, Gondard C, Steeman PAM (1994) J Polym Sci: Part B: Polym Phys 32:111
183. Juinj JA, Gisolf GH, de Jong WA (1973) Kolloid Z Z Polym 251:465
184. Hobson R, Windle AH (1993) Polymer 34:3582
185. Fuoss RM, Kirkwood JG (1941) J Am Chem Soc 63:385
186. Arbe A, Monkenbusch M, Stellbrink J, Richter D, Farago B, Almdahl K, Faust R (2001) Macromolecules 34:1281
187. Arbe A, Colmenero J, Farago B, Monkenbusch M, Buchenau U, Richter D (2003) Chem Phys 292:295

188. Richter D, Zorn R, Farago B, Frick B, Fetters LJ (1992) Phys Rev Lett 68:71
189. Arbe A, Buchenau U, Willner L, Richter D, Farago B, Colmenero J (1996) Phys Rev Lett 76:1872
190. Hofmann A, Alegría A, Colmenero J, Willner L, Buscaglia E, Hadjichristidis N (1996) Macromolecules 29:129
191. Ding Y, Novikov VN, Sokolov AP (2004) J Polym Sci Part B: Polym Phys 42:994
192. Kim EG, Mattice WL (2002) J Chem Phys 117:2389
193. Gee RH, Boyd RH (1994) J Chem Phys 101:8028
194. Vineyard G (1959) Phys Rev 110:999
195. Arbe A, Colmenero J, Frick B, Monkenbusch M, Richter D (1998) Macromolecules 31:4926
196. Slichter WP (1966) J Polym Sci: Part C 14:33
197. Suter UW, Saiz E, Flory PJ (1983) Macromolecules 16:1317
198. Vacatello M, Yoon DY (1992) Macromolecules 25:2502
199. Patterson GD (1977) J Polym Sci: Polym Phys Ed 15:455
200. Williams G (1979) Adv Polymer Sci 33:60
201. Alvarez F, Hofmann A, Alegría A, Colmenero J (1996) J Chem Phys 105:432
202. Bergman R, Alvarez F, Alegría A, Colmenero J (1998) J Chem Phys 109:7546
203. Gómez D, Alegría A, Arbe A, Colmenero J (2001) Macromolecules 287:503
204. Frick B, Farago B, Richter D (1990) Phys Rev Lett 64:2921
205. Gruver JT, Kraus G (1964) J Polym Sci A 2:797
206. Götze W (1999) J Phys: Condens Matter 11:A1
207. Binder K, Baschnagel J, Paul W (2003) Prog Polym Sci 28:115
208. Chong SH, Fuchs M (2002) Phys Rev Lett 88:185702
209. Van Zon A, de Leeuw SW (1999) Phys Rev E 60:6942
210. Russina M, Mezei F, Lechner R, Longeville S, Urban B (2000) Phys Rev Lett 84:3630
211. Flory PJ (1969) Statistical mechanics of chain molecules. Interscience, New York
212. Brückner S (1981) Macromolecules 14:449
213. Allegra G, Ganazzoli F (1981) Macromolecules 14:1110
214. Allegra G, Ganazzoli F (1989) Adv Chem Phys 75:265
215. Harnau L, Winkler RG, Reinecker P (1995) J Chem Phys 102:7750; (1996) J Chem Phys 104:6355
216. Harnau L, Winkler RG, Reinecker P (1997) J Chem Phys 106:2469
217. Richter D, Monkenbusch M, Allgaier J, Arbe A, Colmenero J, Farago B, Cheol Bae Y, Faust R (1999) J Chem Phys 111:6107
218. Zimm BH (1956), J Chem Phys 24:269
219. Richter D, Monkenbusch M, Pyckhout-Hintzen W, Arbe A, Colmenero J (2000) J Chem Phys 113:11398
220. Dubois Violette E, de Gennes PG (1967) Physics (USA) 3:181
221. Richter D, Hayter JB, Mezei F, Ewen B (1978) Phys Rev Lett 41:1484
222. Richter D, Binder K, Ewen B, Stühn B (1984) J Phys Chem 88:661
223. Ewen B, Richter D (1987) Festkörperchemie 27:1
224. Jones AA, Lubianez RP, Hanson MA, Shostak SL (1978) J Polym Sci Polym Phys Ed 16:1685
225. Wang CH, Fischer EW (1985) J Chem Phys 82:632
226. Marvin RS, Aldrich R, Sack HS (1954) J Appl Phys 25:1213
227. Tobolsky AV, Catsiff E (1956) J Polym Sci 19:111
228. Ohl M et al. (unpublished)
229. Arbe A et al. (unpublished)
230. Kob W, Andersen HC (1995) Phys Rev E 51:4626

References

231. Allen G, Higgins JS, Macounachie A, Ghosh R E (1982) Chem Soc Faraday Trans II 78:2117
232. Arbe A (2004) Physica B 350:178
233. de Gennes PG (1959) Physica (Amsterdam) 25:825
234. Sköld K (1967) Phys Rev Lett 19:1023
235. Colmenero et al (unpublished)
236. Karatasos K, Ryckaert J P (2001) Macromolecules 34:7232
237. Chung GC, Kornfield JA, Smith SD (1994) Macromolecules 27:5729
238. Min B, Qiu X, Ediger MD, Pitsikalis M, Hadjichristidis N (2001) Macromolecules 34:4466
239. Arendt BH, Krishnamoori R, Kornfield JA, Smith SD (1997) Macromolecules 30:1127
240. Arendt BH, Kannan RM, Zewail M, Kornfield JA, Smith SD (1994) Rheol Acta 33:322
241. Adams S, Adolf DB (1999) Macromolecules 32:3136
242. Yang X, Halasa A, Hsu WL, Wang SQ (2001) Macromolecules 34:8532
243. Alegría A, Elizetxea C, Cendoya I, Colmenero J (1995) Macromolecules 28:8819
244. Zhang S, Painter PC, Runt J (2002) Macromolecules 35:9403
245. Colby RH (1989) Polymer 30:1275
246. Zetsche A, Fischer EW (1994) Acta Polym 45:168
247. Kumar SK, Colby RH, Anastasiadis SH, Fytas G (1996) J Chem Phys 105:3777
248. Lodge TP, McLeish TCB (2000) Macromolecules 33:5278
249. Leroy E, Alegría A, Colmenero J (2002) Macromolecules 35:5587
250. Lorthioir C, Alegría A, Colmenero J (2003) Phys Rev E 68:031805
251. Leroy E, Alegría A, Colmenero J (2003) Macromolecules 36:7280
252. Mukhopadahyay R, Alegría A, Colmenero J, Frick B (1998) J Non-Cryst Solids 235–237:233
253. Arbe A, Alegría A, Colmenero J, Hoffmann S, Willner L, Richter D (1999) Macromolecules 32:7572
254. Doxastakis M, Kitsiou M, Fytas G, Theodorou DN, Hadjichristidis N, Meier B, Frick B (2000) J Chem Phys 112:8687
255. Götz H, Ewen B, Maschke U, Monkenbusch M, Meier G (2003) e-Polymers 011
256. Vilgis TA (2000) Phys Reports 336:167
257. Muratov CB (1997) Phys Rev Lett 78:3149
258. Koppi KA, Tirrel M, Bates FS (1993) Phys Rev Lett 70:1449
259. Sikka M, Singh N, Karim A, Bates FS, Satija SK, Majkrzak C (1993) Phys Rev Lett 70:307
260. Schwahn D, Frielinghaus H, Mortensen K, Almdal K (1996) Phys Rev Lett 77:3153
261. Leibler L (1980) Macromolecules 13:1602
262. Laradji M, Shi AC, Desai R, Noolandi J (1997) Macromolecules 30:3242
263. Mayes AM, Olvera de la Cruz M (1991) J Chem Phys 95:4670
264. Barrat JL, Frederickson GH (1991) J Chem Phys 95:1281
265. Frederickson GH, Helfand E (1987) J Chem Phys 87:697
266. Gwenza M, Schweizer KS (1998) J Chem Phys 108:1257; 108:1271
267. Genz U, Vilgis, TA (1994) J Chem Phys 101:7111
268. Hoffmann A, Sommer JU, Blumen A (1997) J Chem Phys 106:6709; J Phys A 30:5007
269. Pakula T (1997) Macromolecules 30:8463
270. Akcasu AZ (1993) In: Brown W (ed) Dynamic light scattering, the method and some applications. Clarendon, Oxford
271. Stühn B, Renni AR (1989) Macromolecules 22:2460
272. Tombakoglu M, Akcasu AZ (1992) Polymer 33:1127
273. Borsali R, Duval M, Benmouna, M (1989) Polymer 30:610
274. Borsali R, Benoit H, Legrand JF, Duval M, Picot C, Benmouna M, Farago B (1989) Macromolecules 22:4119

275. Borsali R, Fischer EW, Benmouna M (1991) Phys Rev A 43:5732
276. Boudenne N, Anastasidadis SH, Fytas G, Xenidou M, Hadjichristidis N, Semenov AN, Fleischner G (1996) Phys Rev Lett 77:506
277. Anastasiadis SH, Fytas G, Vogt S, Fischer EW (1993) Phys Rev Lett 70:2415
278. Semenov AN, Anastasiadis SH, Boudenne N, Fytas G, Xenidou M, Hadjichristidis N (1997) Macromolecules 30:6280
279. Hoffmann A, Koch T, Stühn B (1993) Macromolecules 26:7288
280. Lodge TP, Dalvi MC (1995) Phys Rev Lett 75:657
281. Dalvi M, Eastman CE, Lodge TP (1993) Phys Rev Lett 71:2591
282. Zwanzig R (1961) In: Brittin WE, Downs WD, Downs J (eds) Lectures in theoretical physics, vol 3. Wiley, New York, p 106
283. Mori H (1965) Progr Theo Physics 33:423
284. Montes H, Monkenbusch M, Willner L, Rathgeber S, Richter D, Fetters LJ, Farago B (2002) Europhys Lett 58:389
285. Matsen MW, Bates FS (1996) Macromoleules 24:1092
286. Helfand E, Wassermann ZR (1976) Macromolecules 9:879; (1982) In: Goodgman I (ed) Development in block copolymers I. Applied Science, New York
287. Meier G, Fytas G, Momper B, Fleischer G (1993) Macromolecules 26:5310
288. Huang JS, Webb WW (1969) Phys Rev Lett 23:160
289. Semenov AN (1994) Macromolecules 27:2732
290. Hecht AM, Guillermo A, Horkay F, Mallam S, Legrand JF, Geissler E (1992) Macromolecules 25:3677
291. Hecht AM, Horkay F, Schleger P, Geissler E (2002) Macromolecules 35:8552
292. Koizumi S, Monkenbusch M, Richter D, Schwahn D, Annaka M (2001) J Phys Soc Jpn Suppl A70:320
293. Koizumi S, Monkenbusch M, Richter D, Schwahn D, Farago B, Annaka M (2002) Appl Phys A 74 (Suppl):S399
294. Hellweg T, Kratz K, Pouget S, Eimer W (2002) Colloids Surf A: Physicochem Eng Asp 202:223
295. Geissler E, Hecht AM, Horkay F, Legrand JF (1992) Polymer 33:3083
296. Matsuoka H, Yamamoto Y, Nakano M, Endo H, Yamaoka H, Zorn R, Monkenbusch M, Richter D, Seto H, Kawabata Y, Nagao M (2000) Langmuir 16:9177
297. Castelletto V, Hamley IW, Yang Z, Haeussler W (2003) J Chem Phys 119:8158
298. Farago B, Monkenbusch M, Richter D, Huang JS, Fetters LJ, Gast AP (1993) Phys Rev Lett 71:1015
299. Kanaya T, Watanabe H, Matsushita Y, Takeda T, Seto H, Nagao M, Fujii Y, Kaji K (1999) J Phys Chem Solids 60:1367
300. de Gennes PG (1986) C R Acad Sci Paris Ser II 302:765
301. Monkenbusch M, Schneiders D, Richter D, Farago B, Fetters LJ, Huang J (1995) Physica B 213:707
302. Richter D, Stühn B, Ewen B, Nerger D (1987) Phys Rev Lett 58:2462
303. Richter D, Farago B, Huang JS, Fetters LJ, Ewen B (1989) Macromolecules 22:468
304. Stellbrink J, Allgaier J, Monkenbusch M, Richter D, Ehlers G, Schleger P (2002) Appl Phys A-Mater Sci Process 74:S361
305. Chen ZY, Cai C (1999) Macromolecules 32:5423
306. Rathgeber S, Monkenbusch M, Kreitschmann M, Urban V, Brulet A (2002) J Chem Phys 117:4047
307. Funayama K, Imae T, Seto H, Aoi K, Tsutsumiuchi K, Okada M, Nagao M, Furusaka M (2003) J Phys Chem B 107:1353
308. Stark B, Lach C, Farago B, Frey H, Schlenk C, Stühn B (2003) Colloid Polym Sci 281:593

309. Huang JS, Milner ST, Farago B, Richter D (1987) Phys Rev Lett 59:2600
310. Mao G, Fernandez Perera R, Howells WS, Price DL, Saboungi ML (2000) Nature 405:163
311. Mao G, Saboungi ML, Price DL, Armand MB, Mezei F, Pouget S (2000) Macromolecules 35:415
313. Saboungi ML, Price DL, Mao G, Fernandez-Perea R, Borodin O, Smith GD, Armand M, Howells WS (2002) Solid State Ionics 147:225
314. Carlsson P, Zorn R, Andersson D, Farago B, Howells WS, Borjesson L (2001) J Chem Phys 114:9645
315. Carlsson P, Swenson J, Borjesson L, McGreevy RL, Jacobsson P, Torell LM, Howells WS (2000) Electrochim Acta 45:1449
316. Mos B, Verkerk P, Pouget S, van Zon A, Bel JG, de Leeuw SW, Eisenbach CD (2000) J Chem Phys 113:4
317. de Jonge JJ, van Zon A, de Leeuw SW (2002) Solid State Ionics 147:349
318. Mos B, Verkerk P, van Zon A, de Leeuw SW (2000) Physica B 276:351
319. Carlsson P, Mattsson B, Swenson J, Torell LM, Kall M, Borjesson L, McGreevy RL, Mortensen K, Gabrys B (1998) Solid State Ionics Special Issue Dec 115:139
320. Mao G, Saboungi ML, Price DL, Badyal YS, Fischer HE (2001) Europhys Lett 54:347
321. Lonergan MC, Nitzan A, Ratner MA, Shriver DF (1995) J Chem Phys 103:3253
322. Andersson D, Svanberg C, Swenson J, Howells WS, Borjesson L (2001) Physica B 301:44
323. Karlsson C, Best AS, Swenson J, Howells WS, Borjesson L (2003) J Chem. Phys 118:4206
324. Benmouna M, Benoit H, Duval M, Akcasu Z (1987) Macromolecules 20:1107
325. Csiba T, Jannink G, Durand D, Papoular R, Lapp A, Auvray L, Boue F, Cotto JP, Borsali R (1991) J Physique II 1:381–396
326. Colby RH, Rubinstein M (1990) Macromolecules 23:2753
327. Adam M, Lairez D, Raspaud E, Farago B (1996) Phys Rev Lett 77:3673
328. Adam M, Farago B, Schleger P, Raspaud E, Lairez D (1998) Macromolecules 31:9213
329. Adam M, Delsanti M (1985) Macromolecules 18:1760
330. Brulet A, Cotton JP, Lapp A, Jannink G (1996) J Phys II France 6:331
331. Duval M, Picot C, Benoit H, Borsali R, Benmouna M, Lartigue C (1991) Macromolecules 24:3185
332. Longeville S, Doster W, Kali G (2003) Chem Phys 292:413
333. Haeussler W (2003) Chem Phys 292:425
334. Hirai M, Kawai-Hirai M, Iwase H, Hayakawa T, Kawabata T, Takeda T (2002) Appl Phys A (Suppl) 74:S1254
335. Dellerue S, Petrescu A, Smith JC, Longeville S, Bellissent-Funel MC (2000) Physica B 276–278:514
336. Köper I, Bellissent-Funel MC (2002) Appl Phys A (Suppl) 74:S1257
337. Tarek M, Neumann DA, Tobias DJ (2003) Chem Phys 292:435
338. Zanotti J-M, Parello J, Bellissent-Funel MC (2002) Appl Phys A (Suppl) 74:S1277
339. Gall A, Seguin J, Robert B, Bellissent-Funel MC (2002) J Phys Chem B 106:6303
340. Belloni L (1986) J Chem Phys 85:519
341. Hayter JB, Penfold J (1981) Mol Phys 42:109
342. Klein R (1991) In: Chen S-H, Huang JS, Tartaglia P (eds) Structure and dynamics of strongly interacting colloids and supramolecular aggregates in solution, NATO ASI Series C: Mathematical and Physical Sciences, vol 369. Kluver, Dordrecht, p 39
343. Longeville S, Doster W, Diehl M, Gaehler R, Petry W (2003) In: Mezei F, Pappas C, Gutberlet T (eds) Neutron spin echo spectroscopy. Basics, trends and applications, vol 601. Springer, Berlin Heidelberg New York
344. Haeussler W, Wilk A, Gapinski J, Patkowski A (2002) J Chem Phys 117:413

345. Perez J, Zanotti JM, Durand D (1999) Biophys J 77:454
346. Gabel F, Bicout D, Lehnert U, Tehei M, Weik M, Zaccai G (2002) Quart Rev Biophys 35:327
347. Bellissent-Funel MC, Daniel R, Durand D, Ferrand M, Finney JL, Pouget S, Reat V, Smith JC (1998) J Am Chem Soc 120:7347
348. Hinsen K, Petrescu AJ, Dellerue S, Bellissent-Funel MC, Kneller GR (2002) J Mol Liquids 98–99:381
349. Zaccai G (2003) J Phys Condens Matter 15:S1673
350. Wall ME, Gallagher SC, Trewhalla J (2000) Large scale shape changes in proteins and macromolecular complexes. Ann Rev Chem 51:355

Editor: Eugene Terentjev

Abbreviations and Symbols

aPP	Atactic polypropylene
BS	Backscattering
DLS	Dynamic light scattering
DQ-NMR	Double quantum NMR
DS	Dielectric spectroscopy
EISF	Elastic incoherent structure factor
FENE	Finite extensible non-linear elastic potential
ILL	Institut Laue Langevin
KWW	Kohlrausch-William-Watts
MC-simulation	Monte Carlo simulation
MCT	Mode coupling theory
MD-simulation	Molecular dynamics simulation
MSA	Mean spherical approximation
NMR	Nuclear magnetic resonance
NSE	Neutron spin echo
ODT	Order-disorder-temperature
PAMAM	Polyamidoamine
PB	1,4 Polybutadiene
PDMS	Polydimethylsiloxane
PE	Polyethylene
PEE	Polyethylethylene
PEMS	Polyethylmethylsiloxane
PEO	Polyethyleneoxide
PEP	Poylethylenepropylene
PFG-NMR	Pulsed field gradient NMR
PFS	Polyfluorosilicone
PI	Polyisoprene
PIB	Polyisobutylene
PNIPA	Poly(N-isopropylacrylamide)
PS	Polystyrene
PU	Polyurethane
PVC	Polyvinylchloride
PVE	Polyvinylethylene
PVME	Polyvinylmethylether
RPA	Random phase approximation
SANS	Small angle neutron scattering
XPS	X-ray photon correlation
$a_R(T)$	Rheological shift factor
B	Draining parameter
b_i	Neutron scattering length of nucleus "i"
C_∞	Characteristic ratio
$C(q)$	Mode-dependent characteristic ratio
c^*	Overlap concentration
C_{11}	Elastic modulus of longitudinal sound waves
CLF	Contour length fluctuations
CR	Constraint release
d	Tube diameter

d_β	Jump distance in the β-process
$\dfrac{d\sigma}{d\Omega\,dE}$	Double differential cross section
D_c	Collective diffusion coefficient
D_R	Rouse diffusion coefficient
D_{self}	Self-diffusion coefficient
D_z	Zimm translational diffusion coefficient
f_{Debye}	Debye function
$f_x(n,t)$	Component of random force
$f_{Q\max}(T)$	Generalized Debye-Waller factor
$g(E)$	Distribution function of energy barriers
G_N^0	Plateau modulus
$G(t), G(\omega)$	Dynamic shear modulus
$G_{\text{self}}(r,t)$	Self part of the van Hove correlation function
$H(Q)$	Hydrodynamic factor
$\underline{H}(r)$	Oseen Tensor
k	Entropic spring constant
k_B	Boltzmann constant
$k_{i/f}$	Incoming/outgoing neutron wave vector
ℓ	Segment length
ℓ_{mon}	Monomer length
ℓ_0	Bond length
$\tilde{\ell}(q)$	Fourier transformed segment vector
L	Chain contour
M	Chain molecular weight
M_n	Number averaged molecular weight
M_0	Monomer molecular weight
M_w	Weight averaged molecular weight
N	Chain length (number of backbone bonds)
N_e	Number backbone bonds between entanglements
$\text{Prob}(\{\underline{r}_n\})$	Conformational probability function
$\langle r^2(t)\rangle$	Mean square displacement
$\langle R_{\text{cm}}(t)^2\rangle$	Mean square centre of mass displacement
R_E	End-to-end distance
R_g	Radius of gyration
S	Entropy
$S(Q,\omega), S(Q,t)$	Dynamic structure factor
$S_{\text{chain}}(Q,t)$	Single chain dynamic structure factor
$S_{\text{esc}}(Q,t)$	Structure factor for creep motion within the reptation model
$S_{\text{loc}}(Q,t)$	Structure factor for local reptation
$S_{\text{pair}}(Q,t)$	Pair correlation function
$S_{\text{self}}(Q,t)$	Self correlation function
T_g	Glass transition temperature
T_m	Merging temperature of α- and β-relaxation
T_0	Vogel-Fulcher temperature
T_R	Reference temperature
v_1	Flory exponent
$\underline{\underline{v}}(Q,s)$	Excluded volume matrix
W	Elementary Rouse rate

Abbreviations and Symbols

$\tilde{x}(p,t)$	Rouse normal mode
$\underline{\underline{\chi}}(Q,s)$	Dynamic response matrix
χ_{11}	Longitudinal susceptibility
$Z = N/N_e$	Number of entanglements/chain
$\alpha_2(t)$	Second order non-Gaussian parameter
β	Stretching exponent
$\dfrac{d\Sigma}{d\Omega}$	Macroscopic cross section
η	Viscosity
η_s	Solvent viscosity
γ	Surface tension
κ_{ij}	Flory Huggins parameter between monomers "i" and "j"
$\underline{\underline{\mu}}(Q)$	Mobility matrix
$\mu_{DE}(t)$	Tube survival probability
λ	Neutron wavelength
$\underline{\underline{\Omega}}(Q)$	First cumulant matrix
Ω_R	Rouse variable
Π	Osmometic pressure
$\tilde{\sigma}$	Conductivity
τ_α	Characteristic time of the α-process
τ_a	Rotational relaxation time in the Allegra model
τ_β	Characteristic time of β-relaxation
τ_p^b	Characteristic Rouse times for chains with bending elasticity
τ_d	Terminal time for reptation
τ_e	Crossover time Rouse, local reptation
τ_q	Characteristic Rouse times for the all-rotational model
τ_R	Rouse time
$\tau_s(T)$	Characteristic time for structural relaxation
$\tau_{\text{self/pair}}(Q,T)$	KWW-relaxation time of the self/pair correlation function
θ	Scattering angle
$\phi(R,n)$	End to end distribution function of a chain length with n segments
$\phi(t)_{\alpha,\beta}$	Correlator relevant for the glass process (α,β-relaxation)
ξ_n	Hydrodynamic screening length
ξ	Correlation length
ζ	Monomeric friction coefficient
ζ_0	Friction coefficient/bond
$\hbar Q$	Momentum transfer during scattering
$\hbar\omega$	Energy transfer during scattering
Ξ	Inhomogeneity length scale in gels

Author Index Volumes 101–174

Author Index Volumes 1–100 see Volume 100

de, Abajo, J. and *de la Campa, J. G.*: Processable Aromatic Polyimides. Vol. 140, pp. 23–60.
Abetz, V. see Förster, S.: Vol. 166, pp. 173–210.
Adolf, D. B. see Ediger, M. D.: Vol. 116, pp. 73–110.
Aharoni, S. M. and *Edwards, S. F.*: Rigid Polymer Networks. Vol. 118, pp. 1–231.
Albertsson, A.-C. and *Varma, I. K.*: Aliphatic Polyesters: Synthesis, Properties and Applications. Vol. 157, pp. 99–138.
Albertsson, A.-C. see Edlund, U.: Vol. 157, pp. 53–98.
Albertsson, A.-C. see Söderqvist Lindblad, M.: Vol. 157, pp. 139–161.
Albertsson, A.-C. see Stridsberg, K. M.: Vol. 157, pp. 27–51.
Albertsson, A.-C. see Al-Malaika, S.: Vol. 169, pp. 177–199.
Al-Malaika, S.: Perspectives in Stabilisation of Polyolefins. Vol. 169, pp. 121–150.
Améduri, B., Boutevin, B. and *Gramain, P.*: Synthesis of Block Copolymers by Radical Polymerization and Telomerization. Vol. 127, pp. 87–142.
Améduri, B. and *Boutevin, B.*: Synthesis and Properties of Fluorinated Telechelic Monodispersed Compounds. Vol. 102, pp. 133–170.
Amselem, S. see Domb, A. J.: Vol. 107, pp. 93–142.
Andrady, A. L.: Wavelenght Sensitivity in Polymer Photodegradation. Vol. 128, pp. 47–94.
Andreis, M. and *Koenig, J. L.*: Application of Nitrogen-15 NMR to Polymers. Vol. 124, pp. 191–238.
Angiolini, L. see Carlini, C.: Vol. 123, pp. 127–214.
Anjum, N. see Gupta, B.: Vol. 162, pp. 37–63.
Anseth, K. S., Newman, S. M. and *Bowman, C. N.*: Polymeric Dental Composites: Properties and Reaction Behavior of Multimethacrylate Dental Restorations. Vol. 122, pp. 177–218.
Antonietti, M. see Cölfen, H.: Vol. 150, pp. 67–187.
Armitage, B. A. see O'Brien, D. F.: Vol. 126, pp. 53–58.
Arndt, M. see Kaminski, W.: Vol. 127, pp. 143–187.
Arnold Jr., F. E. and *Arnold, F. E.*: Rigid-Rod Polymers and Molecular Composites. Vol. 117, pp. 257–296.
Arora, M. see Kumar, M. N. V. R.: Vol. 160, pp. 45–118.
Arshady, R.: Polymer Synthesis via Activated Esters: A New Dimension of Creativity in Macromolecular Chemistry. Vol. 111, pp. 1–42.
Auer, S. and *Frenkel, D.*: Numerical Simulation of Crystal Nucleation in Colloids. Vol. 173, pp. 149–208.

Bahar, I., Erman, B. and *Monnerie, L.*: Effect of Molecular Structure on Local Chain Dynamics: Analytical Approaches and Computational Methods. Vol. 116, pp. 145–206.
Ballauff, M. see Dingenouts, N.: Vol. 144, pp. 1–48.
Ballauff, M. see Holm, C.: Vol. 166, pp. 1–27.

Ballauff, M. see Rühe, J.: Vol. 165, pp. 79–150.
Baltá-Calleja, F. J., González Arche, A., Ezquerra, T. A., Santa Cruz, C., Batallón, F., Frick, B. and *López Cabarcos, E.*: Structure and Properties of Ferroelectric Copolymers of Poly(vinylidene) Fluoride. Vol. 108, pp. 1–48.
Barnes, M. D. see Otaigbe, J. U.: Vol. 154, pp. 1–86.
Barshtein, G. R. and *Sabsai, O. Y.*: Compositions with Mineralorganic Fillers. Vol. 101, pp. 1–28.
Barton, J. see Hunkeler, D.: Vol. 112, pp. 115–134.
Baschnagel, J., Binder, K., Doruker, P., Gusev, A. A., Hahn, O., Kremer, K., Mattice, W. L., Müller-Plathe, F., Murat, M., Paul, W., Santos, S., Sutter, U. W. and *Tries, V.*: Bridging the Gap Between Atomistic and Coarse-Grained Models of Polymers: Status and Perspectives. Vol. 152, pp. 41–156.
Batallán, F. see Baltá-Calleja, F. J.: Vol. 108, pp. 1–48.
Batog, A. E., Pet'ko, I. P. and *Penczek, P.*: Aliphatic-Cycloaliphatic Epoxy Compounds and Polymers. Vol. 144, pp. 49–114.
Baughman, T. W. and *Wagener, K. B.*: Recent Advances in ADMET Polymerization. Vol 176, pp. 1–42.
Bell, C. L. and *Peppas, N. A.*: Biomedical Membranes from Hydrogels and Interpolymer Complexes. Vol. 122, pp. 125–176.
Bellon-Maurel, A. see Calmon-Decriaud, A.: Vol. 135, pp. 207–226.
Bennett, D. E. see O'Brien, D. F.: Vol. 126, pp. 53–84.
Berry, G. C.: Static and Dynamic Light Scattering on Moderately Concentraded Solutions: Isotropic Solutions of Flexible and Rodlike Chains and Nematic Solutions of Rodlike Chains. Vol. 114, pp. 233–290.
Bershtein, V. A. and *Ryzhov, V. A.*: Far Infrared Spectroscopy of Polymers. Vol. 114, pp. 43–122.
Bhargava R., Wang S.-Q. and *Koenig J. L*: FTIR Microspectroscopy of Polymeric Systems. Vol. 163, pp. 137–191.
Biesalski, M.: see Rühe, J.: Vol. 165, pp. 79–150.
Bigg, D. M.: Thermal Conductivity of Heterophase Polymer Compositions. Vol. 119, pp. 1–30.
Binder, K.: Phase Transitions in Polymer Blends and Block Copolymer Melts: Some Recent Developments. Vol. 112, pp. 115–134.
Binder, K.: Phase Transitions of Polymer Blends and Block Copolymer Melts in Thin Films. Vol. 138, pp. 1–90.
Binder, K. see Baschnagel, J.: Vol. 152, pp. 41–156.
Binder, K., Müller, M., Virnau, P. and *González MacDowell, L.*: Polymer+Solvent Systems: Phase Diagrams, Interface Free Energies, and Nucleation. Vol. 173, pp. 1–104.
Bird, R. B. see Curtiss, C. F.: Vol. 125, pp. 1–102.
Biswas, M. and *Mukherjee, A.*: Synthesis and Evaluation of Metal-Containing Polymers. Vol. 115, pp. 89–124.
Biswas, M. and *Sinha Ray, S.*: Recent Progress in Synthesis and Evaluation of Polymer-Montmorillonite Nanocomposites. Vol. 155, pp. 167–221.
Bogdal, D., Penczek, P., Pielichowski, J. and *Prociak, A.*: Microwave Assisted Synthesis, Crosslinking, and Processing of Polymeric Materials. Vol. 163, pp. 193–263.
Bohrisch, J., Eisenbach, C. D., Jaeger, W., Mori H., Müller A. H. E., Rehahn, M., Schaller, C., Traser, S. and *Wittmeyer, P.*: New Polyelectrolyte Architectures. Vol. 165, pp. 1–41.
Bolze, J. see Dingenouts, N.: Vol. 144, pp. 1–48.
Bosshard, C.: see Gubler, U.: Vol. 158, pp. 123–190.
Boutevin, B. and *Robin, J. J.*: Synthesis and Properties of Fluorinated Diols. Vol. 102. pp. 105–132.

Boutevin, B. see *Amédouri, B.*: Vol. 102, pp. 133–170.
Boutevin, B. see *Améduri, B.*: Vol. 127, pp. 87–142.
Bowman, C. N. see *Anseth, K. S.*: Vol. 122, pp. 177–218.
Boyd, R. H.: Prediction of Polymer Crystal Structures and Properties. Vol. 116, pp. 1–26.
Briber, R. M. see *Hedrick, J. L.*: Vol. 141, pp. 1–44.
Bronnikov, S. V., Vettegren, V. I. and *Frenkel, S. Y.*: Kinetics of Deformation and Relaxation in Highly Oriented Polymers. Vol. 125, pp. 103–146.
Brown, H. R. see *Creton, C.*: Vol. 156, pp. 53–135.
Bruza, K. J. see *Kirchhoff, R. A.*: Vol. 117, pp. 1–66.
Buchmeiser, M. R.: Regioselective Polymerization of 1-Alkynes and Stereoselective Cyclopolymerization of α, ω-Heptadiynes. Vol. 176, pp. 89–119.
Budkowski, A.: Interfacial Phenomena in Thin Polymer Films: Phase Coexistence and Segregation. Vol. 148, pp. 1–112.
Burban, J. H. see *Cussler, E. L.*: Vol. 110, pp. 67–80.
Burchard,W.: Solution Properties of Branched Macromolecules. Vol. 143, pp. 113–194.:
Butté, A. see *Schork, F. J.*: Vol. 175, pp. 129–255.

Calmon-Decriaud, A., Bellon-Maurel, V. and *Silvestre, F.*: Standard Methods for Testing the Aerobic Biodegradation of Polymeric Materials.Vol 135, pp. 207–226.
Cameron, N. R. and *Sherrington, D. C.*: High Internal Phase Emulsions (HIPEs)-Structure, Properties and Use in Polymer Preparation.Vol. 126, pp. 163–214.
de la Campa, J. G. see *de Abajo, J.*: Vol. 140, pp. 23–60.
Candau, F. see *Hunkeler, D.*: Vol. 112, pp. 115–134.
Canelas, D. A. and *DeSimone, J. M.*: Polymerizations in Liquid and Supercritical Carbon Dioxide. Vol. 133, pp. 103–140.
Canva, M. and *Stegeman, G. I.*: Quadratic Parametric Interactions in Organic Waveguides. Vol. 158, pp. 87–121.
Capek, I.: Kinetics of the Free-Radical Emulsion Polymerization of Vinyl Chloride. Vol. 120, pp. 135–206.
Capek, I.: Radical Polymerization of Polyoxyethylene Macromonomers in Disperse Systems. Vol. 145, pp. 1–56.
Capek, I. and *Chern, C.-S.*: Radical Polymerization in Direct Mini-Emulsion Systems. Vol. 155, pp. 101–166.
Cappella, B. see *Munz, M.*: Vol. 164, pp. 87–210.
Carlesso, G. see *Prokop, A.*: Vol. 160, pp. 119–174.
Carlini, C. and *Angiolini, L.*: Polymers as Free Radical Photoinitiators. Vol. 123, pp. 127–214.
Carter, K. R. see *Hedrick, J. L.*: Vol. 141, pp. 1–44.
Casas-Vazquez, J. see *Jou, D.*: Vol. 120, pp. 207–266.
Chandrasekhar, V.: Polymer Solid Electrolytes: Synthesis and Structure. Vol 135, pp. 139–206.
Chang, J. Y. see *Han, M. J.*: Vol. 153, pp. 1–36.
Chang, T.: Recent Advances in Liquid Chromatography Analysis of Synthetic Polymers. Vol. 163, pp. 1–60.
Charleux, B. and *Faust R.*: Synthesis of Branched Polymers by Cationic Polymerization. Vol. 142, pp. 1–70.
Chen, P. see *Jaffe, M.*: Vol. 117, pp. 297–328.
Chern, C.-S. see *Capek, I.*: Vol. 155, pp. 101–166.
Chevolot, Y. see *Mathieu, H. J.*: Vol. 162, pp. 1–35.
Choe, E.-W. see *Jaffe, M.*: Vol. 117, pp. 297–328.

Chow, P. Y. and *Gan, L. M.*: Microemulsion Polymerizations and Reactions. Vol. 175, pp. 257–298.
Chow, T. S.: Glassy State Relaxation and Deformation in Polymers. Vol. 103, pp. 149–190.
Chujo, Y. see Uemura, T.: Vol. 167, pp. 81–106.
Chung, S.-J. see Lin, T.-C.: Vol. 161, pp. 157–193.
Chung, T.-S. see Jaffe, M.: Vol. 117, pp. 297–328.
Cölfen, H. and *Antonietti, M.*: Field-Flow Fractionation Techniques for Polymer and Colloid Analysis. Vol. 150, pp. 67–187.
Colmenero J. see Richter, D.: Vol. 174, pp. 1–221.
Comanita, B. see Roovers, J.: Vol. 142, pp. 179–228.
Connell, J. W. see Hergenrother, P. M.: Vol. 117, pp. 67–110.
Creton, C., Kramer, E. J., Brown, H. R. and *Hui, C.-Y.*: Adhesion and Fracture of Interfaces Between Immiscible Polymers: From the Molecular to the Continuum Scale. Vol. 156, pp. 53–135.
Criado-Sancho, M. see Jou, D.: Vol. 120, pp. 207–266.
Curro, J. G. see Schweizer, K. S.: Vol. 116, pp. 319–378.
Curtiss, C. F. and *Bird, R. B.*: Statistical Mechanics of Transport Phenomena: Polymeric Liquid Mixtures. Vol. 125, pp. 1–102.
Cussler, E. L., Wang, K. L. and *Burban, J. H.*: Hydrogels as Separation Agents. Vol. 110, pp. 67–80.

Dalton, L.: Nonlinear Optical Polymeric Materials: From Chromophore Design to Commercial Applications. Vol. 158, pp. 1–86.
Dautzenberg, H. see Holm, C.: Vol. 166, pp.113–171.
Davidson, J. M. see Prokop, A.: Vol. 160, pp.119–174.
Desai, S. M. and *Singh, R. P.*: Surface Modification of Polyethylene. Vol. 169, pp. 231–293.
DeSimone, J. M. see Canelas D. A.: Vol. 133, pp. 103–140.
DeSimone, J. M. see Kennedy, K. A.: Vol. 175, pp. 329–346.
DiMari, S. see Prokop, A.: Vol. 136, pp. 1–52.
Dimonie, M. V. see Hunkeler, D.: Vol. 112, pp. 115–134.
Dingenouts, N., Bolze, J., Pötschke, D. and *Ballauf, M.*: Analysis of Polymer Latexes by Small-Angle X-Ray Scattering. Vol. 144, pp. 1–48.
Dodd, L. R. and *Theodorou, D. N.*: Atomistic Monte Carlo Simulation and Continuum Mean Field Theory of the Structure and Equation of State Properties of Alkane and Polymer Melts. Vol. 116, pp. 249–282.
Doelker, E.: Cellulose Derivatives. Vol. 107, pp. 199–266.
Dolden, J. G.: Calculation of a Mesogenic Index with Emphasis Upon LC-Polyimides. Vol. 141, pp. 189 –245.
Domb, A. J., Amselem, S., Shah, J. and *Maniar, M.*: Polyanhydrides: Synthesis and Characterization. Vol. 107, pp. 93–142.
Domb, A. J. see Kumar, M. N. V. R.: Vol. 160, pp. 45118.
Doruker, P. see Baschnagel, J.: Vol. 152, pp. 41–156.
Dubois, P. see Mecerreyes, D.: Vol. 147, pp. 1–60.
Dubrovskii, S. A. see Kazanskii, K. S.: Vol. 104, pp. 97–134.
Dunkin, I. R. see Steinke, J.: Vol. 123, pp. 81–126.
Dunson, D. L. see McGrath, J. E.: Vol. 140, pp. 61–106.
Dziezok, P. see Rühe, J.: Vol. 165, pp. 79–150.

Eastmond, G. C.: Poly(ε-caprolactone) Blends. Vol. 149, pp. 59–223.
Economy, J. and *Goranov, K.*: Thermotropic Liquid Crystalline Polymers for High Performance Applications. Vol. 117, pp. 221–256.

Ediger, M. D. and *Adolf, D. B.*: Brownian Dynamics Simulations of Local Polymer Dynamics. Vol. 116, pp. 73–110.
Edlund, U. and *Albertsson, A.-C.*: Degradable Polymer Microspheres for Controlled Drug Delivery. Vol. 157, pp. 53–98.
Edwards, S. F. see Aharoni, S. M.: Vol. 118, pp. 1–231.
Eisenbach, C. D. see Bohrisch, J.: Vol. 165, pp. 1–41.
Endo, T. see Yagci, Y.: Vol. 127, pp. 59–86.
Engelhardt, H. and *Grosche, O.*: Capillary Electrophoresis in Polymer Analysis. Vol.150, pp. 189–217.
Engelhardt, H. and *Martin, H.*: Characterization of Synthetic Polyelectrolytes by Capillary Electrophoretic Methods. Vol. 165, pp. 211–247.
Eriksson, P. see Jacobson, K.: Vol. 169, pp. 151–176.
Erman, B. see Bahar, I.: Vol. 116, pp. 145–206.
Eschner, M. see Spange, S.: Vol. 165, pp. 43–78.
Estel, K. see Spange, S.: Vol. 165, pp. 43–78.
Ewen, B. and *Richter, D.*: Neutron Spin Echo Investigations on the Segmental Dynamics of Polymers in Melts, Networks and Solutions. Vol. 134, pp. 1–130.
Ezquerra, T. A. see Baltá-Calleja, F. J.: Vol. 108, pp. 1–48.

Fatkullin, N. see Kimmich, R.: Vol. 170, pp. 1–113.
Faust, R. see Charleux, B.: Vol. 142, pp. 1–70.
Faust, R. see Kwon, Y.: Vol. 167, pp. 107–135.
Fekete, E. see Pukánszky, B.: Vol. 139, pp. 109–154.
Fendler, J. H.: Membrane-Mimetic Approach to Advanced Materials. Vol. 113, pp. 1–209.
Fetters, L. J. see Xu, Z.: Vol. 120, pp. 1–50.
Fontenot, K. see Schork, F. J.: Vol. 175, pp. 129–255.
Förster, S., Abetz, V. and *Müller, A. H. E.*: Polyelectrolyte Block Copolymer Micelles. Vol. 166, pp. 173–210.
Förster, S. and *Schmidt, M.*: Polyelectrolytes in Solution. Vol. 120, pp. 51–134.
Freire, J. J.: Conformational Properties of Branched Polymers: Theory and Simulations. Vol. 143, pp. 35–112.
Frenkel, S. Y. see Bronnikov, S.V.: Vol. 125, pp. 103–146.
Frick, B. see Baltá-Calleja, F. J.: Vol. 108, pp. 1–48.
Fridman, M. L. see Terent'eva, J. P.: Vol. 101, pp. 29–64.
Fuchs, G. see Trimmel, G.: Vol. 176, pp. 43–87.
Fukui, K. see Otaigbe, J. U.: Vol. 154, pp. 1–86.
Funke, W.: Microgels-Intramolecularly Crosslinked Macromolecules with a Globular Structure. Vol. 136, pp. 137–232.
Furusho, Y. see Takata, T.: Vol. 171, pp. 1–75.

Galina, H.: Mean-Field Kinetic Modeling of Polymerization: The Smoluchowski Coagulation Equation.Vol. 137, pp. 135–172.
Gan, L. M. see Chow, P. Y.: Vol. 175, pp. 257–298.
Ganesh, K. see Kishore, K.: Vol. 121, pp. 81–122.
Gaw, K. O. and *Kakimoto, M.*: Polyimide-Epoxy Composites. Vol. 140, pp. 107–136.
Geckeler, K. E. see Rivas, B.: Vol. 102, pp. 171–188.
Geckeler, K. E.: Soluble Polymer Supports for Liquid-Phase Synthesis. Vol. 121, pp. 31–80.
Gedde, U. W. and *Mattozzi, A.*: Polyethylene Morphology. Vol. 169, pp. 29–73.
Gehrke, S. H.: Synthesis, Equilibrium Swelling, Kinetics Permeability and Applications of Environmentally Responsive Gels. Vol. 110, pp. 81–144.

de Gennes, P.-G.: Flexible Polymers in Nanopores. Vol. 138, pp. 91–106.
Georgiou, S.: Laser Cleaning Methodologies of Polymer Substrates. Vol. 168, pp. 1–49.
Geuss, M. see Munz, M.: Vol. 164, pp. 87–210.
Giannelis, E. P., Krishnamoorti, R. and *Manias, E.*: Polymer-Silicate Nanocomposites: Model Systems for Confined Polymers and Polymer Brushes. Vol. 138, pp. 107–148.
Godovsky, D. Y.: Device Applications of Polymer-Nanocomposites. Vol. 153, pp. 163–205.
Godovsky, D. Y.: Electron Behavior and Magnetic Properties Polymer-Nanocomposites. Vol. 119, pp. 79–122.
González Arche, A. see Baltá-Calleja, F. J.: Vol. 108, pp. 1–48.
Goranov, K. see Economy, J.: Vol. 117, pp. 221–256.
Gramain, P. see Améduri, B.: Vol. 127, pp. 87–142.
Grest, G. S.: Normal and Shear Forces Between Polymer Brushes. Vol. 138, pp. 149–184.
Grigorescu, G. and *Kulicke, W.-M.*: Prediction of Viscoelastic Properties and Shear Stability of Polymers in Solution. Vol. 152, p. 1–40.
Gröhn, F. see Rühe, J.: Vol. 165, pp. 79–150.
Grosberg, A. and *Nechaev, S.*: Polymer Topology. Vol. 106, pp. 1–30.
Grosche, O. see Engelhardt, H.: Vol. 150, pp. 189–217.
Grubbs, R., Risse, W. and *Novac, B.*: The Development of Well-defined Catalysts for Ring-Opening Olefin Metathesis. Vol. 102, pp. 47–72.
Gubler, U. and *Bosshard, C.*: Molecular Design for Third-Order Nonlinear Optics. Vol. 158, pp. 123–190.
van Gunsteren, W. F. see Gusev, A. A.: Vol. 116, pp. 207–248.
Gupta, B. and *Anjum, N.*: Plasma and Radiation-Induced Graft Modification of Polymers for Biomedical Applications. Vol. 162, pp. 37–63.
Gusev, A. A., Müller-Plathe, F., van Gunsteren, W. F. and *Suter, U. W.*: Dynamics of Small Molecules in Bulk Polymers. Vol. 116, pp. 207–248.
Gusev, A. A. see Baschnagel, J.: Vol. 152, pp. 41–156.
Guillot, J. see Hunkeler, D.: Vol. 112, pp. 115–134.
Guyot, A. and *Tauer, K.*: Reactive Surfactants in Emulsion Polymerization. Vol. 111, pp. 43–66.

Hadjichristidis, N., Pispas, S., Pitsikalis, M., Iatrou, H. and *Vlahos, C.*: Asymmetric Star Polymers Synthesis and Properties. Vol. 142, pp. 71–128.
Hadjichristidis, N. see Xu, Z.: Vol. 120, pp. 1–50.
Hadjichristidis, N. see Pitsikalis, M.: Vol. 135, pp. 1–138.
Hahn, O. see Baschnagel, J.: Vol. 152, pp. 41–156.
Hakkarainen, M.: Aliphatic Polyesters: Abiotic and Biotic Degradation and Degradation Products. Vol. 157, pp. 1–26.
Hakkarainen, M. and *Albertsson, A.-C.*: Environmental Degradation of Polyethylene. Vol. 169, pp. 177–199.
Hall, H. K. see Penelle, J.: Vol. 102, pp. 73–104.
Hamley, I. W.: Crystallization in Block Copolymers. Vol. 148, pp. 113–138.
Hammouda, B.: SANS from Homogeneous Polymer Mixtures: A Unified Overview. Vol. 106, pp. 87–134.
Han, M. J. and *Chang, J. Y.*: Polynucleotide Analogues. Vol. 153, pp. 1–36.
Harada, A.: Design and Construction of Supramolecular Architectures Consisting of Cyclodextrins and Polymers. Vol. 133, pp. 141–192.
Haralson, M. A. see Prokop, A.: Vol. 136, pp. 1–52.
Hassan, C. M. and *Peppas, N. A.*: Structure and Applications of Poly(vinyl alcohol) Hydrogels Produced by Conventional Crosslinking or by Freezing/Thawing Methods. Vol. 153, pp. 37–65.

Hawker, C. J.: Dentritic and Hyperbranched Macromolecules Precisely Controlled Macromolecular Architectures. Vol. 147, pp. 113–160.
Hawker, C. J. see Hedrick, J. L.: Vol. 141, pp. 1–44.
He, G. S. see Lin, T.-C.: Vol. 161, pp. 157–193.
Hedrick, J. L., Carter, K. R., Labadie, J. W., Miller, R. D., Volksen, W., Hawker, C. J., Yoon, D. Y., Russell, T. P., McGrath, J. E. and *Briber, R. M.*: Nanoporous Polyimides. Vol. 141, pp. 1–44.
Hedrick, J. L., Labadie, J. W., Volksen, W. and *Hilborn, J. G.*: Nanoscopically Engineered Polyimides. Vol. 147, pp. 61–112.
Hedrick, J. L. see Hergenrother, P. M.: Vol. 117, pp. 67–110.
Hedrick, J. L. see Kiefer, J.: Vol. 147, pp. 161–247.
Hedrick, J. L. see McGrath, J. E.: Vol. 140, pp. 61–106.
Heine, D. R., Grest, G. S. and *Curro, J. G.*: Structure of Polymer Melts and Blends: Comparison of Integral Equation theory and Computer Sumulation. Vol. 173, pp. 209–249.
Heinrich, G. and *Klüppel, M.*: Recent Advances in the Theory of Filler Networking in Elastomers. Vol. 160, pp. 1–44.
Heller, J.: Poly (Ortho Esters). Vol. 107, pp. 41–92.
Helm, C. A.: see Möhwald, H.: Vol. 165, pp. 151–175.
Hemielec, A. A. see Hunkeler, D.: Vol. 112, pp. 115–134.
Hergenrother, P. M., Connell, J. W., Labadie, J. W. and *Hedrick, J. L.*: Poly(arylene ether)s Containing Heterocyclic Units. Vol. 117, pp. 67–110.
Hernández-Barajas, J. see Wandrey, C.: Vol. 145, pp. 123–182.
Hervet, H. see Léger, L.: Vol. 138, pp. 185–226.
Hilborn, J. G. see Hedrick, J. L.: Vol. 147, pp. 61–112.
Hilborn, J. G. see Kiefer, J.: Vol. 147, pp. 161–247.
Hiramatsu, N. see Matsushige, M.: Vol. 125, pp. 147–186.
Hirasa, O. see Suzuki, M.: Vol. 110, pp. 241–262.
Hirotsu, S.: Coexistence of Phases and the Nature of First-Order Transition in Poly-N-isopropylacrylamide Gels. Vol. 110, pp. 1–26.
Höcker, H. see Klee, D.: Vol. 149, pp. 1–57.
Holm, C., Hofmann, T., Joanny, J. F., Kremer, K., Netz, R. R., Reineker, P., Seidel, C., Vilgis, T. A. and *Winkler, R. G.*: Polyelectrolyte Theory. Vol. 166, pp. 67–111.
Holm, C., Rehahn, M., Oppermann, W. and *Ballauff, M.*: Stiff-Chain Polyelectrolytes. Vol. 166, pp. 1–27.
Hornsby, P.: Rheology, Compounding and Processing of Filled Thermoplastics. Vol. 139, pp. 155–216.
Houbenov, N. see Rühe, J.: Vol. 165, pp. 79–150.
Huber, K. see Volk, N.: Vol. 166, pp. 29–65.
Hugenberg, N. see Rühe, J.: Vol. 165, pp. 79–150.
Hui, C.-Y. see Creton, C.: Vol. 156, pp. 53–135.
Hult, A., Johansson, M. and *Malmström, E.*: Hyperbranched Polymers.Vol. 143, pp. 1–34.
Hünenberger, P. H.: Thermostat Algorithms for Molecular-Dynamics Simulations. Vol. 173, pp. 105–147.
Hunkeler, D., Candau, F., Pichot, C., Hemielec, A. E., Xie, T. Y., Barton, J., Vaskova, V., Guillot, J., Dimonie, M. V. and *Reichert, K. H.*: Heterophase Polymerization: A Physical and Kinetic Comparision and Categorization. Vol. 112, pp. 115–134.
Hunkeler, D. see Macko, T.: Vol. 163, pp. 61–136.
Hunkeler, D. see Prokop, A.: Vol. 136, pp. 1–52; 53–74.
Hunkeler, D. see Wandrey, C.: Vol. 145, pp. 123–182.

Iatrou, H. see Hadjichristidis, N.: Vol. 142, pp. 71–128.
Ichikawa, T. see Yoshida, H.: Vol. 105, pp. 3–36.

Ihara, E. see Yasuda, H.: Vol. 133, pp. 53–102.
Ikada, Y. see Uyama,Y.: Vol. 137, pp. 1–40.
Ikehara, T. see Jinnuai, H.: Vol. 170, pp. 115–167.
Ilavsky, M.: Effect on Phase Transition on Swelling and Mechanical Behavior of Synthetic Hydrogels. Vol. 109, pp. 173–206.
Imai, Y.: Rapid Synthesis of Polyimides from Nylon-Salt Monomers. Vol. 140, pp. 1–23.
Inomata, H. see Saito, S.: Vol. 106, pp. 207–232.
Inoue, S. see Sugimoto, H.: Vol. 146, pp. 39–120.
Irie, M.: Stimuli-Responsive Poly(N-isopropylacrylamide), Photo- and Chemical-Induced Phase Transitions. Vol. 110, pp. 49–66.
Ise, N. see Matsuoka, H.: Vol. 114, pp. 187–232.
Ito, H.: Chemical Amplification Resists for Microlithography. Vol. 172, pp. 37–245.
Ito, K. and *Kawaguchi, S.*: Poly(macronomers), Homo- and Copolymerization. Vol. 142, pp. 129–178.
Ito, K. see Kawaguchi, S.: Vol. 175, pp. 299–328.
Ito, Y. see Suginome, M.: Vol. 171, pp. 77–136.
Ivanov, A. E. see Zubov, V. P.: Vol. 104, pp. 135–176.

Jacob, S. and *Kennedy, J.*: Synthesis, Characterization and Properties of OCTA-ARM Polyisobutylene-Based Star Polymers. Vol. 146, pp. 1–38.
Jacobson, K., Eriksson, P., Reitberger, T. and *Stenberg, B.*: Chemiluminescence as a Tool for Polyolefin. Vol. 169, pp. 151–176.
Jaeger, W. see Bohrisch, J.: Vol. 165, pp. 1–41.
Jaffe, M., Chen, P., Choe, E.-W., Chung, T.-S. and *Makhija, S.*: High Performance Polymer Blends. Vol. 117, pp. 297–328.
Jancar, J.: Structure-Property Relationships in Thermoplastic Matrices. Vol. 139, pp. 1–66.
Jen, A. K.-Y. see Kajzar, F.: Vol. 161, pp. 1–85.
Jerome, R. see Mecerreyes, D.: Vol. 147, pp. 1–60.
Jiang, M., Li, M., Xiang, M. and *Zhou, H.*: Interpolymer Complexation and Miscibility and Enhancement by Hydrogen Bonding. Vol. 146, pp. 121–194.
Jin, J. see Shim, H.-K.: Vol. 158, pp. 191–241.
Jinnai, H., Nishikawa, Y., Ikehara, T. and *Nishi, T.*: Emerging Technologies for the 3D Analysis of Polymer Structures. Vol. 170, pp. 115–167.
Jo, W. H. and *Yang, J. S.*: Molecular Simulation Approaches for Multiphase Polymer Systems. Vol. 156, pp. 1–52.
Joanny, J.-F. see Holm, C.: Vol. 166, pp. 67–111.
Joanny, J.-F. see Thünemann, A. F.: Vol. 166, pp. 113–171.
Johannsmann, D. see Rühe, J.: Vol. 165, pp. 79–150.
Johansson, M. see Hult, A.: Vol. 143, pp. 1–34.
Joos-Müller, B. see Funke, W.: Vol. 136, pp. 137–232.
Jou, D., Casas-Vazquez, J. and *Criado-Sancho, M.*: Thermodynamics of Polymer Solutions under Flow: Phase Separation and Polymer Degradation. Vol. 120, pp. 207–266.

Kaetsu, I.: Radiation Synthesis of Polymeric Materials for Biomedical and Biochemical Applications. Vol. 105, pp. 81–98.
Kaji, K. see Kanaya, T.: Vol. 154, pp. 87–141.
Kajzar, F., Lee, K.-S. and *Jen, A. K.-Y.*: Polymeric Materials and their Orientation Techniques for Second-Order Nonlinear Optics. Vol. 161, pp. 1–85.
Kakimoto, M. see Gaw, K. O.: Vol. 140, pp. 107–136.
Kaminski, W. and *Arndt, M.*: Metallocenes for Polymer Catalysis. Vol. 127, pp. 143–187.

Kammer, H. W., Kressler, H. and *Kummerloewe, C.*: Phase Behavior of Polymer Blends – Effects of Thermodynamics and Rheology. Vol. 106, pp. 31–86.

Kanaya, T. and *Kaji, K.*: Dynamcis in the Glassy State and Near the Glass Transition of Amorphous Polymers as Studied by Neutron Scattering. Vol. 154, pp. 87–141.

Kandyrin, L. B. and *Kuleznev, V. N.*: The Dependence of Viscosity on the Composition of Concentrated Dispersions and the Free Volume Concept of Disperse Systems. Vol. 103, pp. 103–148.

Kaneko, M. see Ramaraj, R.: Vol. 123, pp. 215–242.

Kang, E. T., Neoh, K. G. and *Tan, K. L.*: X-Ray Photoelectron Spectroscopic Studies of Electroactive Polymers. Vol. 106, pp. 135–190.

Karlsson, S. see Söderqvist Lindblad, M.: Vol. 157, pp. 139–161.

Karlsson, S.: Recycled Polyolefins. Material Properties and Means for Quality Determination. Vol. 169, pp. 201–229.

Kato, K. see Uyama,Y.: Vol. 137, pp. 1–40.

Kautek, W. see Krüger, J.: Vol. 168, pp. 247–290.

Kawaguchi, S. see Ito, K.: Vol. 142, p 129–178.

Kawaguchi, S. and *Ito, K.*: Dispersion Polymerization. Vol. 175, pp. 299–328.

Kawata, S. see Sun, H.-B.: Vol. 170, pp. 169–273.

Kazanskii, K. S. and *Dubrovskii, S. A.*: Chemistry and Physics of Agricultural Hydrogels. Vol. 104, pp. 97–134.

Kennedy, J. P. see Jacob, S.: Vol. 146, pp. 1–38.

Kennedy, J. P. see Majoros, I.: Vol. 112, pp. 1–113.

Kennedy, K. A., Roberts, G. W. and *DeSimone, J. M.*: Heterogeneous Polymerization of Fluoroolefins in Supercritical Carbon Dioxide. Vol. 175, pp. 329–346.

Khokhlov, A., Starodybtzev, S. and *Vasilevskaya, V.*: Conformational Transitions of Polymer Gels: Theory and Experiment. Vol. 109, pp. 121–172.

Kiefer, J., Hedrick J. L. and *Hiborn, J. G.*: Macroporous Thermosets by Chemically Induced Phase Separation. Vol. 147, pp. 161–247.

Kihara, N. see Takata, T.: Vol. 171, pp. 1–75.

Kilian, H. G. and *Pieper, T.*: Packing of Chain Segments. A Method for Describing X-Ray Patterns of Crystalline, Liquid Crystalline and Non-Crystalline Polymers. Vol. 108, pp. 49–90.

Kim, J. see Quirk, R.P.: Vol. 153, pp. 67–162.

Kim, K.-S. see Lin, T.-C.: Vol. 161, pp. 157–193.

Kimmich, R. and *Fatkullin, N.*: Polymer Chain Dynamics and NMR. Vol. 170, pp. 1–113.

Kippelen, B. and *Peyghambarian, N.*: Photorefractive Polymers and their Applications. Vol. 161, pp. 87–156.

Kirchhoff, R. A. and *Bruza, K. J.*: Polymers from Benzocyclobutenes. Vol. 117, pp. 1–66.

Kishore, K. and *Ganesh, K.*: Polymers Containing Disulfide, Tetrasulfide, Diselenide and Ditelluride Linkages in the Main Chain. Vol. 121, pp. 81–122.

Kitamaru, R.: Phase Structure of Polyethylene and Other Crystalline Polymers by Solid-State 13C/MNR. Vol. 137, pp. 41–102.

Klee, D. and *Höcker, H.*: Polymers for Biomedical Applications: Improvement of the Interface Compatibility. Vol. 149, pp. 1–57.

Klier, J. see Scranton, A. B.: Vol. 122, pp. 1–54.

v. Klitzing, R. and *Tieke, B.*: Polyelectrolyte Membranes. Vol. 165, pp. 177–210.

Klüppel, M.: The Role of Disorder in Filler Reinforcement of Elastomers on Various Length Scales. Vol. 164, pp. 1–86.

Klüppel, M. see Heinrich, G.: Vol. 160, pp. 1–44.

Knuuttila, H., Lehtinen, A. and *Nummila-Pakarinen, A.*: Advanced Polyethylene Technologies – Controlled Material Properties. Vol. 169, pp. 13–27.

Kobayashi, S., Shoda, S. and *Uyama, H.*: Enzymatic Polymerization and Oligomerization. Vol. 121, pp. 1–30.
Köhler, W. and *Schäfer, R.*: Polymer Analysis by Thermal-Diffusion Forced Rayleigh Scattering. Vol. 151, pp. 1–59.
Koenig, J. L. see Bhargava, R.: Vol. 163, pp. 137–191.
Koenig, J. L. see Andreis, M.: Vol. 124, pp. 191–238.
Koike, T.: Viscoelastic Behavior of Epoxy Resins Before Crosslinking. Vol. 148, pp. 139–188.
Kokko, E. see Löfgren, B.: Vol. 169, pp. 1–12.
Kokufuta, E.: Novel Applications for Stimulus-Sensitive Polymer Gels in the Preparation of Functional Immobilized Biocatalysts. Vol. 110, pp. 157–178.
Konno, M. see Saito, S.: Vol. 109, pp. 207–232.
Konradi, R. see Rühe, J.: Vol. 165, pp. 79–150.
Kopecek, J. see Putnam, D.: Vol. 122, pp. 55–124.
Koßmehl, G. see Schopf, G.: Vol. 129, pp. 1–145.
Kozlov, E. see Prokop, A.: Vol. 160, pp. 119–174.
Kramer, E. J. see Creton, C.: Vol. 156, pp. 53–135.
Kremer, K. see Baschnagel, J.: Vol. 152, pp. 41–156.
Kremer, K. see Holm, C.: Vol. 166, pp. 67–111.
Kressler, J. see Kammer, H. W.: Vol. 106, pp. 31–86.
Kricheldorf, H. R.: Liquid-Cristalline Polyimides. Vol. 141, pp. 83–188.
Krishnamoorti, R. see Giannelis, E. P.: Vol. 138, pp. 107–148.
Krüger, J. and *Kautek, W.*: Ultrashort Pulse Laser Interaction with Dielectrics and Polymers, Vol. 168, pp. 247–290.
Kuchanov, S. I.: Modern Aspects of Quantitative Theory of Free-Radical Copolymerization. Vol. 103, pp. 1–102.
Kuchanov, S. I.: Principles of Quantitive Description of Chemical Structure of Synthetic Polymers. Vol. 152, p. 157–202.
Kudaibergennow, S. E.: Recent Advances in Studying of Synthetic Polyampholytes in Solutions. Vol. 144, pp. 115–198.
Kuleznev, V. N. see Kandyrin, L. B.: Vol. 103, pp. 103–148.
Kulichkhin, S. G. see Malkin, A. Y.: Vol. 101, pp. 217–258.
Kulicke, W.-M. see Grigorescu, G.: Vol. 152, p. 1–40.
Kumar, M. N. V. R., Kumar, N., Domb, A. J. and *Arora, M.*: Pharmaceutical Polymeric Controlled Drug Delivery Systems. Vol. 160, pp. 45–118.
Kumar, N. see Kumar M. N. V. R.: Vol. 160, pp. 45–118.
Kummerloewe, C. see Kammer, H. W.: Vol. 106, pp. 31–86.
Kuznetsova, N. P. see Samsonov, G.V.: Vol. 104, pp. 1–50.
Kwon, Y. and *Faust, R.*: Synthesis of Polyisobutylene-Based Block Copolymers with Precisely Controlled Architecture by Living Cationic Polymerization. Vol. 167, pp. 107–135.

Labadie, J. W. see Hergenrother, P. M.: Vol. 117, pp. 67–110.
Labadie, J. W. see Hedrick, J. L.: Vol. 141, pp. 1–44.
Labadie, J. W. see Hedrick, J. L.: Vol. 147, pp. 61–112.
Lamparski, H. G. see O'Brien, D. F.: Vol. 126, pp. 53–84.
Laschewsky, A.: Molecular Concepts, Self-Organisation and Properties of Polysoaps. Vol. 124, pp. 1–86.
Laso, M. see Leontidis, E.: Vol. 116, pp. 283–318.
Lazár, M. and *Rychl, R.*: Oxidation of Hydrocarbon Polymers. Vol. 102, pp. 189–222.
Lechowicz, J. see Galina, H.: Vol. 137, pp. 135–172.

Léger, L., Raphaël, E. and *Hervet, H.*: Surface-Anchored Polymer Chains: Their Role in Adhesion and Friction. Vol. 138, pp. 185–226.
Lenz, R. W.: Biodegradable Polymers. Vol. 107, pp. 1–40.
Leontidis, E., de Pablo, J. J., Laso, M. and *Suter, U. W.*: A Critical Evaluation of Novel Algorithms for the Off-Lattice Monte Carlo Simulation of Condensed Polymer Phases. Vol. 116, pp. 283–318.
Lee, B. see Quirk, R. P.: Vol. 153, pp. 67–162.
Lee, K.-S. see Kajzar, F.: Vol. 161, pp. 1–85.
Lee, Y. see Quirk, R. P: Vol. 153, pp. 67–162.
Lehtinen, A. see Knuuttila, H.: Vol. 169, pp. 13–27.
Leónard, D. see Mathieu, H. J.: Vol. 162, pp. 1–35.
Lesec, J. see Viovy, J.-L.: Vol. 114, pp. 1–42.
Li, M. see Jiang, M.: Vol. 146, pp. 121–194.
Liang, G. L. see Sumpter, B. G.: Vol. 116, pp. 27–72.
Lienert, K.-W.: Poly(ester-imide)s for Industrial Use. Vol. 141, pp. 45–82.
Lin, J. and *Sherrington, D. C.*: Recent Developments in the Synthesis, Thermostability and Liquid Crystal Properties of Aromatic Polyamides. Vol. 111, pp. 177–220.
Lin, T.-C., Chung, S.-J., Kim, K.-S., Wang, X., He, G. S., Swiatkiewicz, J., Pudavar, H. E. and *Prasad, P. N.*: Organics and Polymers with High Two-Photon Activities and their Applications. Vol. 161, pp. 157–193.
Lippert, T.: Laser Application of Polymers. Vol. 168, pp. 51–246.
Liu, Y. see Söderqvist Lindblad, M.: Vol. 157, pp. 139–161.
López Cabarcos, E. see Baltá-Calleja, F. J.: Vol. 108, pp. 1–48.
Löfgren, B., Kokko, E. and *Seppälä, J.*: Specific Structures Enabled by Metallocene Catalysis in Polyethenes. Vol. 169, pp. 1–12.
Löwen, H. see Thünemann, A. F.: Vol. 166, pp. 113–171.
Luo, Y. see Schork, F. J.: Vol. 175, pp. 129–255.

Macko, T. and *Hunkeler, D.*: Liquid Chromatography under Critical and Limiting Conditions: A Survey of Experimental Systems for Synthetic Polymers. Vol. 163, pp. 61–136.
Majoros, I., Nagy, A. and *Kennedy, J. P.*: Conventional and Living Carbocationic Polymerizations United. I. A Comprehensive Model and New Diagnostic Method to Probe the Mechanism of Homopolymerizations. Vol. 112, pp. 1–113.
Makhija, S. see Jaffe, M.: Vol. 117, pp. 297–328.
Malmström, E. see Hult, A.: Vol. 143, pp. 1–34.
Malkin, A. Y. and *Kulichkhin, S. G.*: Rheokinetics of Curing. Vol. 101, pp. 217–258.
Maniar, M. see Domb, A. J.: Vol. 107, pp. 93–142.
Manias, E. see Giannelis, E. P.: Vol. 138, pp. 107–148.
Martin, H. see Engelhardt, H.: Vol. 165, pp. 211–247.
Marty, J. D. and *Mauzac, M.*: Molecular Imprinting: State of the Art and Perspectives. Vol. 172, pp. 1–35.
Mashima, K., Nakayama, Y. and *Nakamura, A.*: Recent Trends in Polymerization of a-Olefins Catalyzed by Organometallic Complexes of Early Transition Metals.Vol. 133, pp. 1–52.
Mathew, D. see Reghunadhan Nair, C.P.: Vol. 155, pp. 1–99.
Mathieu, H. J., Chevolot, Y, Ruiz-Taylor, L. and *Leónard, D.*: Engineering and Characterization of Polymer Surfaces for Biomedical Applications. Vol. 162, pp. 1–35.
Matsumoto, A.: Free-Radical Crosslinking Polymerization and Copolymerization of Multivinyl Compounds. Vol. 123, pp. 41–80.
Matsumoto, A. see Otsu, T.: Vol. 136, pp. 75–138.

Matsuoka, H. and *Ise, N.*: Small-Angle and Ultra-Small Angle Scattering Study of the Ordered Structure in Polyelectrolyte Solutions and Colloidal Dispersions. Vol. 114, pp. 187–232.
Matsushige, K., Hiramatsu, N. and *Okabe, H.*: Ultrasonic Spectroscopy for Polymeric Materials. Vol. 125, pp. 147–186.
Mattice, W. L. see Rehahn, M.: Vol. 131/132, pp. 1–475.
Mattice, W. L. see Baschnagel, J.: Vol. 152, pp. 41–156.
Mattozzi, A. see Gedde, U. W.: Vol. 169, pp. 29–73.
Mauzac, M. see Marty, J. D.: Vol. 172, pp. 1–35.
Mays, W. see Xu, Z.: Vol. 120, pp. 1–50.
Mays, J. W. see Pitsikalis, M.: Vol. 135, pp. 1–138.
McGrath, J. E. see Hedrick, J. L.: Vol. 141, pp. 1–44.
McGrath, J. E., Dunson, D. L. and *Hedrick, J. L.*: Synthesis and Characterization of Segmented Polyimide-Polyorganosiloxane Copolymers. Vol. 140, pp. 61–106.
McLeish, T. C. B. and *Milner, S. T.*: Entangled Dynamics and Melt Flow of Branched Polymers. Vol. 143, pp. 195–256.
Mecerreyes, D., Dubois, P. and *Jerome, R.*: Novel Macromolecular Architectures Based on Aliphatic Polyesters: Relevance of the Coordination-Insertion Ring-Opening Polymerization. Vol. 147, pp. 1–60.
Mecham, S. J. see McGrath, J. E.: Vol. 140, pp. 61–106.
Menzel, H. see Möhwald, H.: Vol. 165, pp. 151–175.
Meyer, T. see Spange, S.: Vol. 165, pp. 43–78.
Mikos, A. G. see Thomson, R. C.: Vol. 122, pp. 245–274.
Milner, S. T. see McLeish, T. C. B.: Vol. 143, pp. 195–256.
Mison, P. and *Sillion, B.*: Thermosetting Oligomers Containing Maleimides and Nadiimides End-Groups. Vol. 140, pp. 137–180.
Miyasaka, K.: PVA-Iodine Complexes: Formation, Structure and Properties. Vol. 108. pp. 91–130.
Miller, R. D. see Hedrick, J. L.: Vol. 141, pp. 1–44.
Minko, S. see Rühe, J.: Vol. 165, pp. 79–150.
Möhwald, H., Menzel, H., Helm, C. A. and *Stamm, M.*: Lipid and Polyampholyte Monolayers to Study Polyelectrolyte Interactions and Structure at Interfaces. Vol. 165, pp. 151–175.
Monkenbusch, M. see Richter, D.: Vol. 174, pp. 1–221.
Monnerie, L. see Bahar, I.: Vol. 116, pp. 145–206.
Mori, H. see Bohrisch, J.: Vol. 165, pp. 1–41.
Morishima, Y.: Photoinduced Electron Transfer in Amphiphilic Polyelectrolyte Systems. Vol. 104, pp. 51–96.
Morton M. see Quirk, R. P: Vol. 153, pp. 67–162.
Motornov, M. see Rühe, J.: Vol. 165, pp. 79–150.
Mours, M. see Winter, H. H.: Vol. 134, pp. 165–234.
Müllen, K. see Scherf, U.: Vol. 123, pp. 1–40.
Müller, A. H. E. see Bohrisch, J.: Vol. 165, pp. 1–41.
Müller, A. H. E. see Förster, S.: Vol. 166, pp. 173–210.
Müller, M. see Thünemann, A. F.: Vol. 166, pp. 113–171.
Müller-Plathe, F. see Gusev, A. A.: Vol. 116, pp. 207–248.
Müller-Plathe, F. see Baschnagel, J.: Vol. 152, p. 41–156.
Mukerherjee, A. see Biswas, M.: Vol. 115, pp. 89–124.
Munz, M., Cappella, B., Sturm, H., Geuss, M. and *Schulz, E.*: Materials Contrasts and Nanolithography Techniques in Scanning Force Microscopy (SFM) and their Application to Polymers and Polymer Composites. Vol. 164, pp. 87–210.

Murat, M. see Baschnagel, J.: Vol. 152, p. 41–156.
Mylnikov, V.: Photoconducting Polymers. Vol. 115, pp. 1–88.

Nagy, A. see Majoros, I.: Vol. 112, pp. 1–11.
Naka, K. see Uemura, T.: Vol. 167, pp. 81–106.
Nakamura, A. see Mashima, K.: Vol. 133, pp. 1–52.
Nakayama, Y. see Mashima, K.: Vol. 133, pp. 1–52.
Narasinham, B. and *Peppas, N. A.*: The Physics of Polymer Dissolution: Modeling Approaches and Experimental Behavior. Vol. 128, pp. 157–208.
Nechaev, S. see Grosberg, A.: Vol. 106, pp. 1–30.
Neoh, K. G. see Kang, E. T.: Vol. 106, pp. 135–190.
Netz, R.R. see Holm, C.: Vol. 166, pp. 67–111.
Netz, R.R. see Rühe, J.: Vol. 165, pp. 79–150.
Newman, S. M. see Anseth, K. S.: Vol. 122, pp. 177–218.
Nijenhuis, K. te: Thermoreversible Networks. Vol. 130, pp. 1–252.
Ninan, K. N. see Reghunadhan Nair, C.P.: Vol. 155, pp. 1–99.
Nishi, T. see Jinnai, H.: Vol. 170, pp. 115–167.
Nishikawa, Y. see Jinnai, H.: Vol. 170, pp. 115–167.
Noid, D. W. see Otaigbe, J. U.: Vol. 154, pp. 1–86.
Noid, D. W. see Sumpter, B. G.: Vol. 116, pp. 27–72.
Nomura, M., Tobita, H. and *Suzuki, K.*: Emulsion Polymerization: Kinetic and Mechanistic Aspects. Vol. 175, pp. 1–128.
Novac, B. see Grubbs, R.: Vol. 102, pp. 47–72.
Novikov, V. V. see Privalko, V. P.: Vol. 119, pp. 31–78.
Nummila-Pakarinen, A. see Knuuttila, H.: Vol. 169, pp. 13–27.

O'Brien, D. F., Armitage, B. A., Bennett, D. E. and *Lamparski, H. G.*: Polymerization and Domain Formation in Lipid Assemblies. Vol. 126, pp. 53–84.
Ogasawara, M.: Application of Pulse Radiolysis to the Study of Polymers and Polymerizations. Vol.105, pp. 37–80.
Okabe, H. see Matsushige, K.: Vol. 125, pp. 147–186.
Okada, M.: Ring-Opening Polymerization of Bicyclic and Spiro Compounds. Reactivities and Polymerization Mechanisms. Vol. 102, pp. 1–46.
Okano, T.: Molecular Design of Temperature-Responsive Polymers as Intelligent Materials. Vol. 110, pp. 179–198.
Okay, O. see Funke, W.: Vol. 136, pp. 137–232.
Onuki, A.: Theory of Phase Transition in Polymer Gels. Vol. 109, pp. 63–120.
Oppermann, W. see Holm, C.: Vol. 166, pp. 1–27.
Oppermann, W. see Volk, N.: Vol. 166, pp. 29–65.
Osad'ko, I. S.: Selective Spectroscopy of Chromophore Doped Polymers and Glasses. Vol. 114, pp. 123–186.
Osakada, K. and *Takeuchi, D.*: Coordination Polymerization of Dienes, Allenes, and Methylenecycloalkanes. Vol. 171, pp. 137–194.
Otaigbe, J. U., Barnes, M. D., Fukui, K., Sumpter, B. G. and *Noid, D. W.*: Generation, Characterization, and Modeling of Polymer Micro- and Nano-Particles. Vol. 154, pp. 1–86.
Otsu, T. and *Matsumoto, A.*: Controlled Synthesis of Polymers Using the Iniferter Technique: Developments in Living Radical Polymerization. Vol. 136, pp. 75–138.

de Pablo, J. J. see Leontidis, E.: Vol. 116, pp. 283–318.
Padias, A. B. see Penelle, J.: Vol. 102, pp. 73–104.

Pascault, J.-P. see Williams, R. J. J.: Vol. 128, pp. 95–156.
Pasch, H.: Analysis of Complex Polymers by Interaction Chromatography. Vol. 128, pp. 1–46.
Pasch, H.: Hyphenated Techniques in Liquid Chromatography of Polymers. Vol. 150, pp. 1–66.
Paul, W. see Baschnagel, J.: Vol. 152, p. 41–156.
Penczek, P. see Batog, A. E.: Vol. 144, pp. 49–114.
Penczek, P. see Bogdal, D.: Vol. 163, pp. 193–263.
Penelle, J., Hall, H. K., Padias, A. B. and *Tanaka, H.*: Captodative Olefins in Polymer Chemistry. Vol. 102, pp. 73–104.
Peppas, N. A. see Bell, C. L.: Vol. 122, pp. 125–176.
Peppas, N. A. see Hassan, C. M.: Vol. 153, pp. 37–65.
Peppas, N. A. see Narasimhan, B.: Vol. 128, pp. 157–208.
Pet'ko, I. P. see Batog, A. E.: Vol. 144, pp. 49–114.
Pheyghambarian, N. see Kippelen, B.: Vol. 161, pp. 87–156.
Pichot, C. see Hunkeler, D.: Vol. 112, pp. 115–134.
Pielichowski, J. see Bogdal, D.: Vol. 163, pp. 193–263.
Pieper, T. see Kilian, H. G.: Vol. 108, pp. 49–90.
Pispas, S. see Pitsikalis, M.: Vol. 135, pp. 1–138.
Pispas, S. see Hadjichristidis, N.: Vol. 142, pp. 71–128.
Pitsikalis, M., Pispas, S., Mays, J. W. and *Hadjichristidis, N.*: Nonlinear Block Copolymer Architectures. Vol. 135, pp. 1–138.
Pitsikalis, M. see Hadjichristidis, N.: Vol. 142, pp. 71–128.
Pleul, D. see Spange, S.: Vol. 165, pp. 43–78.
Plummer, C. J. G.: Microdeformation and Fracture in Bulk Polyolefins. Vol. 169, pp. 75–119.
Pötschke, D. see Dingenouts, N.: Vol 144, pp. 1–48.
Pokrovskii, V. N.: The Mesoscopic Theory of the Slow Relaxation of Linear Macromolecules. Vol. 154, pp. 143–219.
Pospíšil, J.: Functionalized Oligomers and Polymers as Stabilizers for Conventional Polymers. Vol. 101, pp. 65–168.
Pospíšil, J.: Aromatic and Heterocyclic Amines in Polymer Stabilization. Vol. 124, pp. 87–190.
Powers, A. C. see Prokop, A.: Vol. 136, pp. 53–74.
Prasad, P. N. see Lin, T.-C.: Vol. 161, pp. 157–193.
Priddy, D. B.: Recent Advances in Styrene Polymerization. Vol. 111, pp. 67–114.
Priddy, D. B.: Thermal Discoloration Chemistry of Styrene-co-Acrylonitrile. Vol. 121, pp. 123–154.
Privalko, V. P. and *Novikov, V. V.*: Model Treatments of the Heat Conductivity of Heterogeneous Polymers. Vol. 119, pp. 31–78.
Prociak, A. see Bogdal, D.: Vol. 163, pp. 193–263.
Prokop, A., Hunkeler, D., DiMari, S., Haralson, M. A. and *Wang, T. G.*: Water Soluble Polymers for Immunoisolation I: Complex Coacervation and Cytotoxicity. Vol. 136, pp. 1–52.
Prokop, A., Hunkeler, D., Powers, A. C., Whitesell, R. R. and *Wang, T. G.*: Water Soluble Polymers for Immunoisolation II: Evaluation of Multicomponent Microencapsulation Systems. Vol. 136, pp. 53–74.
Prokop, A., Kozlov, E., Carlesso, G. and *Davidsen, J. M.*: Hydrogel-Based Colloidal Polymeric System for Protein and Drug Delivery: Physical and Chemical Characterization, Permeability Control and Applications. Vol. 160, pp. 119–174.
Pruitt, L. A.: The Effects of Radiation on the Structural and Mechanical Properties of Medical Polymers. Vol. 162, pp. 65–95.
Pudavar, H. E. see Lin, T.-C.: Vol. 161, pp. 157–193.

Pukánszky, B. and *Fekete, E.*: Adhesion and Surface Modification. Vol. 139, pp. 109–154.
Putnam, D. and *Kopecek, J.*: Polymer Conjugates with Anticancer Acitivity. Vol. 122, pp. 55–124.

Quirk, R. P., Yoo, T., Lee, Y., M., Kim, J. and *Lee, B.*: Applications of 1,1-Diphenylethylene Chemistry in Anionic Synthesis of Polymers with Controlled Structures. Vol. 153, pp. 67–162.

Ramaraj, R. and *Kaneko, M.*: Metal Complex in Polymer Membrane as a Model for Photosynthetic Oxygen Evolving Center. Vol. 123, pp. 215–242.
Rangarajan, B. see Scranton, A. B.: Vol. 122, pp. 1–54.
Ranucci, E. see Söderqvist Lindblad, M.: Vol. 157, pp. 139–161.
Raphaël, E. see Léger, L.: Vol. 138, pp. 185–226.
Reddinger, J. L. and *Reynolds, J. R.*: Molecular Engineering of p-Conjugated Polymers. Vol. 145, pp. 57–122.
Reghunadhan Nair, C. P., Mathew, D. and *Ninan, K. N.*: Cyanate Ester Resins, Recent Developments. Vol. 155, pp. 1–99.
Reichert, K. H. see Hunkeler, D.: Vol. 112, pp. 115–134.
Rehahn, M., Mattice, W. L. and *Suter, U. W.*: Rotational Isomeric State Models in Macromolecular Systems. Vol. 131/132, pp. 1–475.
Rehahn, M. see Bohrisch, J.: Vol. 165, pp. 1–41.
Rehahn, M. see Holm, C.: Vol. 166, pp. 1–27.
Reineker, P. see Holm, C.: Vol. 166, pp. 67–111.
Reitberger, T. see Jacobson, K.: Vol. 169, pp. 151–176.
Reynolds, J. R. see Reddinger, J. L.: Vol. 145, pp. 57–122.
Richter, D. see Ewen, B.: Vol. 134, pp. 1–130.
Richter, D., Monkenbusch, M. and *Colmenero J.*: Neutron Spin Echo in Polymer Systems. Vol. 174, pp. 1–221.
Riegler, S. see Trimmel, G.: Vol. 176, pp. 43–87.
Risse, W. see Grubbs, R.: Vol. 102, pp. 47–72.
Rivas, B. L. and *Geckeler, K. E.*: Synthesis and Metal Complexation of Poly(ethyleneimine) and Derivatives. Vol. 102, pp. 171–188.
Roberts, G. W. see Kennedy, K. A.: Vol. 175, pp. 329–346.
Robin, J. J.: The Use of Ozone in the Synthesis of New Polymers and the Modification of Polymers. Vol. 167, pp. 35–79.
Robin, J. J. see Boutevin, B.: Vol. 102, pp. 105–132.
Roe, R.-J.: MD Simulation Study of Glass Transition and Short Time Dynamics in Polymer Liquids. Vol. 116, pp. 111–114.
Roovers, J. and *Comanita, B.*: Dendrimers and Dendrimer-Polymer Hybrids. Vol. 142, pp. 179–228.
Rothon, R. N.: Mineral Fillers in Thermoplastics: Filler Manufacture and Characterisation. Vol. 139, pp. 67–108.
Rozenberg, B. A. see Williams, R. J. J.: Vol. 128, pp. 95–156.
Rühe, J., Ballauff, M., Biesalski, M., Dziezok, P., Gröhn, F., Johannsmann, D., Houbenov, N., Hugenberg, N., Konradi, R., Minko, S., Motornov, M., Netz, R. R., Schmidt, M., Seidel, C., Stamm, M., Stephan, T., Usov, D. and *Zhang, H.*: Polyelectrolyte Brushes. Vol. 165, pp. 79–150.
Ruckenstein, E.: Concentrated Emulsion Polymerization. Vol. 127, pp. 1–58.
Ruiz-Taylor, L. see Mathieu, H. J.: Vol. 162, pp. 1–35.
Rusanov, A. L.: Novel Bis (Naphtalic Anhydrides) and Their Polyheteroarylenes with Improved Processability. Vol. 111, pp. 115–176.

Russel, T. P. see Hedrick, J. L.: Vol. 141, pp. 1–44.
Russum, J. P. see Schork, F. J.: Vol. 175, pp. 129–255.
Rychly, J. see Lazár, M.: Vol. 102, pp. 189–222.
Ryner, M. see Stridsberg, K. M.: Vol. 157, pp. 27–51.
Ryzhov, V. A. see Bershtein, V. A.: Vol. 114, pp. 43–122.

Sabsai, O. Y. see Barshtein, G. R.: Vol. 101, pp. 1–28.
Saburov, V. V. see Zubov, V. P.: Vol. 104, pp. 135–176.
Saito, S., Konno, M. and *Inomata, H.*: Volume Phase Transition of N-Alkylacrylamide Gels. Vol. 109, pp. 207–232.
Samsonov, G. V. and *Kuznetsova, N. P.*: Crosslinked Polyelectrolytes in Biology. Vol. 104, pp. 1–50.
Santa Cruz, C. see Baltá-Calleja, F. J.: Vol. 108, pp. 1–48.
Santos, S. see Baschnagel, J.: Vol. 152, p. 41–156.
Sato, T. and *Teramoto, A.*: Concentrated Solutions of Liquid-Christalline Polymers. Vol. 126, pp. 85–162.
Schaller, C. see Bohrisch, J.: Vol. 165, pp. 1–41.
Schäfer R. see Köhler, W.: Vol. 151, pp. 1–59.
Scherf, U. and *Müllen, K.*: The Synthesis of Ladder Polymers. Vol. 123, pp. 1–40.
Schmidt, M. see Förster, S.: Vol. 120, pp. 51–134.
Schmidt, M. see Rühe, J.: Vol. 165, pp. 79–150.
Schmidt, M. see Volk, N.: Vol. 166, pp. 29–65.
Scholz, M.: Effects of Ion Radiation on Cells and Tissues. Vol. 162, pp. 97–158.
Schopf, G. and *Koßmehl, G.*: Polythiophenes – Electrically Conductive Polymers. Vol. 129, pp. 1–145.
Schork, F. J., Luo, Y., Smulders, W., Russum, J. P., Butté, A. and *Fontenot, K.*: Miniemulsion Polymerization. Vol. 175, pp. 127–255.
Schulz, E. see Munz, M.: Vol. 164, pp. 97–210.
Seppälä, J. see Löfgren, B.: Vol. 169, pp. 1–12.
Sturm, H. see Munz, M.: Vol. 164, pp. 87–210.
Schweizer, K. S.: Prism Theory of the Structure, Thermodynamics, and Phase Transitions of Polymer Liquids and Alloys. Vol. 116, pp. 319–378.
Scranton, A. B., Rangarajan, B. and *Klier, J.*: Biomedical Applications of Polyelectrolytes. Vol. 122, pp. 1–54.
Sefton, M. V. and *Stevenson, W. T. K.*: Microencapsulation of Live Animal Cells Using Polycrylates. Vol. 107, pp. 143–198.
Seidel, C. see Holm, C.: Vol. 166, pp. 67–111.
Seidel, C. see Rühe, J.: Vol. 165, pp. 79–150.
Shamanin, V. V.: Bases of the Axiomatic Theory of Addition Polymerization. Vol. 112, pp. 135–180.
Sheiko, S. S.: Imaging of Polymers Using Scanning Force Microscopy: From Superstructures to Individual Molecules. Vol. 151, pp. 61–174.
Sherrington, D. C. see Cameron, N. R., Vol. 126, pp. 163–214.
Sherrington, D. C. see Lin, J.: Vol. 111, pp. 177–220.
Sherrington, D. C. see Steinke, J.: Vol. 123, pp. 81–126.
Shibayama, M. see Tanaka, T.: Vol. 109, pp. 1–62.
Shiga, T.: Deformation and Viscoelastic Behavior of Polymer Gels in Electric Fields. Vol. 134, pp. 131–164.
Shim, H.-K. and *Jin, J.*: Light-Emitting Characteristics of Conjugated Polymers. Vol. 158, pp. 191–241.
Shoda, S. see Kobayashi, S.: Vol. 121, pp. 1–30.

Siegel, R. A.: Hydrophobic Weak Polyelectrolyte Gels: Studies of Swelling Equilibria and Kinetics. Vol. 109, pp. 233–268.
Silvestre, F. see Calmon-Decriaud, A.: Vol. 207, pp. 207–226.
Sillion, B. see Mison, P.: Vol. 140, pp. 137–180.
Simon, F. see Spange, S.: Vol. 165, pp. 43–78.
Singh, R. P. see Sivaram, S.: Vol. 101, pp. 169–216.
Singh, R. P. see Desai, S. M.: Vol. 169, pp. 231–293.
Sinha Ray, S. see Biswas, M: Vol. 155, pp. 167–221.
Sivaram, S. and *Singh, R. P.*: Degradation and Stabilization of Ethylene-Propylene Copolymers and Their Blends: A Critical Review. Vol. 101, pp. 169–216.
Slugovc, C. see Trimmel, G.: Vol. 176, pp. 43–87.
Smulders, W. see Schork, F. J.: Vol. 175, pp. 129–255.
Söderqvist Lindblad, M., Liu, Y., Albertsson, A.-C., Ranucci, E. and *Karlsson, S.*: Polymer from Renewable Resources.Vol. 157, pp. 139–161.
Spange, S., Meyer, T., Voigt, I., Eschner, M., Estel, K., Pleul, D. and *Simon, F.*: Poly(Vinylformamide-co-Vinylamine)/Inorganic Oxid Hybrid Materials. Vol. 165, pp. 43–78.
Stamm, M. see Möhwald, H.: Vol. 165, pp. 151–175.
Stamm, M. see Rühe, J.: Vol. 165, pp. 79–150.
Starodybtzev, S. see Khokhlov, A.: Vol. 109, pp. 121–172.
Stegeman, G. I. see Canva, M.: Vol. 158, pp. 87–121.
Steinke, J., Sherrington, D. C. and *Dunkin, I. R.*: Imprinting of Synthetic Polymers Using Molecular Templates. Vol. 123, pp. 81–126.
Stelzer, F. see Trimmel, G.: Vol. 176, pp. 43–87.
Stenberg, B. see Jacobson, K.: Vol. 169, pp. 151–176.
Stenzenberger, H. D.: Addition Polyimides. Vol. 117, pp. 165–220.
Stephan, T. see Rühe, J.: Vol. 165, pp. 79–150.
Stevenson, W. T. K. see Sefton, M. V.: Vol. 107, pp. 143–198.
Stridsberg, K. M., Ryner, M. and *Albertsson, A.-C.*: Controlled Ring-Opening Polymerization: Polymers with Designed Macromoleculars Architecture. Vol. 157, pp. 27–51.
Sturm, H. see Munz, M.: Vol. 164, pp. 87–210.
Suematsu, K.: Recent Progress of Gel Theory: Ring, Excluded Volume, and Dimension. Vol. 156, pp. 136–214.
Sugimoto, H. and *Inoue, S.*: Polymerization by Metalloporphyrin and Related Complexes. Vol. 146, pp. 39–120.
Suginome, M. and *Ito, Y.*: Transition Metal-Mediated Polymerization of Isocyanides. Vol. 171, pp. 77–136.
Sumpter, B. G., Noid, D. W., Liang, G. L. and *Wunderlich, B.*: Atomistic Dynamics of Macromolecular Crystals. Vol. 116, pp. 27–72.
Sumpter, B. G. see Otaigbe, J. U.: Vol. 154, pp. 1–86.
Sun, H.-B. and *Kawata, S.*: Two-Photon Photopolymerization and 3D Lithographic Microfabrication. Vol. 170, pp. 169–273.
Suter, U. W. see Gusev, A. A.: Vol. 116, pp. 207–248.
Suter, U. W. see Leontidis, E.: Vol. 116, pp. 283–318.
Suter, U. W. see Rehahn, M.: Vol. 131/132, pp. 1–475.
Suter, U. W. see Baschnagel, J.: Vol. 152, p. 41–156.
Suzuki, A.: Phase Transition in Gels of Sub-Millimeter Size Induced by Interaction with Stimuli. Vol. 110, pp. 199–240.
Suzuki, A. and *Hirasa, O.*: An Approach to Artifical Muscle by Polymer Gels due to Micro-Phase Separation. Vol. 110, pp. 241–262.
Suzuki, K. see Nomura, M.: Vol. 175, pp. 1–128.
Swiatkiewicz, J. see Lin, T.-C.: Vol. 161, pp. 157–193.

Tagawa, S.: Radiation Effects on Ion Beams on Polymers. Vol. 105, pp. 99–116.
Takata, T., Kihara, N. and *Furusho, Y.*: Polyrotaxanes and Polycatenanes: Recent Advances in Syntheses and Applications of Polymers Comprising of Interlocked Structures. Vol. 171, pp. 1–75.
Takeuchi, D. see Osakada, K.: Vol. 171, pp. 137–194.
Tan, K. L. see Kang, E. T.: Vol. 106, pp. 135–190.
Tanaka, H. and *Shibayama, M.*: Phase Transition and Related Phenomena of Polymer Gels. Vol. 109, pp. 1–62.
Tanaka, T. see Penelle, J.: Vol. 102, pp. 73–104.
Tauer, K. see Guyot, A.: Vol. 111, pp. 43–66.
Teramoto, A. see Sato, T.: Vol. 126, pp. 85–162.
Terent'eva, J. P. and *Fridman, M. L.*: Compositions Based on Aminoresins. Vol. 101, pp. 29–64.
Theodorou, D. N. see Dodd, L. R.: Vol. 116, pp. 249–282.
Thomson, R. C., Wake, M. C., Yaszemski, M. J. and *Mikos, A. G.*: Biodegradable Polymer Scaffolds to Regenerate Organs. Vol. 122, pp. 245–274.
Thünemann, A. F., Müller, M., Dautzenberg, H., Joanny, J.-F. and *Löwen, H.*: Polyelectrolyte complexes. Vol. 166, pp. 113–171.
Tieke, B. see v. Klitzing, R.: Vol. 165, pp. 177–210.
Tobita, H. see Nomura, M.: Vol. 175, pp. 1–128.
Tokita, M.: Friction Between Polymer Networks of Gels and Solvent. Vol. 110, pp. 27–48.
Traser, S. see Bohrisch, J.: Vol. 165, pp. 1–41.
Tries, V. see Baschnagel, J.: Vol. 152, p. 41–156.
Trimmel, G., Riegler, S., Fuchs, G., Slugovc, C. and *Stelzer, F.*: Liquid Crystalline Polymers by Metathesis Polymerization. Vol. 176, pp. 43–87.
Tsuruta, T.: Contemporary Topics in Polymeric Materials for Biomedical Applications. Vol. 126, pp. 1–52.

Uemura, T., Naka, K. and *Chujo, Y.*: Functional Macromolecules with Electron-Donating Dithiafulvene Unit. Vol. 167, pp. 81–106.
Usov, D. see Rühe, J.: Vol. 165, pp. 79–150.
Uyama, H. see Kobayashi, S.: Vol. 121, pp. 1–30.
Uyama, Y: Surface Modification of Polymers by Grafting. Vol. 137, pp. 1–40.

Varma, I. K. see Albertsson, A.-C.: Vol. 157, pp. 99–138.
Vasilevskaya, V. see Khokhlov, A.: Vol. 109, pp. 121–172.
Vaskova, V. see Hunkeler, D.: Vol.: 112, pp. 115–134.
Verdugo, P.: Polymer Gel Phase Transition in Condensation-Decondensation of Secretory Products. Vol. 110, pp. 145–156.
Vettegren, V. I. see Bronnikov, S. V.: Vol. 125, pp. 103–146.
Vilgis, T. A. see Holm, C.: Vol. 166, pp. 67–111.
Viovy, J.-L. and *Lesec, J.*: Separation of Macromolecules in Gels: Permeation Chromatography and Electrophoresis. Vol. 114, pp. 1–42.
Vlahos, C. see Hadjichristidis, N.: Vol. 142, pp. 71–128.
Voigt, I. see Spange, S.: Vol. 165, pp. 43–78.
Volk, N., Vollmer, D., Schmidt, M., Oppermann, W. and *Huber, K.*: Conformation and Phase Diagrams of Flexible Polyelectrolytes. Vol. 166, pp. 29–65.
Volksen, W.: Condensation Polyimides: Synthesis, Solution Behavior, and Imidization Characteristics. Vol. 117, pp. 111–164.
Volksen, W. see Hedrick, J. L.: Vol. 141, pp. 1–44.
Volksen, W. see Hedrick, J. L.: Vol. 147, pp. 61–112.
Vollmer, D. see Volk, N.: Vol. 166, pp. 29–65.

Wagener, K. B. see Baughman, T. W.: Vol 176, pp. 1–42.
Wake, M. C. see Thomson, R. C.: Vol. 122, pp. 245–274.
Wandrey C., Hernández-Barajas, J. and *Hunkeler, D.*: Diallyldimethylammonium Chloride and its Polymers. Vol. 145, pp. 123–182.
Wang, K. L. see Cussler, E. L.: Vol. 110, pp. 67–80.
Wang, S. Q.: Molecular Transitions and Dynamics at Polymer/Wall Interfaces: Origins of Flow Instabilities and Wall Slip. Vol. 138, pp. 227–276.
Wang, S.-Q. see Bhargava, R.: Vol. 163, pp. 137–191.
Wang, T. G. see Prokop, A.: Vol. 136, pp. 1–52; 53–74.
Wang, X. see Lin, T.-C.: Vol. 161, pp. 157–193.
Webster, O. W.: Group Transfer Polymerization: Mechanism and Comparison with Other Methods of Controlled Polymerization of Acrylic Monomers. Vol. 167, pp. 1–34.
Whitesell, R. R. see Prokop, A.: Vol. 136, pp. 53–74.
Williams, R. J. J., Rozenberg, B. A. and *Pascault, J.-P.*: Reaction Induced Phase Separation in Modified Thermosetting Polymers. Vol. 128, pp. 95–156.
Winkler, R. G. see Holm, C.: Vol. 166, pp. 67–111.
Winter, H. H. and *Mours, M.*: Rheology of Polymers Near Liquid-Solid Transitions. Vol. 134, pp. 165–234.
Wittmeyer, P. see Bohrisch, J.: Vol. 165, pp. 1–41.
Wu, C.: Laser Light Scattering Characterization of Special Intractable Macromolecules in Solution. Vol 137, pp. 103–134.
Wunderlich, B. see Sumpter, B. G.: Vol. 116, pp. 27–72.

Xiang, M. see Jiang, M.: Vol. 146, pp. 121–194.
Xie, T. Y. see Hunkeler, D.: Vol. 112, pp. 115–134.
Xu, Z., Hadjichristidis, N., Fetters, L. J. and *Mays, J. W.*: Structure/Chain-Flexibility Relationships of Polymers. Vol. 120, pp. 1–50.

Yagci, Y. and *Endo, T.*: N-Benzyl and N-Alkoxy Pyridium Salts as Thermal and Photochemical Initiators for Cationic Polymerization. Vol. 127, pp. 59–86.
Yannas, I. V.: Tissue Regeneration Templates Based on Collagen-Glycosaminoglycan Copolymers. Vol. 122, pp. 219–244.
Yang, J. S. see Jo, W. H.: Vol. 156, pp. 1–52.
Yamaoka, H.: Polymer Materials for Fusion Reactors. Vol. 105, pp. 117–144.
Yasuda, H. and *Ihara, E.*: Rare Earth Metal-Initiated Living Polymerizations of Polar and Nonpolar Monomers. Vol. 133, pp. 53–102.
Yaszemski, M. J. see Thomson, R. C.: Vol. 122, pp. 245–274.
Yoo, T. see Quirk, R. P.: Vol. 153, pp. 67–162.
Yoon, D. Y. see Hedrick, J. L.: Vol. 141, pp. 1–44.
Yoshida, H. and *Ichikawa, T.*: Electron Spin Studies of Free Radicals in Irradiated Polymers. Vol. 105, pp. 3–36.

Zhang, H. see Rühe, J.: Vol. 165, pp. 79–150.
Zhang, Y.: Synchrotron Radiation Direct Photo Etching of Polymers. Vol. 168, pp. 291–340.
Zhou, H. see Jiang, M.: Vol. 146, pp. 121–194.
Zubov, V . P., Ivanov, A. E. and *Saburov, V. V.*: Polymer-Coated Adsorbents for the Separation of Biopolymers and Particles. Vol. 104, pp. 135–176.

Subject Index

All rotational state model (ARS)   118, 120, 124
Apoferitin   201

Backscattering   82, 84, 90, 91, 109, 155, 200
Bending forces   120
Biomolecules   200
Block equation   16
Breathing mode   172
BS experiments   84
BS instrument   91

Chain, worm like   120
Chain rigidity   117
Chain statistics, mode description   118
Chain stiffness   126
Characteristic ratio   25, 118, 126
Coherent scattering   10
Conformational entropy   26
Constraint release (CR)   63
Contour length fluctuation (CLF)   63, 66
Contrast variation   11
Copolymers, diblock   162
Cross section, double differential   9

Debye-Waller factor (DWF)   83
Dendrimers   184, 186
Diblock copolymers   162
Displacement, mean square   11, 32, 42, 45, 54, 56, 119
Double differential cross section   9
Draining parameter   124
Dynamic light scattering   21
Dynamic modulus   41, 47, 55

Elastic incoherent structure factor (EISF)   100, 110, 111
End-to-end-distance   6
Energy transfer   9

Entanglement   49, 50, 57
Entropic forces   24
Entropy, conformational   26
Entropy-driven dynamics   25
Ergodicity parameter, non-   113–115, 141

First cumulant matrix   165, 170
Flipper   13, 16, 19
Flory Huggins interaction parameter   164, 166, 168, 172
Fresnel coil   17
Friction coefficient   26, 35, 36, 166, 170, 173

Gaussian approximation   10, 39, 45, 54, 87–89
Gaussian chain   25, 26, 34, 195
Gaussian parameter, non-   89, 90
Gel   182
Glass transition   67
Glass transition temperature   67
Glutamate dehydrogenase   204

Hydrodynamic interaction   123, 193, 200

Incoherent scattering   10, 15, 53, 82
Interface   169

Kohlrausch-William-Watts (KWW) function   69, 73, 82, 87, 88, 97, 138, 142

Langevin equation   26
Larmor frequency   13
Length scale dynamics   147
Light scattering, dynamic   21

Matrix, first cumulant   165, 170
MDS   94

Mean spherical approximation (MSA) 201
Mean square displacement 11, 32, 42, 45, 54, 56, 119
Merging, $\alpha\beta$- 107
MIEZE technique 20
Mobility matrix 166
Mode coupling theory (MCT) 69, 88, 112, 141, 142, 146
Molecular dynamics simulation 37, 38, 56, 58, 68, 90–94, 142, 146, 150, 190, 204, 205
Modulus, dynamic 41, 47, 55
–, plateau 41
Momentum transfer 9
Monte Carlo simulation 56, 61
Motions, collective 136
–, self- 142
Multiple scattering 15, 138
Myoglobin 201, 205

Neutron spin echo (NSE) spectroscopy 1
– – –, resonance 20
NMR, pulsed field gradient 21

Order disorder transition (ODT) 162, 169, 173

Pair correlation function 29, 32
Phenoxy 84
Photon correlation spectroscopy 81
C-Phycocyanin 204
PIB 6, 74, 77–81, 84, 85, 95, 107, 108, 116, 122–125, 130, 133, 136, 149
–, intrachain viscosity 133
Plateau modulus 41
Poly(fluorosilicone) (PFS) 182
Poly(N-isopropyl-acrylamide) (PNIPA) 6, 181, 183
Polyamidoamine (PAMAM) 6, 186
Polybutadiene (PB) 6, 36, 39, 72–84, 96, 107, 114
Polydimethylsiloxane (PDMS) 6, 31, 36, 125, 129, 131, 161, 181, 196
Polyethylene (PE) 6, 36, 37, 48–54, 59–62, 136, 185
Polyethyleneoxide (PEO) 6, 189
Polyethylene-propylene (PEP) 6, 32, 36, 37, 51–54, 59, 60, 167, 185
Polyethylethylene (PEE) 6, 32, 36, 167
Polyethylmethylsiloxane (PEMS) 161

Polyisobutylene (PIB) 6, 74, 77–81, 84, 85, 95, 107, 108, 116, 122–125, 130, 133, 136, 149
Polyisoprene (PI) 6, 36, 77–95, 114, 115, 140, 146, 147, 156, 185, 198
Polymer blend dynamics 153
– – –, neutron scattering 155
Polymer dynamics 1
Polymer solutions 130
Polymeric electrolytes 188
Polymeric micelles 184
Polymers 6
Polypropylene (aPP) 6, 74–77, 80
Polypropyleneoxide (PPO) 6, 188
Polystyrene 185, 198
Polyurethane (PU) 6, 74–80
Polyvinylchloride (PVC) 6, 74–77, 95
Polyvinylethylene (PVE) 6, 36, 74–77, 84, 93, 136, 140, 143, 156
Polyvinylmethylether (PVME) 6, 84, 90
Protons, self-atomic motions 82

Random phase approximation (RPA) 162, 163, 199
Relaxation, glassy 1
α-Relaxation 67–69, 72, 82, 117, 144, 153, 155
β-Relaxation 67, 70, 96, 99, 111, 155
Reptation model 1, 24, 42
Resolution 14
Resonance NSE 20
Rheology 35, 55
Rotational isomeric state 118
Rotational transitions 117
Rouse diffusion coefficient 28, 42, 175
Rouse model 24–26, 30–35, 38, 117, 119, 142, 193, 200
– –, generalized 47
Rouse time 27
RPA 162, 163, 199

Scattering, backscattering 82
–, coherent 10
–, incoherent 82
–, multiple 15, 138
Scattering function, intermediate 10
– –, static 165
Scattering length 9
Segment length 25
Self correlation function 28, 31, 45
Self-motion 142

Semidilute solutions 195
θ-Solvent 197
Stars 184, 185
Stiffness, chains 126
Stochastic force 26, 27
Structure factor, dynamic 165
– –, static 167
Surface tension 181

Terminal time 44
Tube diameter 41, 51, 54, 63, 65

Undulation 179

Viscosity 35, 42, 61, 123
Viscosity effects, internal 123, 128

Worm-like chain 120

X-ray photon correlation spectroscopy 23

Zimm model 123, 130, 182, 193
Zwanzig-Mori projection operator formalism 165

Printing: Krips bv, Meppel
Binding: Litges & Dopf, Heppenheim